MAN·AND·THE NATURAL·WORLD

BY THE SAME AUTHOR

Religion and the Decline of Magic
There Was a Man of Our Town

MAN·AND·THE NATURAL·WORLD

A HISTORY OF THE MODERN SENSIBILITY

KEITH THOMAS

PANTHEON BOOKS, NEW YORK

 Published in the United States by Pantheon Books, a division of Random House, Inc., New York. Originally published in Great Britain by Allen Lane, Penguin Books, London, as *Man and the Natural World: Changing Attitudes in England 1500–1800.*

Library of Congress Cataloging in Publication Data

Thomas, Keith Vivian.
Man and the natural world.
Bibliography: p.
Includes index.
1. Man–Influence on nature–History. I. Title.
GF75.T47 1983 304.2 82–14384
ISBN 0–394–49945–X

Manufactured in the United States of America

First American Edition

TO VALERIE

CONTENTS

CONTENTS

PREFACE

This book is an expanded version of the George Macaulay Trevelyan Lectures delivered in the University of Cambridge in Lent Term 1979. I have included a good deal of additional material, but the essential form of the lectures has been preserved.

I am glad to have the chance of thanking the Electors, who by inviting me to give the Trevelyan Lectures conferred upon me the greatest honour to which any Oxford historian can aspire. G. M. Trevelyan is not a fashionable historian today, but to my generation, at school in the late 1940s, he was far and away the best known and most accessible of all historical writers. No one could be wholly immune to the poetic appeal of his writing; and, in my case, his moving evocation (in *Clio, a Muse*) of the garden front of St John's, witness of the last days of the courtiers of Charles I, shaped my perception of my present college long before I ever set eyes on it. The theme of my lectures can, I hope, be associated with his name. In the introduction to his *English Social History* he listed among the components of social history 'the attitude of man to nature'; and his writings abound in reflections on the subject.

I must also thank those (some, alas, no longer living) whose kindness and warm hospitality made my weekly visits to Cambridge so memorable. I am particularly indebted to the Master and Fellows of Sidney Sussex College, who most generously put a set of rooms at my disposal throughout the term. Finally, I am deeply grateful to the friendly and tolerant audience who endured these lectures in conditions so wintry as to remind all present that man's conquest of the natural world is still far from complete.

ACKNOWLEDGEMENTS

Many of the individual themes discussed in this book have been treated by previous writers and although I have sometimes departed from their interpretations my debt to them is very great. I have received particular stimulus from the pioneering works of Dix Harwood, Arthur O. Lovejoy and Charles E. Raven, as well as from the more recent writings of John Passmore, David Elliston Allen and John Barrell.

For permission to quote from Seamus Heaney's *Death of a Naturalist* (USA title *Poems 1965–1975*) thanks are due to Faber and Faber (London) and Farrar, Straus & Giroux, Inc. (New York).

I am grateful for permission to cite from the unpublished theses of Drs J. Addy, B. G. Reay, M. J. Ingram, P. J. Lineham and J. A. Sharpe. I have also used references kindly supplied me by Jonathan Barry, Peter Clark, James Cockburn, Howard Colvin, Patricia Crawford, Julie Crossley, George Gallop, Brian Harrison, Arnold Harvey, William Lamont, Jill Lewis, Stephen Logan, Alan Macfarlane, Sara Mendelson, Victor Morgan, Geoffrey Nuttall, John Prest, Paul Slack and John Walsh. Many other people have sent me helpful suggestions (often unsolicited) and I thank them all. At the final stage I benefited greatly from the constructive advice of Sir John Plumb.

I am deeply grateful to the staff of St John's College Library, the Codrington Library and, above all, the Bodleian Library. I have made heavy (and often unreasonable) demands on them all and have received nothing but kindness in return.

I also wish to thank the archivists of the county record offices of Clwyd, Essex, Kent and Northamptonshire for supplying copies of documents in their custody; Miss B. Atkinson and Mrs A. Hill for help with the typing; Susan Rose-Smith for help with the illustrations; Judy Nairn for scrupulous copy-editing; and my publisher Peter Carson for his patience.

The chief burden of this book has been borne by my family. My son Edmund has lent me cheerful assistance. My daughter Emily has given up much time to read proofs and check references. My wife Valerie has given me helpful advice, trenchant criticism and constant support.

St John's College, Oxford K.T.

INTRODUCTION

Unlike most historians, who are a sedentary lot, George Macaulay Trevelyan was passionately devoted to the open air and the countryside. In later life he was notorious for taking unsuspecting guests on short strolls after lunch which would turn out to be thirty-mile walks; and as a young man his tastes were notably strenuous. 'A run on foot after the fox among the moss-hags, on the very top of Great Cheviot itself, on a frosty morning, with both kingdoms full in view,' he wrote, 'is no ill way to begin the year.'[1] At Wallington, the Trevelyan family home in Northumberland, now the property of the National Trust, former servants of the family remember him because he used to come over each year for the shooting.

In the early twentieth century a devotion to rural pursuits was characteristic of the English upper classes to whom Trevelyan belonged. It was shared by that devoted fisherman and bird-watcher Lord Grey of Fallodon, whose biography he wrote; and it was affected by the former Midlands industrialist Stanley Baldwin, who appointed him to the Regius Chair at Cambridge. For centuries the English aristocracy had been country-based, because a heavily capitalized agriculture was the foundation of their wealth. Farming and estate management were the central interests of the gentry. They pursued country sports; they took an obsessive interest in dogs and horses; they were often knowledgeable about natural history; and they self-consciously designed a rural landscape which would provide for both profit and recreation. Wallington itself was built out of the coal fortunes of the Blacketts, but it reflects the nature-loving preoccupations of its owners. It does not merely contain a fine library and the two desks on which Trevelyan's great-uncle, Thomas Babington Macaulay, wrote his *History of England* and his father, George Otto Trevelyan, composed his history of *The American Revolution*. It also has grounds laid out in the style of Charles Bridgeman, a lake by Capability Brown, and a flower garden full of exotics. There are stuffed birds, books on natural history and paintings of flowers.

This feeling for the countryside, real or imagined, was not confined to the upper classes, but was common to many members of the first industrial nation. Already in the late eighteenth century it had begun to produce the characteristic home-sickness of English travellers abroad, like William Beckford, lying in his Portuguese hotel in 1787, 'haunted all night with rural ideas of England'.[2] As the factories multiplied, the nostalgia of the town-dweller was reflected in his little bit of garden, his pets, his holidays in Scotland or the Lake District, his taste for wild flowers and bird-watching, and his dream of a weekend cottage in the country. Today it can be seen in the enduring popularity of those self-consciously 'rural' writers who, from Izaak Walton in the seventeenth century to James Herriot in the twentieth, have sustained the myth of a country arcadia. Whether or not the preoccupation with nature and rural life is in reality peculiarly English, it is certainly something which the English townsman has for a long time liked to think of as such; and much of the country's literature and intellectual life has displayed a profoundly anti-urban bias.[3]

In G. M. Trevelyan the feeling for wild nature went much deeper than these vaguely rural longings. Of all his writings none is more eloquent than his Rickman Godlee Lecture of 1931 on *The Call and Claims of Natural Beauty*, in which he laments the erosion of rural England and, in terms echoing his beloved Wordsworth, proclaims the importance of natural scenery for man's spiritual life. Trevelyan's view was deeply pessimistic. Until the end of the eighteenth century, he held, the works of man had only added to the beauties of nature; thereafter deterioration had been rapid. Beauty was no longer produced by ordinary economic circumstances, and the only remaining hope lay in the preservation of what had not already been destroyed. As an active member and benefactor of the National Trust, Trevelyan looked to that body for the preservation of 'all that is lovely and solitary in Britain'. Most of the land then acquired by the Trust was wholly uncultivated; and his view was that it should be maintained in that natural condition. Without access to wild nature the English would spiritually perish.[4]

Only a few hundred years earlier the idea that human cultivation was something to be resisted rather than encouraged would have been unintelligible. For how had civilization progressed, if not by the clearance of the forests, the cultivation of the soil and the conversion of wild landscape into human settlement? Kings and great landowners might set aside forests and parks as reserves for hunting and timber, but in Tudor England the artificial preservation of uncultivated hilltops would have seemed as absurd as the creation of sanctuaries for wild birds or animals which could not be eaten or hunted. Man's task, in the words of Genesis

(i. 28), was to 'replenish the earth and subdue it': to level the woods, till the soil, drive off the predators, kill the vermin, plough up the bracken, drain the fens. Agriculture stood to land as did cooking to raw meat. It converted nature into culture. Uncultivated land meant uncultivated men; and, when seventeenth-century Englishmen moved to Massachusetts, part of their case for occupying Indian territory would be that those who did not themselves subdue and cultivate the land had no right to prevent others from doing so.[5]*

Trevelyan's passion to preserve wild scenery and his faith in the healing powers of unexploited nature would have been inconceivable without the profound shift in sensibilities (some aspects of which are explored in this book) which occurred in England between the sixteenth and late eighteenth centuries. Concern for the natural environment and worries about man's relationship to other species are normally regarded as relatively recent phenomena. Lord Ashby, for example, remarks that man's attitude to nature has changed imperceptibly over the last hundred years, while another commentator calls it 'the most important revolution of feeling since the Second World War'.[6] Nowadays one cannot open a newspaper without encountering some impassioned debate about culling grey seals or cutting down trees in Hampton Court or saving an endangered species of wild animal. But to understand these present-day sensibilities we must go back to the early modern period. For it was between 1500 and 1800 that there occurred a whole cluster of changes in the way in which men and women, at all social levels, perceived and classified the natural world around them. In the process some long-established dogmas about man's place in nature were discarded. New sensibilities arose towards animals, plants and landscape. The relationship of man to other species was redefined; and his right to exploit those species for his own advantage was sharply challenged. It was these centuries which generated both an intense interest in the natural world and those doubts and anxieties about man's relationship to it which we have inherited in magnified form.

An attempt to chart some of these developments may look like a Whiggish search for the intellectual origins of the National Trust, the Council for the Protection of Rural England, Animal Liberation and the Friends of the Earth. But the aim of this book is not just to explain the present; it also attempts to reconstruct an earlier mental world in its

* The same argument was employed in the case of Ireland. The native Irish, thought Sir John Davies in 1610, had failed to exploit the land; 'therefore it stands neither with Christian policy nor conscience to suffer so good and fruitful country to lie waste like a wilderness'; *Historical Tracts by Sir John Davies* (1786), 288.

own right. It seeks to expose the assumptions, some barely articulated, which underlay the perceptions, reasonings and feelings of inhabitants of early modern England towards the animals, birds, vegetation and physical landscape amongst which they spent their lives, often in conditions of proximity which are now difficult for us to appreciate.

Unfortunately, the subject is so vast and the material so abundant that no single author can hope to encompass it, least of all in a relatively short book. The present work is merely an attempt to sketch out some of the topic's more obvious implications. It is confined to England, though many of its themes can be closely paralleled in the history of Wales, Scotland and Ireland, as well as in Europe and North America. It also makes heavy, though unrepentant, use of literary sources of a kind not currently fashionable among historians. For all the defects of imaginative literature as a historical source, there is nothing to surpass it as a guide to the thoughts and feelings of at least the more articulate sections of the population. The book is thus intended to do something to reunite the studies of history and of literature in the way G. M. Trevelyan continually urged.

It also seeks to persuade its readers that its subject-matter deserves more serious historical treatment than it has yet received. Man's ascendancy over the animal and vegetable world has, after all, been a basic precondition of human history. The way in which he has rationalized and questioned that ascendancy is a large and daunting theme which in recent years has received a good deal of attention from philosophers, theologians, geographers and literary critics. The subject also has much to offer historians, for it is impossible to disentangle what the people of the past thought about plants and animals from what they thought about themselves.

I

HUMAN ASCENDANCY

Ask any one of the undistinguished mass of people, for what purpose every thing exists? The general answer is, that every thing was created for our practical use and accommodation!... In short, the whole magnificent scene of things is daily and confidently asserted to be ultimately intended for the peculiar convenience of mankind. Thus do the bulk of the human species vauntingly elevate themselves above the innumerable existences that surround them.

G. H. Toulmin, *The Antiquity and Duration of the World* (1780; 1824 edn), 51–2.

i. THEOLOGICAL FOUNDATIONS

In Tudor and Stuart England the long-established view was that the world had been created for man's sake and that other species were meant to be subordinate to his wishes and needs. This assumption underlay the actions of that vast majority of men who never paused to reflect upon the matter. But those theologians and intellectuals who felt the need to justify it could readily appeal to the classical philosophers and the Bible. Nature made nothing in vain, said Aristotle, and everything had a purpose. Plants were created for the sake of animals and animals for the sake of men. Domestic animals were there to labour, wild ones to be hunted. The Stoics had taught the same: nature existed solely to serve man's interests.[1]

It was in this spirit that Tudor commentators interpreted the biblical account of creation. Although modern scholars discern conflicting accounts embodied in the narrative of Genesis, the theologians of the early modern period usually had no difficulty in arriving at a generally accepted synthesis.[2] The Garden of Eden, they said, was a paradise prepared for man in which Adam had God-given dominion over all living things (Genesis, i. 28). At first, man and beast had cohabited peacefully. The humans were probably not carnivorous and the animals were tame. But with the Fall the relationship changed. By rebelling against God, man forfeited his easy dominance over other species. The earth degenerated. Thorns and thistles grew up where there had been only fruits and flowers (Genesis, iii. 18). The soil became stony and less fertile, making arduous labour necessary for its cultivation. There appeared fleas, gnats and other odious pests. Many animals cast off the yoke, becoming fierce, warring with each other and attacking men. Even domestic animals had now to be coerced into submission.

Then, after the Flood, God renewed man's authority over the animal creation:

> The fear of you and the dread of you shall be upon every beast of the earth, and upon every fowl of the air, upon all that moveth upon the earth, and upon all the fishes of the sea; into your hand are they delivered. Every moving thing that liveth shall be meat for you (Genesis, ix. 2–3).

Henceforth men were carnivorous and animals might lawfully be killed and eaten, subject only to the prevailing dietary restrictions. This was the Old Testament charter upon which human rule over nature was founded.[3] It was further reinforced by the coming of Christ, who was taken by some commentators to have reconfirmed man's rights over the natural world; though it was now possible to argue that such rights were legitimately enjoyed only by truly regenerate Christians.[4]

It is difficult nowadays to recapture the breathtakingly anthropocentric spirit in which Tudor and Stuart preachers interpreted the biblical story. For they did not hesitate to represent the world's physical attributes as a direct response to Adam's sin: 'Cursed is the ground for thy sake' (Genesis, iii. 17). It was only because of the Fall that wild animals were fierce, that obnoxious reptiles existed, and that domestic animals had to undergo blows in misery. 'The creatures were not made for themselves, but for the use and service of man,' said a Jacobean bishop. 'Whatsoever change for the worse is come upon them is not their punishment, but a part of ours.'[5]

Human ascendancy was, therefore, central to the Divine plan. Man was the end of all God's works, declared Jeremiah Burroughes in 1657; 'He made others for man, and man for himself.' 'All things,' agreed Richard Bentley in 1692, were created 'principally for the benefit and pleasure of man.' 'Man, if we look to final causes, may be regarded as the centre of the world,' mused Francis Bacon, 'insomuch that if man were taken away from the world, the rest would seem to be all astray, without aim or purpose.'[6] Some divines thought that after the Day of Judgement the world would be annihilated; it had been made only to accommodate humanity and would have no further use.[7]

Despite the Fall, therefore, man's right to rule remained intact. He was still 'the Vicegerent and Deputy of Almighty God'. 'All the creatures were made for man, subjected to his government and appointed for his use.'[8] The animals were less docile than they had been, but they had not all forgotten their duty. As Andrew Willet observed in 1605, there still remained 'a natural instinct of obedience in those creatures which are for man's use, as the ox, ass, horse'.[9] 'Sometimes,' said Jeremiah Burroughes in 1643, 'you may see a little child driving before him a hundred oxen or kine this way or that way as he pleaseth; it showeth that God hath preserved somewhat of man's dominion over the creatures.'[10] The instinct which brought fish in shoals to the sea-shore, noted the nonconformist divine Philip Doddridge a century later, 'seems an intimation that they are intended for human use'. The only purpose of animals, declared Thomas Wilcox, an Elizabethan, was to minister to man, 'for whose sake all the creatures were made that are made'.[11]

It was with human needs in mind that the animals had been carefully designed and distributed. Camels, observed a preacher in 1696, had been sensibly allotted to Arabia, where there was no water, and savage beasts 'sent to deserts, where they may do less harm'. It was a sign of God's providence that fierce animals were less prolific than domestic ones and that they lived in dens by day, usually coming out only at night, when men were in bed.[12] Moreover, whereas members of wild species all looked alike, cows, horses and other domestic animals had been conveniently variegated in colour and shape, in order 'that mankind may the more readily distinguish and claim their respective property'. The physician George Cheyne in 1705 explained that the Creator made the horse's excrement smell sweet, because he knew that men would often be in its vicinity.[13]

Every animal was thus intended to serve some human purpose, if not practical, then moral or aesthetic. Savage beasts were necessary instruments of God's wrath, left among us 'to be our schoolmasters', thought James Pilkington, the Elizabethan bishop; they fostered human courage and provided useful training for war.[14] Horse-flies, guessed the Virginian gentleman William Byrd in 1728, had been created so 'that men should exercise their wits and industry to guard themselves against them'. Apes and parrots had been ordained 'for man's mirth'. Singing-birds were devised 'on purpose to entertain and delight mankind'.[15] The lobster, observed the Elizabethan George Owen, served several purposes in one: it provided men with food, for they could eat its flesh; with exercise, for they had first to crack its legs and claws; and with an object of contemplation, for they could behold its wonderful suit of armour, with its 'tases, vaunthraces [vamplates], pouldrons, coushes [cuisses], gauntlets and gorgets curiously wrought and forged by the most admirable workman of the world'.* As for cattle and sheep, Henry More in 1653 was convinced that they had only been given life in the first place so as to keep their meat fresh 'till we shall have need to eat them'. As late as the 1830s the authors of the Bridgewater Treatises on 'God's goodness as manifested in the Creation' were still maintaining that all inferior species had been made to serve man's purpose. God created the ox and the horse to labour in our service, said the naturalist William Swainson; the dog to display affectionate attachment, and the chicken to show 'perfect contentment in a state of partial confinement'. The louse was indispensable, explained the Rev. William Kirby, because it provided a powerful incentive to habits of cleanliness.[16]

Vegetables and minerals were regarded in the same way. Henry More thought that their only purpose was to enhance human life. Without wood, men's houses would have been merely 'a bigger sort of beehives or birds' nests, made of contemptible sticks and straw and dirty mortar'; and, without metals, men would have been deprived of the 'glory and pomp' of war, fought with swords, guns and trumpets; instead there would have been 'nothing but howlings and shoutings of poor naked men belabouring one another ... with sticks or dully falling together by the ears at fisticuffs'. Even weeds and poisons had their essential uses, noted a herbalist: for they exercised 'the industry of man to weed them out ... Had he nothing to struggle with, the fire of his spirit would be half extinguished.'[17]

It was in the later seventeenth and early eighteenth centuries that these arguments about the perfection of the Creator's design reached their most ingenious and fanciful. In the century after the Reformation, by contrast, the tendency of theologians was to lay greater stress on the Fall. They emphasized the wretched, decaying state of the natural world and the obstacles God had put in man's way; and they seldom claimed that all was as it ideally ought to be. But from the mid seventeenth century there was an increasing disposition to play down the Fall and to stress, not the decay of nature, but its benevolent design. All was for the best; there were no real disharmonies between man's need and those of subordinate creation; and apparent conflicts of interest were purely superficial.

In the eighteenth century it was widely urged that domestication was *good* for animals; it civilized them and increased their numbers: 'we multiply life, sensation and enjoyment'.[18] Cows and sheep were better off in man's care than left to the mercy of wild predators. To butcher them for meat might seem cruel, said Thomas Robinson in 1709, but, 'when

* Tasses and cuisses were pieces of armour protecting the thighs; vamplates were hand-guards; pouldrons covered the shoulders and gorgets the throat.

more closely enquired into,' it proved 'a kindness rather than cruelty'; their despatch was quick and they were spared the sufferings of old age. There was no injustice about killing oxen to provide food for 'a more noble animal', thought Archbishop King; it was only on that condition that the beasts had been given life in the first place. Besides, added William Wollaston, the sufferings of brutes were not like the sufferings of men. They had no conception of the future and lost nothing by being deprived of life. It was therefore 'best for the beasts that they should be under man'.[19]

Man's authority over the natural world was thus virtually unlimited. He might use it as he pleased, said John Day in 1620: 'for his profit or for his pleasure'.[20] Vegetables obviously had no rights, for they were destitute of sense and therefore incapable of injury.[21] Animals had no rights either. They 'can have no right of society with us,' said Lancelot Andrewes, 'because they want reason.' They could not own land, for God had given the Earth to men, not to sheep or deer.[22] Unlike men, beasts had no divine authority for their dominion over the creatures they consumed. They did not even own their own lives. 'They have no right or propriety in anything,' emphasized Samuel Gott, 'no, not in themselves.' 'We may put them to any kind of death that the necessity either of our food or physic will require,' declared Bishop Hopkins.[23] When animals grew troublesome, agreed Henry More, then man had the right to curb them, 'for there is no question but we are more worth than they'. Vivisection, thought Isaac Barrow, was 'a most innocent cruelty, and easily excusable ferocity'. Even Thomas Hobbes, who rejected scriptural sanctions for man's ascendancy, agreed that there could be no obligations to animals, because 'to make covenants with brute beasts is impossible'.[24]

So when travellers came back with reports of how Eastern religions held a totally different view, and how Jains, Buddhists and Hindus respected the lives of animals, even of insects, the general reaction was of baffled contempt. It was 'unaccountable folly' of the Hindus, thought one seventeenth-century observer, to let a widow burn on her husband's pyre and yet be so sparing of the lives of insignificant creatures, 'as if the life of a man were of less consequence and consideration than that of a beast'.[25]* Remnants of a similar outlook in the West were equally

*At Shipton-under-Wychwood, Oxfordshire, in 1615 the preacher Henry Mills illustrated the profaneness of the Turkish religion to his rural audience by telling them 'of a woman that, travelling a long journey, should make water in her hand and give it her dog that fainted, to restore him, and how that this woman in the Turkish religion was taken up to heaven for the same deed'; Bodleian Library, Oxford Diocesan Papers, c 25, fol. 266.

condemned. 'I cannot approve,' said a preacher in 1612, 'of Pythagoras's too, too pitiful philosophy, which would not allow that the life of ... either plant or beast should be violated.' Lancelot Andrewes followed St Augustine and Aquinas in scornfully rejecting the Manichaean doctrine that man had no right to kill other creatures. The sixth commandment against murder, he explained, did not apply to non-humans.[26] In the same spirit a seventeenth-century bee-keeper refuted the 'pitiful humour' of those who thought it a pity to kill the bees so as to get at their honey (which was normal practice at this period): 'Hath not God given all creatures unto us for our benefit,' he asked, 'and to be used accordingly as may seem good unto us for our good? ... Is it not lawful for us to use these silly creatures in such sort as they may be most for our benefit, which I take to be the right use of them and the very end of their creation?' In the eighteenth century Philip Doddridge considered that, because animals were 'capable of but small degrees of happiness in comparison with man,' it was 'fit that their interests should give way to that of the human species whenever in any considerable article they come in competition with each other'.[27]

Even those who wished to kill animals for pleasure could, as Thomas Fuller observed in 1642, cite 'man's charter of dominion over the creatures'. Of bear-baiting and cock-fighting they could say: 'Christianity gives us a placard to use these sports.' Conventional wisdom about man's authority over animals was well summed up in 1735 by the hunting poet William Somervile:

> The brute creation are his property,
> Subservient to his will, and for him made.
> As hurtful these he kills, as useful those
> Preserves; their sole and arbitrary king.[28]

Contemporary theology thus provided the moral underpinnings for that ascendancy of man over nature which had by the early modern period become the accepted goal of human endeavour. The dominant religious tradition had no truck with that 'veneration' of nature which many Eastern religions still retained and which the scientist Robert Boyle correctly recognized as 'a discouraging impediment to the empire of man over the inferior creatures'. Since Anglo-Saxon times the Christian Church in England had stood out against the worship of wells and rivers. The pagan divinities of grove, stream and mountain had been expelled, leaving behind them a disenchanted world, to be shaped, moulded and dominated.[29]

In 1967 the American historian Lynn White, Jr, described Christianity, especially in its Western form, as 'the most anthropocentric religion the world has seen'; and his brief article blaming the medieval Church for the horrors of modern pollution became almost a sacred text for modern ecologists.[30]* Professor White was not the first to attribute the Western exploitation of nature to Europe's distinctive religious inheritance.† But, like his predecessors, he almost certainly overrated the extent to which human actions have been determined by official religion alone. In the 1680s the English sectary Thomas Tryon had also contrasted the moderate demands made on nature by the North American Indians with the ruthlessly manipulative approach of the European invaders.‡ But he recognized that it was new commercial incentives that had made the difference: it was less the replacement of pagan animism by Christianity, than the pressure of the international fur trade which led to overhunting and the unprecedented onslaught on Canadian wild life. As Karl Marx would note, it was not their religion, but the coming of private property and a money economy, which led Christians to exploit the natural world in a way the Jews had never done; it was what he called the 'great civilising influence of capital' which finally ended the 'deification of nature'.[31] More recent critics of Professor White's thesis have observed that the ancient Romans exploited natural resources in the pre-Christian world more effectively than did their Christian medieval successors; and that in modern times the Japanese worship of nature has not prevented

* He also urged that St Francis be recognized as the patron saint of ecology, a suggestion which was adopted by the Pope in April 1980.

† Over a century earlier, the German philosopher Arthur Schopenhauer had denounced the view that men have no duties to animals as 'one of revolting coarseness, a barbarism of the West, whose source is Judaism'; *The Basis of Morality*, trans. A. B. Bullock (1903), 218. In 1900 Wilfrid Scawen Blunt blamed Christianity for 'the atrocious doctrine that beasts and birds were made solely for man's use and pleasure, and that he has no duties towards them'; *My Diaries* (1932 edn), 343. The Oxford theologian Hastings Rashdall also attributed the indifference of philosophers to the issue of cruelty to animals to 'prejudices of theological origin'; *The Theory of Good and Evil* (2nd edn, 1924), i. 214.

‡ In *Keepers of the Game. Indian–Animal Relationships and the Fur Trade* (1978), Calvin Martin has ingeniously argued that the religion of the Eastern Algonkin Indians, which taught the existence of a tacit contract between men and animals, deterred them from over-exploiting the wild life of the area, but that, with the arrival of the Europeans (who brought new diseases for which the Indian shamans had no remedy), they became convinced that the animals had broken faith and therefore declared war upon them. His interpretation seems to take insufficient account of the new incentives for hunting provided by the Western market, a state of affairs for which the notion that the animals had broken their contract surely provided a rationalization. See the comments of Cornelius J. Jaenen in *Jnl of Interdisciplinary History*, x (1979), 376.

the industrial pollution of Japan. Ecological problems are not peculiar to the West, for soil erosion, deforestation and the extinction of species have occurred in parts of the world where the Judaeo-Christian tradition has had no influence. The Maya, the Chinese and the people of the Near East were all capable of destroying their environment without the aid of Christianity. Indeed, Christian teaching was less idiosyncratic than Professor White suggested, for there were other, non-Christian religions which also had their myths about man's God-given authority to dominate the natural world. It was reported in 1632 of the American Indians, for example, that 'they have it amongst them by tradition that God made one man and one woman and bade them live together, and get children, kill deer, beasts, birds, fish and fowl and what they would at their pleasure.'[32] Anthropocentrism was not peculiar to Western Europe.

Besides, the Judaeo-Christian inheritance was deeply ambivalent. Side by side with the emphasis on man's right to exploit the inferior species went a distinctive doctrine of human stewardship and responsibility for God's creatures. The English theologians who have been quoted so far tended to disregard those sections of the Old Testament which suggest that man has a duty to act responsibly towards God's creation. They passed quickly over the embarrassing passage in Proverbs (xii. 10) which taught that a good man regarded the life of his beast, and the section in Hosea (ii. 18) which implied that animals were members of God's covenant. 'That this expression is figurative,' said an Oxford professor in 1685 in his commentary on Hosea, 'cannot be doubted, seeing the things here named are not fit parties for making a covenant.' 'Many learned men of great judgement', therefore, took the passage to be a mere renewal of the league by which animals were subjected to Adam.[33] As for Proverbs, the commentators gratefully quoted St Paul's question in his first Epistle to the Corinthians (ix. 9.): 'Doth God take care for oxen?' – which they took to mean, perhaps wrongly, that he didn't.[34]

It can indeed be argued that Greek and Stoic influence distorted the Jewish legacy so as to make the religion of the New Testament much more man-centred than that of the Old; Christianity, it can be said, teaches, in a way that Judaism has never done, that the whole world is subordinate to man's purposes.[35] Fortunately, modern theological argument about the actual meaning of the Bible is irrelevant to our present purpose. It is not necessary here to determine whether or not Christianity is in itself intrinsically anthropocentric. The point is that in the early modern period its leading English exponents, the preachers and commentators, undoubtedly were. In due course, Christian doctrines would be drawn upon to buttress an altogether different view of man's relationship to animals. But at the start of our period exploitation, not stewardship, was the dominant theme. A reader who came fresh to the moral and theological writings of the sixteenth and seventeenth centuries could be forgiven for inferring that their main purpose was to define the special status of man and to justify his rule over other creatures.

ii. THE SUBJUGATION OF THE NATURAL WORLD

Human civilization indeed was virtually synonymous with the conquest of nature. The vegetable world had always been the source of food and fuel; and the West was by this time distinctive for its exceptionally high reliance upon animal resources, whether for labour, for food, for clothing or for transport. The civilization of medieval Europe would have been inconceivable without the ox and the horse.* Indeed it has been estimated that the use of animals for draught and burden gave the fifteenth-century European a motor power five times that of his Chinese counterpart. Like the Chinese, the Aztec and Inca societies of America had fewer animals than their European conquerors; it was the Spaniards who introduced horses, cattle, sheep and pigs to the New World.[1]

* The thirty or so copies of the Gutenberg Bible printed on vellum in 1456 used the skins of some five thousand calves; Curt F. Bühler, *The Fifteenth-Century Book* (Philadelphia, 1960), 41–2.

Europeans, moreover, were exceptionally carnivorous by comparison with the vegetable-eating peoples of the East.[2]

Nowhere in Europe was this dependence upon animals greater than in England, which, certainly by the eighteenth century and probably much earlier, had a higher ratio of domestic beasts per cultivated acre and per man than any other country, save the Netherlands.[3] The early modern period in England saw a striking expansion in the use of horses for draught, thus gradually releasing oxen to be used for human food.[4] Foreign visitors marvelled to see so many butchers' shops and so much meat-eating. 'Our shambles,' declared the Elizabethan Thomas Muffett, are 'the wonder of Europe, yea, verily, rather of the whole world.' London, thought Henry Peacham, 'eateth more good beef and mutton in one month than all Spain, Italy and a part of France in a whole year.'[5] Of course, meat was a relative luxury. 'Our poor country-people,' remarked a theologian in 1608, 'feed for the most part upon hard cheese, milk and roots'; and at the end of the seventeenth century Gregory King calculated that a quarter of the population could afford to eat meat only two days in seven, and another quarter not more than once a week. But this still left half the inhabitants eating meat regularly and averaging 147½ lb. of it per annum (and King excluded imported Dutch beef and Westphalian bacon from his calculations). Eighteenth-century British seamen were allowed 208 lb. of beef a year and 104 lb. of pork.[6] In 1726 it was estimated that in London alone there were killed annually some 100,000 beeves, 100,000 calves and 600,000 sheep. In 1748 the Swedish visitor Pehr Kalm noticed that England was different from other countries, in that butcher's meat formed the greater part of the main meal of the day: 'I do not believe that any Englishman who is his own master has ever eaten a dinner without meat.'[7] Everyone's ideal was a heavy meat diet, since flesh, particularly beef, was, according to the doctors, 'of all food ... most agreeable to the nature of man and breedeth most abundant nourishment to the body'; and it was thought to make men virile and aggressive.[8] A Scottish theologian even felt it necessary to explain that the Jewish doctrine that a man who ate more than a pound of meat at a time was a glutton, did not apply 'in these cold countries', where it was emphatically *not* gluttonous to consume over a pound of flesh at a sitting.[9]

From the sixteenth to the eighteenth centuries, accordingly, the roast beef of England was a national symbol.[10] It was no accident that carving meat at the table was so important a social accomplishment, or that it was associated with a lordly (and distinctly sadistic) vocabulary:

> Break that deer; ... rear that goose; lift that swan; sauce that chicken; unbrace that mallard; unlace that cony; dismember that heron; display that crane; disfigure that peacock; unjoint that bittern; ... mince that plover; ... splay that bream; ... tame that crab ...[11]

Meanwhile, the scientists and economic projectors of the seventeenth century anticipated yet further triumphs over the inferior species. For Bacon, the purpose of science was to restore to man that dominion over the creation which he had partly lost at the Fall, while Robert Boyle was egged on by his correspondent John Beale to establish what Beale called 'the empire of mankind'. To scientists reared in this tradition, the whole purpose of studying the natural world was 'that, Nature being known, it may be master'd, managed, and used in the services of human life'.[12] As William Forsyth remarked in 1802, in a plea for the observation of caterpillars: 'it would be of great service to get acquainted as much as possible with the economy and natural history of all these insects, as we might thereby be enabled to find out the most certain method of destroying them.'[13] The initial motive for the study of natural history was practical and utilitarian. Botany began as an attempt to identify the 'uses and virtues' of plants, primarily for medicine, but also for cooking and manufacture. It was the conviction that every part of the plant world had been designed to serve a human purpose which led Sir John Colbatch in 1719 to discover the medical use of mistletoe: 'It immediately enter'd into my mind that there must be something extraordinary in that uncommon beautiful plant; that the Almighty had design'd it for farther and more noble uses than barely to feed thrushes or to be hung up superstitiously in houses to drive away evil spirits ... I concluded, *a priori*, that it was ... very likely to subdue ... epilepsy.'[14]

Zoology was equally practical in its intentions. The Royal Society encouraged the study of animals with a view to determining 'whether they may be of any advantage to mankind, as food or physic; and whether those or any other uses of them can be further improved'. ''Tis no slight point of philosophy,' thought their secretary Henry Oldenburg, 'to know ... what animals may be tamed for human use and what commixtures with other animals may be advanced.'[15] Centuries of selective breeding had already refined the stock of domestic animals, cows, sheep, chickens and pigeons, but many new possibilities were yet to be explored. Pigs, urged Sir William Petty, could be taught to labour; and, if their diet were changed, the flesh of domestic stock could be improved. One of Samuel Hartlib's correspondents suggested that more exotic animals could be introduced into English agriculture: if turkeys, then why not elephants, buffaloes or mules?* 'I pray God for the introduction of new creatures into this island,' Christopher Smart would sing, 'For I pray God for the ostriches of Salisbury Plain, the beavers of the Medway and silver fish of Thames.' In the nineteenth century the official purpose of the Zoological Society would be to acclimatize and breed new domestic animals.[16] Animals, as the Rev. William Kirby put it in 1835, were of the deepest interest to everyone, because of their diversity, their beauty, 'but above all, their pre-eminent utility to mankind'.[17]

Plants were equally malleable. A large range of cultivated plants had been inherited from remote antiquity, but continuous breeding and experimentation opened new vistas. Agricultural writers described the great improvements which could be made by 'altering the species of such vegetables that are naturally produced, totally suppressing the one, and propagating another in its place'.[18] A gardener declared in 1734 that man now had the power 'to govern the vegetable world to a much greater improvement, satisfaction and pleasure than ever was known in the former ages of the world'. An infinity of exotic trees, flowers, fruits, vegetables and industrial crops was waiting to be introduced.[19] It was a plastic world, ready to be shaped and moulded.

In the conjectural history which became increasingly popular during the European Enlightenment of the eighteenth century, man's victory over other species was made the central theme. The true origin of human society, it was said, lay in the combination of men to defend themselves against wild beasts.[20] Then came hunting and domestication. Man's crucial act, thought Buffon, was the taming of the dog. It led, agreed Thomas Bewick, to the conquest and peaceable possession of the earth.[21] Without the camel, thought Herder, the deserts of Africa and Arabia would have been inaccessible, and, without the horse, the Europeans could never have conquered America. Lord Kames noted that, without the reindeer, Lapland would have been impenetrable. Adam Smith observed that crops and herds were the earliest forms of private property. 'Our toil is lessened,' pronounced Edward Gibbon, 'and our wealth is increased by our dominion over the useful animals.'[22]

Today, when our ascendancy over nature seems nearly complete, there are plenty of commentators ready to look back with nostalgia at earlier periods when a more even balance obtained. But in the Tudor and Stuart age the characteristic attitude was one of exaltation in hard-won human

* A popular writer opined that 'riding upon an ostrich may one day become the favourite, as it most certainly is the swiftest, mode of conveyance'; (anon.), *The Natural History of Remarkable Birds* (Dublin, 1821), 11.

dominance. Man's dominion over nature was the self-consciously proclaimed ideal of early modern scientists. Yet, despite the aggressively despotic imagery explicit in their talk of 'mastery', 'conquest' and 'dominion', they saw their task, thanks to generations of Christian teaching, as morally innocent. 'It never harmed any man,' said Francis Bacon, 'never burdened a conscience with remorse.' The husbandman, sang Abraham Cowley, confined his craft to 'innocent wars, on beasts and birds alone', and William Somervile agreed: though 'bloody' in deed, hunting was 'yet without guilt'.[23]

In the 'country-house' poems of the early seventeenth century birds and beasts find their fulfilment in yielding themselves up to be eaten by man.

> The pheasant, partridge and the lark
> Flew to thy house, as to the Ark.
> The willing ox of himself came
> Home to the slaughter, with the lamb;
> And every beast did thither bring
> Himself to be an offering.[24]

The rituals of contemporary hunters betray uninhibited delight in the capture and killing of wild animals. When James I hunted the stag he would personally cut its throat and daub the faces of his courtiers with blood, which they were not permitted to wash off; and it remained customary 'for ladies and women of quality after the hunting of a deer to stand by until they are ripped up, that they might wash their hands in the blood, supposing it will make them white'.[25] In the equestrian manuals, horse-riding was not just a convenient mode of transport. In spirit it was more like a sideshow at Bartholomew Fair, displaying 'wild beasts made tame';[26] it symbolized the human triumph; it was reason mastering the animal passions. The spectacle of a gentleman thus 'daunting a fierce and cruel beast' created 'a majesty and dread to inferior persons, beholding him above the common course of other men', declared Sir Thomas Elyot. As he curvetted and bounded, galloped and turned, or skilfully shuffled sideways, the rider of the great horse proclaimed both his social superiority and his conquest of the animal creation.[27] Contemporary horses differed in social nuance as much as do motor-cars today. But any horse was better than none and the number of horses multiplied. A foreign visitor noted in 1557 that in England no person would walk if he could possibly help it; and a century later Sir William Petty could adduce no better proof that the Irish had benefited from English rule than that 'the poorest now in Ireland ride on horseback, when heretofore the best ran on foot like animals'. In the early eighteenth century one of the supposed attractions of life in South Carolina was that horses were so plentiful that one seldom saw anyone there travel on foot, except negroes, and even they often went on horseback.[28]

In the same spirit, the colonists in Virginia began the task of converting the Indians by offering them a cow for every eight wolves they could kill, an exchange which neatly symbolized their view of the uses to which the natural world should be put. To give the Indians cattle, it was urged, would be 'a step to civilizing them and to making them Christians'.[29] It was for the same reason that, two hundred years later at the Great Exhibition of 1851, a booth displayed monkey skins from Africa. It was painful, wrote a sensitive contemporary, to think of the sufferings the creatures must have undergone. But there was a silver lining: 'the work of catching these monkeys is civilizing the African.'[30]

iii. HUMAN UNIQUENESS

Inhibitions about the treatment of other species were dispelled by the reminder that there was a fundamental difference in kind between humanity and other forms of life. The justification for this belief went back beyond Christianity to the Greeks. According to Aristotle, the soul comprised three elements: the nutritive soul, which was shared by man with vegetables; the sensitive soul, which was shared by animals; and the intellectual or rational soul, which was peculiar to man.[1] This doctrine had been taken over by the medieval scholastics and fused with the Judaeo-Christian teaching that man was made in the image of God (Genesis i. 27). Instead of representing man as merely a superior animal, it elevated him to a wholly different status, halfway between the beasts and the angels. In the early modern period it was accompanied by a great deal of self-congratulation.

Man, it was said, was more beautiful, more perfectly formed than any of the other animals. He had 'more of divine majesty in his countenance' and 'a more exquisite symmetry of parts'.[2] Jeremiah Burroughes reminded his congregation that, when God saw his other works, he only said that they were 'good', whereas when he had made man he said '*very* good': 'Observe, it is never said "very good" till the last day, till man is made.'[3]

Even so, there was a marked lack of agreement as to just where man's unique superiority lay. The search for this elusive attribute has been one of the most enduring pursuits of Western philosophers, most of whom have tended to fix on one feature and emphasize it out of all proportion, sometimes to the point of absurdity. Thus man has been described as a political animal (Aristotle); a laughing animal (Thomas Willis); a tool-making animal (Benjamin Franklin); a religious animal (Edmund Burke); and a cooking animal (James Boswell, anticipating Lévi-Strauss). As the novelist Peacock's Mr Cranium observes, man has at one time or another been defined as a featherless biped, an animal which forms opinions and an animal which carries a stick.[4] What all such definitions have in common is that they assume a polarity between the categories 'man' and 'animal' and that they invariably regard the animal as the inferior. In practice, of course, the aim of such definitions has often been less to distinguish men from animals than to propound some ideal of human behaviour, as when Martin Luther in 1530 and Pope Leo XIII in 1891 each declared that the possession of private property was an essential difference between men and beasts.[5]

By Tudor times the amount of inherited law on the subject was already enormous. Since Plato a great deal had been made of man's erect posture: beasts looked down, but he looked up to Heaven.[6] Aristotle had developed the theme, adding that men laughed, that their hair went grey, and that they alone couldn't wiggle their ears.[7] In the early modern period differences in anatomy continued to impress. According to one early Stuart doctor:

> Man is of a far different structure in his guts from ravenous creatures as dogs, wolves, etc., who, minding only their belly, have their guts descending almost straight down from their ventricle or stomach to

> the fundament: whereas in this noble microcosm man, there are in these intestinal parts many anfractuous circumvolutions, windings and turnings, whereby, longer retention of his food being procured, he might so much the better attend upon sublime speculations, and profitable employments in Church and Commonwealth.[8]

In the late eighteenth century the aesthete Uvedale Price drew special attention to the nose. 'Man is, I believe, the only animal that has a marked projection in the middle of the face.'[9]

Three other human attributes were particularly stressed. The first was speech, a quality which John Ray described as 'so peculiar to man that no beast could ever attain to it'. It was through speech, said Ben Jonson, that man expressed his superiority to other creatures. Without it, agreed Bishop Wilkins, man would be 'a very mean creature'. Because beasts lacked language, explained the eighteenth-century economist James Anderson, their experience could not be transmitted to their posterity: man progressed, but every animal species had 'the same powers and propensities ... that they had at the earliest period they were known'.[10]

The second distinguishing quality was reason. Man, as Bishop Cumberland put it, was 'an animal endowed with a mind'. Whether the difference was of kind or only of degree was a matter of debate. Some regarded animals as utterly irrational. Robert Lovell in 1661 divided the whole animal creation into two categories, 'rational' and 'irrational', putting only man in the former class. Gervase Markham reported the 'strongly held opinions' of 'many farriers' that horses had no brains at all; he himself had cut up the skulls of many dead horses and found nothing inside.[11] But most thought animals had elementary powers of understanding, albeit highly inferior ones. They had some practical intelligence, taught Aristotle, but they lacked the capacity for deliberation or speculative reason. From man's vast intellectual superiority, it was agreed, came his superior memory, his greater imagination, his curiosity, his sense of time, his sharper concept of the future, his use of numbers, his sense of beauty, his capacity for progress.[12] Above all, man could choose, whereas animals were prisoners of their instinct, guided only by appetite and incapable of free will.[13]

This distinctive human capacity for free agency and moral responsibility led on to the third, and, in the theologians' view, most decisive difference. This was not reason, which was, after all, shared to some extent by inferior creatures, but religion. Unlike animals, man had a conscience and a religious instinct.[14] He also had an immortal soul, whereas beasts perished and were incapable of an afterlife. This was no matter for regret: 'The life of a beast,' as a seventeenth-century preacher put it, was quite 'long enough for a beast-like life'. To suggest that animals might be immortal, said another in 1695, was an 'offensive absurdity'. Belief in the posthumous extinction of beasts was very important, he explained. It preserved the dignity of human nature, by showing an essential difference between the spirit of man and the souls of animals.[15]

In the seventeenth century the most remarkable attempt to magnify this difference was a doctrine originally formulated by a Spanish physician, Gomez Pereira, in 1554, but independently developed and made famous by René Descartes from the 1630s onwards. This was the view that animals were mere machines or automata, like clocks, capable of complex behaviour, but wholly incapable of speech, reasoning, or, on some interpretations, even sensation. For Descartes, the human body was also an automaton; after all, it performed many unconscious functions, like that of digestion. But the difference was that within the human machine there was a mind and therefore a separable soul, whereas brutes were automata without minds or souls. Only man combined both matter and intellect.[16]

This doctrine anticipated much later mechanistic psychology and contained the germs of the materialism of La Mettrie and other eighteenth-century thinkers. In due course, it would make it possible for scientists to argue that consciousness could be explained mechanically and that the whole of an individual's psychic life was the product of his physical organization. What Descartes said of animals would one day be said of man.[17] In the meantime, however, the Cartesian doctrine had the effect of further downgrading animals by comparison with human beings. Descartes himself seems to have modified his doctrine in later years and was unwilling to conclude that brutes were wholly incapable of sensation; for him the essential point was that they lacked the faculty of cogitation. He denied souls to animals because they exhibited no behaviour which could not be accounted for in terms of mere natural impulse.[18] But his supporters went further. Animals, they declared, did not feel pain; the cry of a beaten dog was no more evidence of the brute's suffering than was the sound of an organ proof that the instrument felt pain when struck.[19] Animal howls and writhings were merely external reflexes, unconnected with any inner sensation.

Today, this doctrine may seem to fly in the face of common sense. But it is not surprising that Cartesianism had its supporters at the time. An age accustomed to a host of mechanical marvels – clocks, watches, moving figures and automata of every kind – was well prepared to believe that animals were also machines, though made by God, not man.[20]* Besides, Descartes was only sharpening a distinction already implicit in scholastic teaching. Aquinas, after all, had taught that the so-called prudence of animals was no more than divinely implanted instinct.[21] Moreover, Cartesianism seemed an excellent way of safeguarding religion. Its opponents, by contrast, could be made to seem theologically suspect, for when they conceded to beasts the powers of perception, memory and reflection, they were implicitly attributing to animals all the ingredients of an immortal soul, which was absurd; and if they denied that they had an immortal soul, even though they had such powers, they were by implication questioning whether man had an immortal soul either.[22] Cartesianism was a way of escaping both of these unequally unacceptable alternatives. It denied that animals had souls and it maintained that men were something more than mere machines. It was, thought Leibniz, an opinion into which its supporters had foolishly rushed 'because it seemed necessary either to ascribe immortal souls to beasts or to admit that the soul of man could be mortal'.[23]

But the most powerful argument for the Cartesian position was that it was the best possible rationalization for the way man actually treated animals. The alternative view had left room for human guilt by conceding that animals could and did suffer; and it aroused worries about the motives of a God who could allow beasts to undergo undeserved miseries on such a scale. Cartesianism, by contrast, absolved God from the charge of unjustly causing pain to innocent beasts by permitting humans to ill-treat them; it also justified the ascendancy of men, by freeing them, as Descartes put it, from 'any suspicion of crime, however often they may eat or kill animals'.[24] By denying the immortality of beasts, it removed any lingering doubts about the human right to exploit the brute creation. For, as the Cartesians observed, if animals really had an immortal element, the liberties men took with them would be impossible to justify; and to concede that animals had sensation was to make human behaviour seem intolerably cruel.[25] The suggestion that a beast could feel or possess an immaterial soul, commented John Locke, had so worried some men that they 'had rather thought fit to conclude all beasts perfect machines rather than allow their souls immortality'.[26] Descartes's explicit aim had been to make men 'lords and possessors of nature'.[27] It fitted in well with his intention that he should have portrayed other species as inert and lacking any spiritual dimension. In so doing he

* The Cartesians cited the Chinese ruler who, when shown a watch, was supposed to have mistaken it for a living creature; Sir Kenelm Digby, *Two Treatises* (1645), i. 400.

created an absolute break between man and the rest of nature, thus clearing the way very satisfactorily for the uninhibited exercise of human rule.

The Cartesian view of animal souls generated a vast learned literature, and it is no exaggeration to describe it as a central preoccupation of seventeenth- and eighteenth-century European intellectuals.[28] Yet, though Descartes's work was disseminated in England, the country threw up only half a dozen or so explicit defenders of the Cartesian position. They included the virtuoso Sir Kenelm Digby, who did not hesitate to declare that birds were machines, and that their motions when building their nests and feeding their young were no different from the striking of a clock or the ringing of an alarm.[29] Many physiologists agreed that the body had its mechanical and involuntary movements. But the theologian Henry More was more representative of English opinion when he bluntly told Descartes in 1648 that he thought his a 'murderous' doctrine.[30] Most later English intellectuals felt with Locke and Ray that the whole idea of beast-machine was 'against all evidence of sense and reason' and 'contrary to the commonsense of mankind'. As Bolingbroke remarked, the plain man would persist in believing that there was a difference between the town bull and the parish clock. The nonconformist divine John Howe could understand Descartes's anxiety to distinguish between men and animals, 'lest any prejudice should be done to the doctrine of the human soul's immortality'. But he thought his formulation 'a great deal more pious than ... cogent'.[31]

Yet Descartes had only pushed the European emphasis on the gulf between man and beast to its logical conclusion. A transcendent God, outside his creation, symbolized the separation between spirit and nature. Man stood to animal as did heaven to earth, soul to body, culture to nature. There was a total qualitative difference between man and brute. In England the doctrine of human uniqueness was propounded from every pulpit. John Evelyn heard a sermon in 1659 on how man was 'a creature of different composure from the rest of animals; as both to soul and body; [and] how the one was to be the subject to the other.' In 1683 the Dean of Winchester conceded that animals had some human qualities, albeit 'in an inferior manner', but he denounced the idea that animals and men were therefore the same as a 'dangerous imagination'.[32] Throughout the eighteenth century the theme was reiterated. 'In the ascent from brutes to man,' declared Oliver Goldsmith, 'the line is strongly drawn, well marked, and unpassable.' 'How slender so ever it may sometimes appear,' wrote the naturalist William Bingley, 'the barrier which separates men from brutes is fixed and immutable.' The practical advantages of this distinction were clear, even if its theoretical rationale was elusive. 'Animals, whom we have made our slaves,' Charles Darwin would write, 'we do not like to consider our equal.'[33]

iv. MAINTAINING THE BOUNDARIES

Of course most of these learned disquisitions passed far above the heads of ordinary people. But consciously or unconsciously, the fundamental distinction between man and animals underlay everyone's behaviour. What, for example, were religion and morality, if not attempts to curb the supposedly animal aspects of human nature, what Plato called 'the wild beast within us'?[1] As Richard Baxter put it, 'he that hath well learned ... wherein a man doth differ from a brute, hath laid such a foundation for a holy life, as all the reason in the world is never able to overthrow.' If a man's mind was not pure, said Oliver Cromwell, there was no difference between him and a beast. The eighteenth-century Evangelical John Fletcher explained that regeneration meant passing from nature to grace: 'He was an animal man; in being born again he becomes a spiritual man.'[2] It was no accident that the symbol of Anti-Christ was the Beast, or that the Devil was regularly portrayed as a mixture of man and animal. When people saw what they thought were evil spirits, it was usually in the guise of some animal: a dog, a cat or a rat; one diarist noted the case of a man 'heaved into the water by one in the shape of a bull'.[3]

Like morals and religion, polite education, 'civility' and refinement were also intended to raise men above the animals. England was not one of those societies, like Bali,[4] where the consumption of food was regarded as a disgusting operation, best carried out in private. But people cooked their meat, rather than eating it raw like animals, and they thought gluttony a 'beastly' vice. ('[I was] a little swinish at dinner,' writes the eighteenth-century Irish Quaker John Rutty in his spiritual diary.)[5] His contemporary Oliver Goldsmith considered that 'of all other animals we spend the least time in eating; this is one of the great distinctions between us and the brute creation; and [he added piously] eating is a pleasure of so low a kind that none but such as are nearly allied to the quadruped desire its prolongation.' (Goldsmith had been understandably upset by the contemporary case of the young man from Bristol who came of a ruminating family and, a quarter of an hour after each meal, began to chew his meat all over again, declaring it tasted better the second time round.)[6]

Long before Goldsmith, Erasmus's decisively influential textbook on civility had made differentiation from animals the very essence of good table manners, more so even than differentiation from 'rustics'. Don't smack your lips, like a horse, he warned; don't swallow your meat without chewing, like a stork; don't gnaw the bones, like a dog; don't lick the dish, like a cat. (Even so, the Venetian ambassador in 1618 was shocked to discover that Londoners shamelessly munched fruit in the street, 'like so many goats'.)[7] Erasmus's rules for bodily comportment show the same preoccupation: don't shake your hair like a colt; don't neigh when you laugh, like a horse, or show your teeth, like a dog; don't move your whole body when speaking, like a wagtail; don't speak through your nose: 'It is the property of crows and elephants.' In the eighteenth century Henry Fielding remarked that it was 'those great polishers of our manners', the dancing-masters, who were 'by some thought to teach what principally distinguishes us from the brute creation'.[8]

Since all the bodily functions had undesirable animal associations, some commentators thought that it was physical modesty, even more than reason, which distinguished men from beasts.[9] There is an instructive passage in the diary of the New England clergyman Cotton Mather for 1700:

> I was once emptying the cistern of nature, and making water at the wall. At the same time, there came a dog, who did so too, before me. Thought I; 'What mean and vile things are the children of men ... How much do our natural necessities abase us, and place us ... on the same level with the very dogs!'
>
> My thought proceeded. 'Yet I will be a more noble creature; and at the very time when my natural necessities debase me into the condition of the beast, my spirit shall (I say *at that very time*!) rise and soar ...
>
> Accordingly, I resolved that it should be my ordinary practice, whenever I step to answer the one or other necessity of nature to make it an opportunity of shaping in my mind some holy, noble, divine thought ...

It thus became in 1711 his firm resolution to use the occasion of the usual evacuations of nature to form

> some thoughts of piety wherein I may differ from the brutes (which in the actions themselves I do very little).[10]

Not everyone reached so exquisite a level of self-consciousness. But most people were taught to regard their bodily impulses as 'animal' ones, needing to be subdued. The alternative would be 'beastly' or 'brutish'.[11] Lust, in particular, was synonymous with the animal condition, for the sexual connotations of such terms as 'brute', 'bestial' and 'beastly' were much stronger than they are today.[12] Lust, said a sixteenth-century moralist, made men 'like ... swine, goats, dogs and the most savage and brutish beasts in the world'.[13] In the bestiaries and emblem books a remarkably high proportion of the animals which appear are meant to symbolize lasciviousness or sexual infidelity. For Gerrard Winstanley, sexual freedom was 'the freedom of wanton unreasonable beasts'. For Jeremy Collier, the loose morality of the Restoration stage broke down 'the distinctions between man and beast. Goats and monkeys, if they could speak, would express their brutality in such language as this.' The sexual impulse in man was usually conceived of as thrusting upwards from below.[14]

Wherever we look in early modern England, we find anxiety, latent or explicit, about any form of behaviour which threatened to transgress the fragile boundaries between man and the animal creation. Physical cleanliness was necessary because, as John Stuart Mill would put it, its absence, 'more than of anything else, renders man bestial'.[15] Nakedness was bestial, for clothes, like cooking, were a distinctively human attribute.[16] It was bestial for men to have unduly long hair: 'Beasts are more hairy than men,' wrote Bacon, 'and savage men more than civil.'[17] It was bestial to work at night, for the same reason that burglary was a worse crime than daylight robbery; the night, as Sir Edward Coke explained, was 'the time wherein man is to rest, and wherein beasts run about seeking their prey'.[18] It was even bestial to go swimming, for, apart from being in many Puritan eyes a dangerous form of semi-suicide, it was essentially a non-human method of progression. As a Cambridge divine observed in 1600: men walked; birds flew; only fish swam.[19] One commentator even thought that the reason some Red Indians coloured their teeth black was that they supposed it 'essential to men to differ from the brutes in every respect, and therefore it was necessary not even to have teeth of the same colour'.[20]

Even to pretend to be an animal for purposes of ritual or entertainment was unacceptable. William Prynne declared it immoral to dress as a beast on the stage because to do so obliterated man's glorious image. Many moralists shared his objection to animal disguises; and in the early seventeenth century the hobby horse seems to have largely disappeared from the morris dance. Other ways of dressing as animals also became uncommon until they were revived by folklorists in modern times. At the same time traditional tales about the metamorphosis of humans into animals were condemned as either poetical fancies or diabolical fictions.[21] One of the reasons that monstrous births caused such horror was that they threatened the firm dividing-line between men and animals.

Close relations with animals were also frowned upon. When in 1667 Dr Edmund King planned the transfusion of a lamb's blood into the veins of a man, the experiment was at first held up because of 'some considerations of a moral nature'; and in the nineteenth century one of the great arguments against vaccination would be that inoculation with fluid from cows would result in the 'animalization' of human beings.[22] Bestiality, accordingly, was the worst of sexual crimes because, as one Stuart moralist put it, 'it turns man into a very beast, makes a man a member of a brute creature'.[23] The sin was the sin of confusion; it was immoral to mix the categories.* Injunctions against 'buggery with beasts' were standard in seventeenth-century moral literature, though occasionally the topic was passed over, 'the fact being more filthy than to be spoken of'.[24] Bestiality became a capital offence in 1534 and, with one brief interval, remained so until 1861. Incest, by contrast, was not a secular crime at all until the twentieth century.[25]

* It is revealing of the extent to which sensibilities on this point have changed that a learned modern commentator should find the legal prohibition of bestiality 'pointless'; Tony Honoré, *Sex Law* (1978), 176.

In early modern England even animal pets were morally suspect, especially if admitted to the table and fed better than the servants. It was against the rules of civility to handle dogs at the table, ruled Erasmus. 'Over-familiar usage of any brute creature is to be abhorred,' said a moralist in 1633.[26] An unconventional pet – a toad or a fly or weasel – could be identified as a witch's familiar, while for gentlewomen to cherish pet monkeys in their bosoms was, as Helkiah Crooke ruled in 1631, 'a very wicked and inhumane thing'. The godly remembered the story of the pious Elizabethan Katherine Stubbes, who, on her deathbed, caught sight of her favourite little puppy.

> She had no sooner espied her, but she beat her away, and calling her husband to her, said 'Good husband, you and I have offended God grievously in receiving many a time this bitch into our bed; we would have been loathe to have received a Christian soul ... into our bed, and to have nourished him in our bosoms, and to have fed him at our table, as we have done this filthy cur many times. The Lord give us grace to repent it' ... and afterwards she could not abide to look upon the bitch any more.[27]

It was during these centuries that most farmers finally moved the animals out of their houses into separate accommodation.*

Sentiments about animals, say the anthropologists, are usually projections of attitudes to man.[28] In early modern England the official concept of the animal was a negative one, helping to define, by contrast, what was supposedly distinctive and admirable about the human species. By embodying the antithesis of all that was valued and esteemed, the idea of the brute was as indispensable a prop to established human values as were the equally unrealistic notions held by contemporaries about witches or Papists. 'The meaning of order,' it has been well said, 'could only be grasped by exploring its antithesis or "contrary".'[29] Animal analogies came particularly readily to the lips of those who saw more of animals, wild and domestic, than do most people today. The brute creation provided the most readily-available point of reference for the continuous process of human self-definition. Neither the same as humans, nor wholly dissimilar, the animals offered an almost inexhaustible fund of symbolic meaning.

Yet there was little objective justification for the way in which the beasts were perceived. 'As drunk as a dog,' the proverb said. But who has ever seen a drunken dog?[30] Men attributed to animals the natural

* See below, p. 95.

impulses they most feared in themselves – ferocity, gluttony, sexuality – even though it was men, not beasts, who made war on their own species, ate more than was good for them and were sexually active all the year round. It was as a comment on *human* nature that the concept of 'animality' was devised. As S. T. Coleridge would observe, to call human vices 'bestial' was to libel the animals.[31]

V. INFERIOR HUMANS

In drawing a firm line between man and beast, the main purpose of early modern theorists was to justify hunting, domestication, meat-eating, vivisection (which became common scientific practice in the late seventeenth century) and the wholesale extermination of vermin and predators. But this abiding urge to distinguish the human from the animal also had important consequences for relations between men. For, if the essence of humanity was defined as consisting in some specific quality, then it followed that any man who did not display that quality was subhuman, semi-animal. 'In each constructed world of nature,' writes one modern anthropologist, 'the contrast between man and not-man provides an analogy for the contrast between the member of the human society and the outsider.' It is common, says another, for tribes to appropriate for themselves the arrogant title of 'man', referring to other peoples as 'monkeys'.[1]

In early modern England there were exclusive groups, like the Family of Love, of whom it was said in Elizabethan times that 'whosoever is not of their sect they account him as a beast that hath no soul'.[2] But the same exclusive attitude was more widely displayed towards those 'primitive' peoples who lacked the same attributes as those in which the animals were deficient: technology, intelligible language, Christianity. Many of the early explorers would have agreed with Gibbon that 'the human brute, without arts or laws, ... is poorly distinguished from the rest of the animal creation'.[3] Culture was as necessary to man as was domestication for plants and animals. Robert Gray declared in 1609 that 'the greater part' of the earth was 'possessed and wrongfully usurped by wild beasts ... or by brutish savages, which by reason of their godless ignorance, and blasphemous idolatry, are worse than those beasts'. The Earl of Clarendon agreed: 'the greatest part of the world is yet inhabited by men as savage as the beasts who inhabit with them.' 'Their words are sounded rather like that of apes than men,' reported Sir Thomas Herbert of the inhabitants of the Cape of Good Hope; 'I doubt that many of them have no better predecessors than monkeys.' 'The Hottentots,' thought a Jacobean clergyman, were 'beasts in the skin of man' and their speech 'an articulate noise rather than language, like the clucking of hens or gabbling of turkeys.' They were 'filthy animals', said a later traveller, who 'hardly deserve the name of rational creatures'. The seventeenth and eighteenth centuries saw many discourses on the animal nature of the negroes, their beastlike sexuality and their brutish nature.[4]

The American Indians were not normally thought of in this way, but occasionally they too were described in similar language. Frobisher found them living in caves and hunting their prey, 'even as the bear or other wild beasts do'. Robert Johnson saw them wandering 'up and down in troops, like herds of deer in a forest'. In Virginia they were observed 'creeping upon all fours ... like bears'; their houses were said to be like a 'den or hog-sty', and they themselves, 'more brutish than the beasts they hunt'. In 1689 Edmund Hickeringill, an English cleric who had once been to the West Indies, spoke disdainfully of 'the poor silly naked Indians' as 'just one degree (if they be so much) remov'd from a monkey'.[5]

Many saw the Irish in a similar light. They lived 'like beasts', thought the Elizabethan Barnaby Rich; 'in a brutish, nasty condition', said Sir William Petty. They ate raw flesh and drank hot blood from their cows.[6] The Irishman's animal nature had been discovered long before those Victorian caricatures which depicted him with simian features. In the 1650s a captain in General Ireton's regiment told how, when an Irish garrison was slaughtered at Cashel in 1647, they found among the bodies of the dead 'divers that had tails near a quarter of a yard long'; and when the story was challenged, forty soldiers came forward to testify on oath that they personally had seen them.[7]

There were other animals nearer home. 'What is an infant,' asked a Jacobean writer, 'but a brute beast in the shape of a man? And what is a young youth but (as it were) a wild untamed ass-colt unbridled?' Small children were not in control of their actions and the language of infancy was 'no little, if at all, better than the sounds the most sagacious brute animals make to each other'.[8] Young men, being still unable to control their passions, were only a little better. They were 'like wild asses and wild heifers', said George Fox; like young colts, thought Gerrard Winstanley.[9]

Women were also near the animal state. Over many centuries theologians had debated, half frivolously, half seriously, whether or not the female sex had souls, a discussion which closely paralleled the debate about animals and was sometimes echoed at a popular level. At Witley in Surrey in 1570, one Nicholas Woodies allegedly asserted that women had no souls; at Earls Colne, Essex, in 1588 the minister himself said the same; and in the diocese of Peterborough in 1614 a local wit was reported for 'avowing and obstinately defending that women have no souls, but their shoe soles'. The Quaker George Fox met a group of people who held that women had 'no souls, no more than a goose'.[10] Contemporary gynaecologists laid heavy emphasis on the animal aspects of child-bearing. A pregnant woman was commonly said to be 'breeding'; and one pre-Civil-War clergyman in the pulpit compared women to sows. Puritan opponents of the churching ceremony sometimes did the same, referring to the mother as a sow with her piglets following her.[11] Until the eighteenth century, the suckling of babies was usually regarded by the upper classes as a debasing activity, to be avoided if possible by putting infants out to the wet-nurse. Jane Austen was in a long tradition when she described her sex as 'poor animals', worn out by annual childbearing.[12]

Still more beastlike were the poor – ignorant, irreligious, squalid in their living conditions and notably lacking in some of the accomplishments supposed to be distinctively human: letters, numbers, manners and a developed sense of time. Intellectuals had long been in the habit of regarding the uneducated as subhuman.[13] In the early modern period the attitude lingered. 'The numerous rabble that seem to have the signatures of man in their faces,' explained Sir Thomas Pope Blount in 1693, 'are but brutes in their understanding ... 'tis by the favour of a metaphor we call them men, for at the best they are but Descartes's automata, moving frames and figures of men, and have nothing but their outsides to justify their titles to rationality.'[14] To other observers, the poor were 'the vile and brutish part of mankind'; their occupations were 'bestial' and they 'toiled like their horses'.[15] In his answer to the Lincolnshire rebels of 1536, King Henry VIII described the commonalty of Lincolnshire as 'one of the most brute and beastly of the whole realm'. In Elizabethan Pembrokeshire George Owen saw young people herding cattle, burned black by the sun, 'their skins all chapped like elephants'. The villagers of Tottington, Lancashire, were 'mere brutes', thought the local incumbent in 1696. In the Essex marshes in 1700 there were 'people of so abject and sordid a temper that they seem almost to have undergone poor Nebuchadnezzar's fate, and by conversing continually with the beasts to have learned their manners'. At Madeley, Shropshire, the vicar, John Fletcher, mused in 1772 on the condition of the bargemen:

> Fastened to their lines as horses to their traces, wherein do they differ from the laborious brutes? Not in an erect posture of the body, for, in the intenseness of their toil, they bend forward, their head is foremost, and their hands upon the ground. If there is any difference, it consists in this: horses are indulged with a collar to save their breasts; and these, as if theirs were not worth saving, draw without one; the beasts tug in silent patience and mutual ritual harmony; but the men with loud contention and horrible imprecations.[16]

Most beastlike of all were those on the margins of human society: the mad, who seemed to have been taken over by the wild beast within; and the vagrants, who followed no calling, but lived what the Puritan William Perkins called 'the life of a beast'.[17] It has been truly said that the image of animality haunted the madhouse.[18] The same image runs through contemporary indictments of the vagrants, who did not 'range themselves into families, but consorted together as beasts'. Beggars were also like brutes because they spent all day seeking food.[19]

Once perceived as beasts, people were liable to be treated accordingly. The ethic of human domination removed animals from the sphere of human concern. But it also legitimized the ill-treatment of those humans who were in a supposedly animal condition. In the colonies, slavery, with its markets, its brandings and its constant labour, was one way of dealing with men thought to be beastlike. The Portuguese, reported an English traveller, marked slaves 'as we do sheep, with a hot iron', and at the slave market in Constantinople, Fynes Moryson saw the buyers taking their slaves indoors to inspect them naked, handling them 'as we handle beasts, to know their fatness and strength'.[20] Slaves were often given names of the kind normally reserved for dogs and horses.[21] One eighteenth-century London goldsmith advertised 'silver padlocks for blacks or dogs'; and English advertisements for runaway negroes show that they often had collars around their necks.[22] Historians now think that black slavery preceded assertions of the negro's semi-animal status. Fully-developed theories of racial inferiority came later.[23]* But it is hard to believe that the system would ever have been tolerated if negroes had been credited with fully human attributes. Their dehumanization was a necessary precondition of their maltreatment.

At home animal domestication furnished many of the techniques for dealing with delinquency: bridles for scolding women; cages, chains and straw for madmen; halters for wives sold by auction in the market, in the widely accepted informal ritual of divorce.[24] The training of youth was frequently compared to the breaking of horses; and it was no accident that the emergence in the seventeenth and eighteenth centuries of more humane methods of horse-breaking would coincide with a reaction against the use of corporal punishment in education.†

Above all, the common people were repeatedly portrayed as animals who needed to be forcibly restrained if they were not to break out and become dangerous. The best way to deal with them, thought Timothy Nourse in 1700,

> will be to bridle them, and to make them feel the spur too, when they begin to play their tricks and kick. The saying of an English gentleman was much to the purpose, that three things ought always to be kept under: a mastiff dog, a stone horse [i.e. a stallion] and a clown; and really I think a snarling, cross-grained clown to be the most unlucky beast of [the] three.[25]

This was not typical of the way in which all members of the well-to-do classes regarded their inferiors. But neither was it unique; and a recently-discovered letter by the gentle Charles Lamb, undated but evidently written at a time of agricultural unrest,‡ reminds us of how long it survived.

> It was never good times in England since the poor began to speculate upon their condition. Formerly they jogged on with as little reflection

* See below, pp. 135–6.

† Cf. below, p. 189.

‡ Perhaps in 1822. Cf. *John Constable's Correspondence*, ed. R. B. Beckett (Suffolk Records Soc., 1962–8), vi. 88 ('never a night without seeing fires').

> as horses. The whistling ploughman went cheek by jowl with his brother that neighed. Now the biped carries a box of phosphorus in his leather breeches ... and half a county is grinning with new fires.[26]

Some anthropologists believe that it was the management of herds of domestic animals which first gave rise to an interventionist and manipulative conception of political life. Inhabitants of societies which, like those of Polynesia, lived by vegetable-gardening and growing crops which require relatively little human intervention seem to have taken a relatively unambitious view of the ruler's function. They believed that nature should be left to take its course and that men could be trusted to fend for themselves without regulation from above. But the domestication of animals generated a more authoritarian attitude.[27] In early modern England human rule over the lower creatures provided the mental analogue on which many political and social arrangements were based. Moreover, the two kinds of rule reinforced each other. The 'dominion' which God gave Adam over the animals, explained a Jacobean commentator, meant 'such a prevailing and possessing as a master hath over servants'.[28] Men enjoyed dominion over the lower creatures, but not all men. As a familiar proverb had it, 'The wisest of men saw it to be a great evil that servants should ride on horses.'[29]

Domestication thus became the archetypal pattern for other kinds of social subordination. The model was a paternal one, with the ruler a good shepherd, like the bishop with his pastoral staff. Loyal, docile animals obeying a considerate master were an example to all employees.

> Their faculties of mind are ... proportioned to this state of subjection [wrote an observer in 1758] ... they have knowledge peculiar to their several spheres, and sufficient for the underpart they are to act ... if they had a higher degree of knowledge ... they would be the plagues of mankind; they would repine, resent ... combine, rebel ... they would no longer endure their present necessary, and much happier, state of subordination.

This was not an eighteenth-century politician resisting a proposal for the education of the poor. It was a naturalist (William Borlase) discussing the lower animals. As Oliver Goldsmith wrote of the mole: 'A small degree of vision is sufficient for a creature that is ever destined to live in darkness. A more extensive sight would only have served to show the horrors of its prison.'[30]

The ideal of human ascendancy, therefore, had implications for men's relations to each other, no less than for their treatment of the natural world. Some men were seen as useful beasts, to be curbed, domesticated and kept docile; others were vermin and predators, to be eliminated. 'Let him bear the wolf's head,' they said in the thirteenth century of an outlaw. 'They act like wolves and are to be dealt withal as wolves,' remarked a clergyman in 1703 of the Indians in New England, in justification of their being hunted with dogs.[31] In Jacobean Scotland a Campbell chieftain offered the same reward for the head of a MacGregor as for the head of a wolf; and in Cromwellian Ireland the Tories were frequently compared to ravening wolves.[32] John Locke thought that an aggressor who ignored the dictates of human reason rendered himself liable to be destroyed like a beast. So in 1657 did the inhabitants of the Essex village of Great Horkesley: when, confronted by one Samuel Warner, 'the most dangerous, bloody villain in the county', who was said to have killed one man and assaulted another, they requested the authorities 'to tie him up, as they used to deal with savage beasts'.[33] It was in the same decade that, when the Quaker Edward Billing was attacked by a mob, 'a great one' said, 'trouble not a magistrate with him. Dash out his brains ... they are like dogs in time of plague. They are to be killed as they go up and down the streets, that they do not infect.'[34]

It was, therefore, a serious matter when controversialists tried to dehumanize their opponents, as when John Milton compared his enemies to 'owls and cuckoos, asses, apes and dogs'; or when the godly Nehemiah Wallington described the Royalists as 'tigers and bears for cruelty ... boars for waste and destruction ... swine for drunkenness ... wolves for greediness'.[35] From the medieval schoolman Albert the Great, who accused his opponents of blaspheming 'like brute beasts', to Karl Marx, who called Malthus a 'baboon', such language has been part of the tradition of European learned controversy.[36] In early modern England it was a regular weapon of religious and political polemic, whether used by Thomas More, who called William Tyndale 'a bold beast' and stigmatized his writings as 'a poisoned stinking tale of some stinking serpent', or by the Puritans, who denounced non-preaching clergymen as 'dumb dogs'. The bishops, said the anonymous author of the Marprelate Tracts, were 'hogs, dogs, wolves, foxes'.[37] Animal analogies were equally conspicuous in popular satire and abuse. Opponents of the Church's ceremonies frequently staged mock baptisms or funerals of cows, pigs, cats, dogs and horses.[38] And sometimes the Church's supporters retaliated in kind: in 1643 a Puritan triumphantly recorded the birth of a monstrous child to two Popish parents, a judgement on the grandmother, who some years earlier 'out of an inveterate malignity ... and in devilish derision' towards the famous victims of Archbishop Laud had named her three cats 'Bastwick', 'Burton' and 'Prynne', cutting off their ears, 'in desperate disdain of their glorious sufferings'.[39]

Animal insults remain a feature of human discourse today. But they have lost the force they possessed in an age when beasts enjoyed no claim to moral consideration. For to describe a man as a beast was to imply that he should be treated as such. The story of religious persecution in the early modern period makes it abundantly clear that, for those who committed acts of bloody atrocity, the dehumanization of their victims by reclassifying them as animals was often a necessary mental preliminary.[40]

Yet nearly all the protests which were made on behalf of the poor and oppressed in the early modern period were couched in terms of the very same ideology of human domination that was used to justify their oppression. Slavery was attacked because it confused the categories of beast and man,[41] while political tyranny was denounced on the grounds that it was wrong that human beings should be treated as if they were animals. In 1596 the rioters in Oxfordshire protested that servants were being 'held in and kept like dogs'. James Harrington thought the people in Scotland were oppressed because they were 'little better than the cattle of the nobility', while for Edmund Ludlow the main question in dispute during the Civil War was whether the King should govern his people by law, or rule them by force 'like beasts'.[42] 'Men,' as one opponent of monarchy put it in 1654, 'are not like sheep under a shepherd, where the dignity of the kind may justly challenge superiority and dominion over the inferior kind, in regard of the great difference of the species.' Reason ruled men, pronounced John Locke. Force was only for brutes.[43]

The common people themselves were always extremely sensitive to the suggestion that they were to be equated with their animal inferiors. When the first Duke of Buckingham took to being carried round in a sedan chair, there were objections to the immorality of his 'employing his fellow-creatures to do the service of beasts'; and in Victorian times it was said to be a 'barbarous practice' that at hiring-fairs 'men and women should stand in droves, like cattle, for inspection'. Parents dreaded lest their infants should die unbaptized and be committed to the earth 'like dogs'; hence the distress in 1539 when the French foster-parents refused to bury an English nurse-child who had died in their care, and returned it to Calais 'as if it had been a dead calf'.[44]

Much popular protest during the period consequently took the form of demanding that everyone should be admitted to share in that ascendancy over the lower creation which God had bestowed on mankind. 'All the land, trees, beasts, fish, fowl, etc.,' complained the radical author of *Light Shining in Buckinghamshire* (1648), 'are enclosed into a few mercenary hands.'[45] Many objected to the medieval forest law which restricted access to the royal game reserves; and they hated the statutes which, from the fourteenth century onwards, had confined the right to hunt game to those above a certain social level ('It is not fit,' James I explained, 'that clowns should have these sports.')[46] The poor too wanted to be able to kill the deer and shoot the birds. They accepted the private ownership of domestic animals, but they held to the old common-law view that there was no property in wild ones (*ferae naturae*) until they had been killed or tamed.

European jurists, like Grotius or Pufendorf, would devote much energy to refuting the belief of 'ignorant persons' that every man was entitled to a share in the God-given dominion over nature, but ordinary people remained unconvinced.[47] During the Civil War a group of Parliamentary troopers, quartered in Leamington, did much damage to Baron Trevor's dove-house. When their captain remonstrated,

> they answered him that pigeons were fowls of the air given to the sons of men, and all men had a common right in them that could get them, and they were as much theirs as the Baron's, and therefore they would kill them ... and not part from their right; upon which ... the Captain said he was so convinced with their arguments he could not answer them, and so came away, letting them do as they would.

In the eighteenth century the great lawyer William Blackstone confirmed that forest and game laws were 'both founded on the same unreasonable notions of permanent property in wild creatures'.[48] It is not surprising that poachers were often unrepentant when apprehended: 'the deer were wild beasts,' said a convicted deer-stealer in 1722, 'and ... the poor, as well as the rich, might lawfully use them.' Wild animals, birds and fish were God's gift to all men, 'everyone's property'.[49]

The main dispute during the period, therefore, was between those who held that all humanity had dominion over the creatures, and those who believed that human rights over inferior creatures should be confined to a privileged group. Disputes about the game laws did not lead on to doubts about man's right to hunt birds and animals, because the lower classes were as committed to the idea of human domination as anyone else.* After all, even labourers ruled over domestic animals, whom they could kick and curse when things went wrong. Farm animals were a sort of inferior class, reassuring the humblest rural worker that he was not at the absolute bottom of the social scale, a consolation which his industrial successor was to lack. The ox, as the Greeks used to say, was the poor man's slave; and even the poorest tinker had a dog at his heels on which to bestow the kick which indicated his superiority. The lower classes, thought Mary Wollstonecraft in 1792, domineered over animals 'to revenge the insults that they are obliged to bear from their superiors'.[50]

Yet the uncompromisingly aggressive view of man's place in the natural world which has been sketched out in this chapter was by no means representative of all opinion in early modern England. Not everyone thought that the world was made exclusively for man, that nature was to be feared and subjugated, that the inferior species had no rights or that the differences between man and beast were unbridgeable. On the contrary, reality was much more complicated than that. If we look below the surface we shall find many traces of guilt, unease and defensiveness about the treatment of animals; and many of the official attitudes which have been described so far were remote from the actual practice of many people. The rest of this book will try to do justice to other, more ambiguous, modes of thought and action.

* In the nineteenth century Fourier, Saint-Simon, Engels and other Socialists explicitly aimed at the full separation of man from the animal kingdom and his complete lordship over inferior species; the exploitation of man by man was to be replaced by the exploitation by man of nature. See Frank E. Manuel and Fritzie P. Manuel, *Utopian Thought in the Western World* (Oxford, 1979), 517, 604, 665–6, 707; Karl Marx and Frederick Engels, *Selected Works* (Moscow, 1951), ii. 140.

II
NATURAL HISTORY AND VULGAR ERRORS

This is a general and main error, running through all the conceptions of mankind, unless great heed be taken to prevent it, that what subject soever they speculate upon, whether it be of substances that have a superior nature to theirs, or whether it be of creatures inferior to them, they are still apt to bring them to their own standard, and to frame such conceptions of them as they would do of themselves.

Sir Kenelm Digby, *Two Treatises* (1645), i. 419.

i. CLASSIFICATION

It has been suggested so far that in early modern England it was conventional to regard the world as made for man and all other species as subordinate to his wishes. We must now see how this assumption was gradually eroded by a combination of developments, some of them already in operation when the period started, others emerging as time went on. Of these developments, the first was the growth of natural history, the scientific study of animals, birds and vegetation.

There was, of course, nothing new about the realization that the natural world had a life of its own, independent of human needs. The detached observation of nature had achieved an astonishing maturity in the work of Aristotle, and, although the literary culture of the subsequent fifteen hundred years added very little to them, so far as botany and zoology were concerned, there were plenty of people in medieval England who observed the natural world very carefully. In the late twelfth century Gerald of Wales recorded extremely accurate descriptions of fish and birds, while in the fifteenth century the antiquary William of Worcester gave the nesting habits of birds similarly close attention. Striking evidence of the direct perception of nature is to be found in the meticulous depictions of foliage in the thirteenth-century sculpture of Southwell Minster, the careful representation of flowers in medieval embroidery, and the marvellous fourteenth-century coloured drawings of birds in the Bird Psalter at the Fitzwilliam Museum and the sketchbook in the Pepysian Library.[1] In 1753, when the Society of Antiquaries was shown an illustrated fifteenth-century missal, its members concluded that 'the insects and flowers were as if painted by a profess'd naturalist'.[2]

Yet the naturalism of English medieval art was only sporadic. Usually the intentions of the artists were purely emblematic; and older models or stereotypes often prove to lie behind even the most 'realistic' of their achievements.[3] Official book-learning was very largely unaffected by direct experience of the natural world. It was only in Tudor times that there began an unbroken succession of active field naturalists, running from William Turner (born in 1508) to John Ray (who died in 1705). Members of a wider European scientific fraternity, it was they who, by their cumulative labours, searching for plants, listing and describing wild creatures, and corresponding with continental naturalists, laid the foundations of modern botany, zoology, ornithology, and the other life sciences.

There is no need to re-tell that well-known story.[4] But it is necessary to emphasize its relevance to the theme of this book. For all observation of the natural world involves the use of mental categories with which we, the observers, classify and order the otherwise incomprehensible mass of phenomena around us; and it is notorious that, once these categories have been learned, it is very difficult for us to see the world in any other way. The prevailing system of classification takes possession of us, shaping our perception and thereby our behaviour. What was important about the early modern naturalists was that they developed a novel way of looking at things, a new system of classification and one which was more detached, more objective, less man-centred than that of the past. In turn, this new way of looking at nature had a marked impact upon the perceptions of non-scientists and ordinary people, ultimately proving destructive of many popular assumptions. By 1800 it had become possible to regard plants and animals in a light which was very different from the anthropocentric vision of earlier times.

Yet the development was a very gradual one. For, at the start of the early modern period, even the naturalists themselves regarded the world from an essentially human viewpoint and tended to classify it less according to its intrinsic qualities than according to its relationship to man. Plants, for example, were studied primarily for the sake of their human uses, and perceived accordingly. There were seven kinds of herb, thought William Coles in 1656: pot herbs; medical herbs; corn; pulse; flowers; grass; and weeds.[5] The illustrated herbals of the Tudor and Stuart period were meant to facilitate the instant recognition of any plant. Sometimes their authors merely listed the plants alphabetically, but often they followed the classical writers Theophrastus, Dioscorides and Pliny in distinguishing them according to their taste, their smell, their edibility and above all their medicinal value, often subdividing them according to the part of the body they could heal. John Parkinson in 1640 used such categories as 'sweet-smelling', 'purging', 'venomous, sleepy and hurtful' and 'strange and outlandish', while *The Grete Herball* of 1526 briskly divided mushrooms into two kinds: 'one ... is deadly and slayeth them that eateth of them; ... and the other doth not.'[6] As in Anglo-Saxon times, the main impulse to botanical study was medical. It was no accident that nearly all the early botanists were doctors or apothecaries, preoccupied with the plants' uses and 'virtues'. The opening up of the New World intensified the search for plants which were medically useful; and it was in so-called 'physic gardens' that the new species were cultivated. The practical utility of the plant world to mankind long provided the botanists with their essential organizing principle.

Animals were often perceived in the same way. Early modern zoologists inherited from Aristotle the practice of classifying beasts according to their anatomical structure, their habitat and their mode of reproduction. But they also considered their utility to man, and their value as food and medicine and as moral symbols. The most 'natural' order for arranging animals, thought the great eighteenth-century French naturalist Buffon, was according to the degree of their relationship to man; and he, like his predecessors the Zürich naturalist Konrad Gesner and the Bolognese Ulisse Aldrovandi, devoted most attention to those animals which, like the dog and the horse, were of greatest interest to humans. When in 1607 Edward Topsell, the divine, published his *Historie of Foure-Footed Beasts* he explained unashamedly that his main purpose was to show which beasts were the friends of man; which could be trusted; and which could be eaten.[7]

Essentially, there were three categories for animals: edible and inedible; wild and tame; useful and useless. Of these, the first was the most fundamental, though the least discussed. Early modern England had no equivalent to the formal dietary rules, set out in Leviticus, which had regulated the practice of ancient Israel. In the eighth century the Church had issued prohibitions against eating jackdaws, crows, storks, hares, beavers and horses, the latter because of their association with a pagan cult.[8] But in the Reformation era it became axiomatic that all meats were lawful and that dietary habits were not a matter of religion. The distinction between clean and unclean foods had been abrogated by the coming of Christ; to the pure all things were pure.[9] Only a few Lollards and sectaries maintained the doctrine that it was sinful to eat pork, either because of the Old Testament prohibition or because Jesus had sent evil spirits into swine.[10]*

Implicit prohibitions of equal force, however, underlay the everyday eating habits of the English people; and they were reflected in the classifications of the naturalists. Most often, it was the nature of the particular creature's diet which determined its edible status. The animals most generally eaten were the vegetarians, feeding on grass or berries, whereas carnivorous beasts or scavengers, who ate carrion or excrement, were rejected as unclean (hence perhaps the enduring uncertainty about pigs).[11] So also were those birds which, like woodpeckers or swallows, lived wholly on insects. A normally carnivorous creature when fed on a vegetarian diet could change its status and become edible, like young rooks, which, being reared on grain, could be put into pies, or even rats, which Jamaican planters were prepared to cook when they had been fed on sugar in the canefields. Captain Cook and his companions found that the South Sea dog, being fed on vegetables, was, when cooked, little inferior to English lamb.[12]

The animal's diet was not the only consideration, however. There had long been a prejudice against killing indispensable working animals for food: horses, dogs and, in parts of Europe, oxen, all came under this prohibition. In the seventeenth century the objection to eating oxen survived in many Mediterranean countries and also in Ireland, where, it was said, 'the common sort' would never kill a cow. The rise of the cult of the roast beef of England closely paralleled the decline of the ox as a working animal.[13] The ancient Britons had held it wrong to eat hares, cocks or geese because they were creatures kept for pleasure and amusement; and doubts about the edibility of hares long continued, it being widely asserted that a pregnant woman who ate hare's flesh would bear a child with a hare-lip.[14]

There was an equal distaste for meat which was in appearance too reminiscent of human flesh: many travellers reported that monkeys were delicious, but, as the naturalist William Bingley observed, there was 'something extremely disgusting in the idea of eating what appears, when skinned and dressed, so like a child. The skull, the paws, and indeed every part of them remind us, much too strongly, of the idea of

* It was notorious in the seventeenth century that the Highland Scots would not eat pig's flesh, and the taboo survived in some parts of Scotland until very modern times; Walter Blith, *The English Improver Improved* (1652), 145–6; [John Dunton], *The Athenian Oracle* (1703–4), ii. 360; T. C. Smout, *A History of the Scottish People, 1560–1830* (1969), 132.

devouring a fellow-creature.'[15] Finally, there was an objection to eating those creatures which were thought to have been engendered out of putrefaction: what an early seventeenth-century doctor called 'the excrements of the earth, the slime and scum of the water, the superfluity of the woods and the putrefaction of the sea: to wit ... frogs, snails, mushrooms and oysters'. The lack of inhibition displayed by the French and Italians towards these commodities was an enduring subject of contemptuous comment.[16]

Each of these prohibitions was justified in terms of the prevailing humoral physiology. Topsell, for example, explained that rapacious creatures engendered melancholy in those who ate them; he condemned the 'wanton eating of frogs' as 'perilous to life and health'. Others rationalized the prejudice against horse-meat by saying that it was bad for the digestion.[17] In practice, however, the categories were far from rigid and the classification of what was or was not edible shifted slightly during the early modern period. As the eighteenth-century naturalist Thomas Pennant remarked, 'it would be curious to trace the revolutions of fashion in the article of eatables. What epicure first rejected the seagull and heron? And what delicate stomach first nauseated the greasy flesh of the porpoise?' A nineteenth-century editor of the household regulations of the fifth Earl of Northumberland (1512) observed that 'in the lists of birds here served up to the table are many fowls which are now discarded as little better than rank carrion'.[18] Herons and cranes, once the dish of kings, were by this time regarded as fit only for the cat. Yet in the seventeenth and eighteenth centuries the list of creatures which at least some authorities regarded as suitable for human consumption included squirrels, badgers, seals, owls, hedgehogs, otters and tortoises; and when Pennant's contemporary Thomas Gray drew up a list of birds which it was customary to eat, he included rooks, lapwings, puffins and curlews.[19] In the last resort, poverty would break down most inhibitions. 'Of fleshes,' wrote Francis Bacon, 'some are edible; some, except it be in famine, are not.' In times of scarcity the eighteenth-century Cornish would eat seals, just as English travellers noted that the Italians, being short of food, ate starlings and other 'birds which we esteem unwholesome'.[20]

The second way of classifying animals was in terms of their general utility to human beings. Thus Dr John Caius in his book *Of English Dogges* (translated and expanded by Abraham Fleming in 1576) divides dogs into three categories: a 'generous' kind, used in hunting or by fine ladies; a 'rustic' kind, used for necessary tasks; and a 'degenerate', currish kind, used as turnspits and for other menial purposes. The hunting kind is then subdivided according to whether it excels at smelling, spying, speed or subtlety; and whether it is better against beasts or birds and, if against birds, whether water birds or land birds. The rustic kind is subdivided into sheepdogs, watchdogs, etc., watchdogs being further subdivided according to whether they bark or whether they bite, or, if they do both, whether they bark before they bite or bite before they bark. This way of classifying dogs by their human uses would survive even in the work of the great Swedish naturalist Linnaeus, who in the mid eighteenth century distinguished the faithful dog (*Canis familiaris*) from the wolf and the wild dog, and subdivided him into such varieties as the sheepdog (*Canis domesticus*) and the turnspit (*Canis vertegus*).[21]

The third distinction was that between 'wild' and 'tame'. This criterion did not have the unequivocal authority of Aristotle since, in his opinion, its application would have subdivided some species, notably that of man.[22] But it was popular with the naturalists. In 1661, for example, Robert Lovell divided his class of viviparous *digitales* (mammals with toes) into the 'wild' (e.g. tigers and wolves), the 'wildish' (e.g. foxes, apes) and the 'domestic' (cats and dogs).[23] Bees, observed Thomas Muffett in his book on insects, were 'neither wild nor tame', but 'of a middle nature'.[24] As has already been seen, this distinction was important for lawyers, who ruled that no property was possible in animals which were wild (*ferae naturae*) unless they had been killed or tamed by human industry.* In practice, only those animals domesticated for draught or for food were deemed private property; there could, by contrast, be no property in a dog because, as a judge put it in 1521, a dog was 'vermin and savage by nature'. This view, however, was not taken by his colleagues at the time; and the case of *Ireland v. Higgins* (1588) confirmed that there could be property in greyhounds and other dogs. Sir Edward Coke, who seems to have been more reluctant to concede property in animals than were some of his contemporaries, reasserted the traditional doctrine, to the extent of denying that there could be larceny of mastiffs, bloodhounds or other dogs; they were, he held, so 'base' that no man should lose his life or limb for stealing them. He confined full property in animals to those creatures which could be tamed for food or were intrinsically 'noble and generous', like falcons. The more general view, however, was that beasts, even monkeys or muskrats, could, when tamed, become the property of men, but that they ceased to be property if they reverted

* Above, p. 49.

to their original wild state.[25] Even today, English lawyers impose human criteria upon animals by dividing species into 'dangerous' and 'non dangerous', though admitting that 'such a division is not to be found in nature'.[26]

In the early modern period animals were further categorized according to whether or not men found them physically handsome. Thus Topsell held that 'the body of an ape is ridiculous', 'by reason of an indecent likeness and imitation of man'. The frog, said Bishop Babington, is, 'as we all know, a foul and filthy creature, abiding in foul places, as bogs and miry plashes, all the day long, and at night peeping out with the head above the water, making a hateful noise with many others of his sort till the day appear again'. The cormorant, explained Thomas Pennant, 'has the rankest and most disagreeable smell of any bird. Its form is disgusting, its voice hoarse and croaking, and its qualities base.'[27] The toad, the swine, the rat were all of 'an hateful aspect', agreed Henry More. Spiders and caterpillars were so 'loathsome' that some ladies would scream at the sight of them.[28] A seventeenth-century writer remarked of the bullhead (or miller's thumb) that it was 'an excellent fish for taste, but of so ill a shape that many women care not for dressing him, he so much resembles a toad'.[29]

Reptiles, insects and amphibians were especially detested, though the reasons for this loathing were seldom clearly articulated.[30] Modern anthropologists suggest that the explanation lies in their anomalous status. Fish were creatures which lived only in water; birds flew in the air, had two legs and laid eggs; beasts had four legs and lived on land. But many reptiles and insects moved ambiguously between earth, air and water, while snakes, though land animals, laid eggs and had no legs.[31]

Whatever the cause of the popular prejudice, it was one from which even the naturalists found it hard to emancipate themselves. John Ray never really liked snakes: 'I have such a natural abhorrency of that sort of animal that I was not very inquisitive after them.' In 1721 Richard Bradley, later Professor of Botany at Cambridge, thought the female frog a 'despicable, loathsome, little animal'. 'Many persons, of which number I am one,' confessed Oliver Goldsmith in his *History of ... Animated Nature*, 'have an invincible aversion to caterpillars and worms of every species; there is something disagreeable in their slow crawling motion for which the variety of their colouring can never compensate.' Thomas Pennant had seen frogs displayed in French markets, but his 'strong dislike to these reptiles prevented a close examination'. At the end of the eighteenth century the popular zoologist

William Bingley was still using this kind of language. 'When we examine the human louse with the microscope, its external deformity excites disgust.' As for the egret monkey, 'few animals are more dirty, ugly, or loathsome'.[32] Sometimes these sensibilities were shared by the animals themselves. The elephant, reported a Stuart preacher, was unwilling to enter water and, if forced in, would deliberately muddy the stream so as to be unable to discern the reflection of his deformity.[33]

Other species were praised for their fine appearance. John Hill in his *Natural History* (1751) commended the little owl, 'an extremely pretty little bird', and the red charr, 'a very elegant fish'. Even the jackal was 'a very beautiful creature', by contrast with the bear, 'an uncouth and unsightly animal'.[34] In the seventeenth century Robert Lovell divided birds into 'melodious' and 'not melodious'; over a hundred years later Bingley elaborated on this criterion in a complicated table which assessed the song of each British species under the respective heads of 'melodiousness of tone', 'sprightliness', 'plaintiveness', 'compassion' and 'execution', giving so many marks out of twenty for each (rather like guides to vintages of claret: '9 = very good; 0 = no good').[35]

Prevailing classifications also reflected the social assumptions of the time. Aristotle had popularized the notion that animals differed in character, some being noble and others mean. The naturalists preserved this distinction, even the pioneer seventeenth-century ornithologist Francis Willoughby dividing landfowl into the 'more generous', like eagles, and the 'more cowardly and sluggish', like buzzards.[36] It was only among the vulgar, explained a seventeenth-century writer, that blackbirds were esteemed as singing-birds; their song was loud and coarse, their speech 'boorish' and their condition 'rustic'.[37] The law of the land, as we have seen, similarly distinguished between animals 'of a base nature', like ferrets, mastiffs and cats, and those which were 'noble and generous', like falcons.* It also divided wild animals into 'beasts of forest, chase and warren', which were to be hunted in a ritual and socially exclusive manner, and 'beasts of prey' or 'vermin', which could be exterminated by anyone, anywhere. As Oliver St John said during the Earl of Strafford's trial in 1641: 'we give law [i.e. a chance to escape] to hares and deers, because they be beasts of chase.' But 'it was never accounted either cruelty or foul play to knock foxes and wolves on the head, as they can be found; because these be beasts of prey'.[38] In fact, he was wrong, because wolves and foxes were both

* Above, p. 56.

privileged beasts at this period.* But his meaning was plain enough.

The selective breeding of domestic stock had long established a similarly hierarchical view of cattle and horses. In the later fourteenth century a horse might be worth anything from two shillings to £50, depending on its breed and condition.[39] By the later seventeenth century the range was even greater: as the historian of the subject remarks, there was now 'a highly differentiated demand for highly differentiated animals at a wide range of prices'.[40] Horses were used in war, in hunting, in agriculture and in industry. They pulled carts, wagons and (from the 1560s onwards) private coaches. They were also indispensable means of private transport. For each of these tasks different qualities were required. The demands of the economy led inexorably to the development of a specialized system of breeding horses for use in farming, industry and trade, while the government, from the reign of Henry VIII onwards, issued many instructions designed to encourage the breeding of horses for war. In pursuit of these different ends there was much experiment with imported breeds from Europe, North Africa and the Levant. The most effective stimulus to careful horse-breeding proved to be the rise of organized horse-racing, in which the gentry participated with increasing enthusiasm from the late Elizabethan period onwards.† By the later seventeenth century the thoroughbred racehorse had become an aristocratic obsession. Its strength, speed and courage symbolized the superior status of its owner; and a noble family's stud books and pedigrees were maintained with a precision which would have done credit to the College of Arms and probably exceeded that bestowed upon many parish registers.[41] To their owners, the great oriental progenitors Byerley Turk (in action by 1690), Darley Arabian (imported 1704) and Godolphin Arab (1724–53)‡ were a kind of equine Adam, Noah or William the Conqueror. One gentleman, Nathaniel Harley, said in 1715 that he would not sell

* According to the Elizabethan lawyer John Manwood, the beasts of forest were red deer, boars and wolves; beasts of chase were fallow deer, foxes, martens and roes; and the beasts and fowls of warren were hares, rabbits, pheasants and partridges; *A Treatise and Discourse of the Lawes of the Forrest* (1598), fols. 21–2ᵛ. Sir Edward Coke included foxes among the beasts of the forest; *Institutes of the Laws of England* (1794–1817 edn), iv, chap. 73. St John's error was noticed by Sir Peter Pett, *The Happy Future State of England* (1688), 8.

† In 1613 the English envoy to Venice, Sir Henry Wotton, could best convey the character of the Venetian regatta by describing it as 'a horse-race of boats'; John Walter Stoye, *English Travellers Abroad, 1604–1667* (1952), 142n.

‡ Respectively great-great-grandfather of Herod (b. 1758), great-great-grandfather of Eclipse (b. 1764), and grandfather of Matchem (b. 1748).

his Dun Arabian imported from Aleppo for a thousand guineas.[42]

By the eighteenth century, cattle, sheep, foxhounds and even pigeons* were being bred with comparable attention.[43] The outlook was ruthlessly eugenic: 'As soon as the bitch hath littered,' explained a seventeenth-century handbook, 'it is requisite to choose them you intend to preserve, and throw away the rest'; the kennel book of a Yorkshire dog-breeder (1691–1720) contains such laconic entries as: 'three of this litter given to Br. Thornhill, the rest hanged, because not liked.'[44] Emphasis on breeding, stock and ancestry thus generated an extremely hierarchical attitude to domestic animals. As early as 1609 the Bishop of Coventry could rule that he wanted no 'riff-raff horses' in his market, while a later writer observed that a country gentleman

> will tell you, 'that his best-bred creatures, and of the truest race, are ever the noblest and most generous in their natures; that it is this chiefly which makes a difference between the horses of good blood and the errant jade of a base breed; between the game cock and the dunghill craven; ... and between the right mastiff, hound or spaniel, and the very mongrel.'[45]

Blood was important; there was a social hierarchy among animals no less than men, the one reinforcing the other.[46]

The whole natural world indeed was conventionally assumed to be ordered in a hierarchical scale, moving up from man to the angels and descending from him in what were regarded as diminishing degrees of perfection.[47] In popular imagery the arrangement was monarchical, with the lion, the eagle and the whale standing at the head of each respective order of being, though late-seventeenth-century scientists spent much time discussing whether the elephant or the ape or the beaver or the dolphin came at the top of the animal hierarchy.[48] One of the aims of zoology, thought an early-nineteenth-century writer, was to determine 'the station which an animal holds in the scale of creation'.[49] Such a picture of the natural world had obvious implications for human society, and there was at least some symbolic truth in the story that King Henry VII once ordered the execution of all mastiffs, after they had baited a lion, 'being deeply displeased ... that an ill-

* Richard Atherton, a Lancashire gentleman who was 'a very complete judge of a pigeon', planned to build 'a stately house ... on the top of which he design'd to have four turrets in which his pigeons were to be dispos'd according to the nearness of relation between the different species'; he died in 1726 before completing the work. See John Moore, *Columbarium: or, the Pigeon-House. Being an Introduction to a Natural History of Tame Pigeons* (1735), viii.

favoured rascal cur should with such violent villainy assault the valiant lion, king of all beasts'.[50]

The work of many anthropologists suggests that it is an enduring tendency of human thought to project upon the natural world (and particularly the animal kingdom) categories and values derived from human society and then to serve them back as a critique or reinforcement of the human order, justifying some particular social or political arrangement on the grounds that it is somehow more 'natural' than any alternative.[51] The diversity of animal species has been used on innumerable occasions to provide conceptual support for social differentiation among humans; and there can have been few societies where 'nature' has never been appealed to for legitimation and justification. It would surely be an exaggeration to maintain, as do some Marxists, that '*all* statements about nature ... express aspects of the social order',[52] but some of them certainly do, and never more than in the early modern period, when the universal belief in analogy and correspondence made it normal to discern in the animal world a mirror image of human social and political organization. For it was not merely the hierarchy of natural species which was invoked to justify social inequalities with the human species. Even within individual natural species there were believed to be social and political divisions closely paralleling those in the human world.

Ants, for example, were known to live in a tightly-governed commonwealth, in which the whole community joined together for the common interest.[53] Cranes, noted Henry More, had a captain who watched over them when the others slept.[54] Rooks had a 'parliament', which used to order the execution of delinquents.* Many birds, thought Oliver Goldsmith, observed 'general laws' and had 'a kind of republican form of government established among them'. Storks were allegedly so attached to republicanism that, when one was shot on the Norfolk coast, 'some there were who took it for an evil omen, saying, "If storks come over into England, God send that a commonwealth doth not come after."'[55] Beavers too exemplified the 'model of a pure and perfect republic'.[56] Even the baboons of Africa were reported to live in 'a kind of commonwealth'.[57] Among the peewits of Shebben Pool in Staffordshire, there was observed

* As late as 1917 the author of a letter to the *Morning Post* reported the annual meeting of starlings in the plane trees of the Temple in London: 'Year after year they gather in the middle of September, and sometimes they have resolutions and amendments which cause the meeting to be adjourned for a couple of days'; cited by T. S. Hawkins, *The Soul of an Animal* (n.d.), 20.

> a certain old one that seems to be somewhat more concern'd than the rest, being clamorous and striking down upon the very heads of the men [who tried to catch them]; which has given ground of suspicion that they have some government amongst them, and that this is their prince, that is so much concern'd for its subjects.[58]

The occasional existence of lawless animals only reinforced the point. The wild dogs of Puerto Rico, thought an English traveller, were 'a notable instruction to man ... how easily he may grow wild, if once he begin to like better of licentious anarchy than of wholesome obedience'. Another bad example was provided by the Cornish chough: 'an incendiary thievish bird, often setting fire to houses and stealing and hiding small money.'[59]

The most impressive monarchical community was that of the bees. The ancient parallel between human society and the beehive was never more popular than in the Stuart period, when numerous published treatises on bee-keeping gave as much attention to the insects' political virtues as to their practical utility. 'The subject here treated of,' said Joseph Warder in one of them, 'is of Princes and Potentates, Kingdoms and Territories, Prerogative and Property, Dominion and Loyalty, War and Peace.'[60] Bees, according to one authority, governed themselves by 'a regal power or civil discipline answerable to our martial laws'. They were ruled, noticed another, by 'a fair and stately bee, having a majestic gait and aspect'.[61] Writers laid heavy emphasis on the hive's monarchical structure,[62] though the embarrassing discovery that their monarch was not a king, as had always been assumed, but a queen, remained controversial until the 1740s. 'A Queen-Bee,' explained an encyclopedia in 1753, was the 'term given by late writers to what used to be called the King-Bee.'[63]* One book on bees came out in 1657, the

* Aristotle (*Hist. An.*, 553) and Pliny (*Nat. Hist.*, ii. 5) had both said that the large bee was male. This view was reiterated by English writers until the mid eighteenth century, though the belief that the bees' ruler was a queen was occasionally held on intuitive grounds, e.g. by Charles Butler, *The Feminine Monarchie* (1609), and Samuel Purchas, *A Theatre of Politicall Flying-Insects* (1657). It was the Dutch entomologist Swammerdam (d. 1685) who conclusively proved the large bee's female sex, but his work was not published until 1740 and then only in Dutch and Latin. His findings were disseminated by René Antoine Ferchault de Réaumur in 1741 and in an English translation of Gilles Augustin Bazin, *The Natural History of the Bees* (1744). When John Thorley upheld the new view in 1744, he had to defend it against the opinions of some 'modern writers'; *ΜΕΛΙΣΣΗΛΟΓΙΑ or, the Female Monarchy* (1744), 75.

year when Cromwell was offered the Crown;[64] another appeared in 1679, the first year of the Exclusion Crisis. It was by Moses Rusden, the royal bee-keeper, and it offered a salutary warning to 'such as look with a malicious eye upon kingly government as being the effect of necessity and force, and not of a natural inclination'.[65] In every hive, said an early Hanoverian clergyman, 'there are a queen, the nobles and the commonalty, acting all in their several places, and the meanest doing their duty with as much cheerfulness as the greatest. There are no murmurers nor complainers amongst them, no schismatics nor separatists ... Would to God we men were but as wise.' Throughout the eighteenth century this hierarchy in nature was invoked to defend hierarchy in human society.[66]

Social insects like bees were particularly popular with the commentators because of their intricate political order. But creatures of every kind could exemplify accepted morality; and as that morality changed so did men's perception of them. First, the animals were looked to for their political organization. 'God endued not only men, but all creatures, with their natural propensity to monarchy,' thought Sir Robert Filmer;[67] levelling principles, declared Joseph Caryl, could be confuted by an appeal to the hierarchy in nature.[68] Then came increasing emphasis on the bourgeois virtues of hard work, diligence and frugality; hence the seventeenth-century obsession with bees and ants.[69] With the sentimentalizing of family relationships, more attention was paid to the domestic virtues: in wild animals commentators saw maternal affection, marital fidelity, piety to aged parents.[70] A growing interest in cleanliness brought into favour such creatures as the badger, who, observed Pennant, always left its quarters when obeying the call of nature.[71] By 1807 new standards of schoolboy honour made it possible for an observer to overhear a tame raven say to three birds fighting: 'Fair play, gentlemen! Fair play! For God's sake gentlemen, fair play. Two against one will never do.'[72] Early-nineteenth-century naturalists studied birds for their reassuring family relationships: 'the sober, domestic attachments of the hedge-sparrow'; the 'maternal care and intelligence' of the titmouse. The popular Victorian writer Mrs Brightwen describes the hedge-sparrow as a 'quiet, gentle bird, the type of a very homely person, always in the path of duty and never interfering with other people'.[73]* In such ways did the world of wild life

* One of her contemporaries, however, saw sparrows as 'the Irishmen of birds, with their noise and their squabbles, their boldness and ubiquity'; Charlotte M. Yonge, *An Old Woman's Outlook in a Hampshire Village* (1896), 16.

provide a mirror-image of human relationships. Descriptions of nature constantly involved the use of metaphors derived from the social organization of the day.

This tendency to see in each species some socially relevant human quality was very ancient, for men had always looked to animals to provide categories with which to describe themselves.[74] The characters attributed to individual beasts were usually stereotypes, based less on observation than literary inheritance. They were derived from Greek, Roman and medieval compilations, not from scrutiny of life in the fields and the woods. For centuries the fox continued to be cunning, the goat lustful and the ant provident. In the works of Goldsmith and other eighteenth-century popularizers of natural history, the pig was still filthy and 'disgusting', the tiger 'cruel', the snake 'treacherous' and the weasel 'cruel, voracious and cowardly'.[75]

In heraldry, pageantry and artistic symbolism, the creatures thus continued to provide a vocabulary and a set of categories with which human qualities could be described and classified. Many preachers held that this was precisely why they had been created: to teach moral lessons to man. 'God would have a lively image of virtues and vices to be in the creatures,' explained the Elizabethan preacher Thomas Wilcox, 'that even in them we might be provoked to virtue and deterred from vice.' 'A good natural philosopher,' agreed the Hanoverian physician George Cheyne, 'might show with great reason and probability that there is scarce beast, bird, reptile nor insect that does not in each particular climate instruct and admonish mankind of some necessary truth for their happiness either in body or mind.'[76] In the medieval bestiaries the animal kingdom had been a collection of types and symbols conveying Christian doctrine; and this mode of emblematic thinking was kept alive in the seventeenth century by the great vogue for emblem books, with their allegorical pictures and accompanying verses.[77] The world was a cryptogram full of hidden meanings for man, but awaiting decipherment. Thus the fly was a reminder of the shortness of life, and the glow-worm of the light of the Holy Spirit. The mole symbolized the blind Papist, unable to see his way out of error; and the caterpillar was an emblem of the Resurrection.[78] The wilderness of America, thought Roger Williams, founder of Rhode Island, was 'a clear resemblance of the world, where greedy and furious men persecute and devour the harmless and innocent as the wild beasts pursue and devour the hinds and roes'. It was in this spirit that the diarist Ralph Josselin thought of Satan as 'the lapwing crying before me with ... temptation or vanity to draw my mind from ...

God'.[79] The contemporary capacity to invest the natural world with symbolic meaning for human life was almost infinite.

Against this long-established tendency of men to see animals and plants as mere symbols of themselves, we should place the search for new and more objective principles of classification which dominated the scientific botany and zoology of the early modern period. After first endeavouring to identify the modern counterparts of the plants described by Dioscorides and other classical authorities, European botanists embarked upon a more ambitious attempt to range the whole plant world in order.[80] What is notable about their work is that they tried increasingly to group plants, not alphabetically or according to their human uses, but by relation to their intrinsic structural characteristics. Most of these systems were 'artificial' in the sense that they arbitrarily concentrated on one external visible feature, whether the character of the leaves (Mathias de l'Obel, 1538–1616) or the fruit (Andrea Cesalpino, 1519–1603) or the flower (A. Q. Bachmann (Rivinus), 1652–1723), instead of attempting a 'natural' classification, based on the overall similarities between plants. But they showed a growing awareness of the natural affinities between species and they made diminishing use of criteria based on the plant's uses or relationship to man.

Each of these classificatory schemes represented an ambitious attempt to impose a new form of intellectual order upon the natural world, to reduce 'all kinds of animals and vegetables into method', as a contemporary put it.[81] Most of them involved a principle of identification whereby any individual specimen could be recognized, and some offered a table of classes, orders, genera, species and varieties into which it could be fitted. The most important English contribution to this European effort was made by John Ray in the late seventeenth century. Ray was indebted to the Italian Andrea Cesalpino, who had, a century earlier, tried to devise an artificial system which concentrated on the plant's parts of fructification. Ray's own system was a natural one, which, though starting with the seed, attempted to take into account the plant's whole structure.[82] It stimulated a new bout of European system-making, but ultimately prevailed in England until superseded by that of the Swede Linnaeus. The Linnaean system was developed from 1735, and accepted in this country in the early 1760s.[83] It was an artificial one, based, as far as plants were concerned, on the number, situation and proportion of the parts of fructification, the stamens and pistils; and its heavy sexual emphasis excited many prudish objections to its supposedly 'licentious' character. Botany seemed a doubtful recreation for young ladies when it involved so close a scrutiny of the 'private parts' of wild flowers.[84]* From about 1810 or so the Linnaean system in turn gave way to other, more 'natural' schemes.

These competing classifications did not by any means abandon the old analogy between the human and the natural worlds. All of them had inescapably hierarchical implications; and there was an obvious parallel between the descending categories of scientific taxonomy and the diminishing units of human society. The Linnaean system, as propounded in late-eighteenth-century England, pursued the parallel very closely indeed. The 'Vegetable Kingdom' was divided into 'Tribes' and 'Nations', the latter bearing titles which were more sociological than botanical: the grasses were 'plebeians' – 'the more they are taxed and trod upon, the more they multiply'; the lilies were 'patricians' – 'they amuse the eye and adorn the vegetable kingdom with the splendour of courts'; the mosses were 'servants', who 'collect for the benefit of others the daedal soil'; the flags were 'slaves' – 'squalid, revivescent, abstemious, almost naked'; and the funguses were 'vagabonds' – 'barbarous, naked, putrescent, rapacious, voracious'.[85] The assimilation of the natural world to human society could hardly have been closer.

Nevertheless, for all their anthropomorphic tendency, these new modes of classification represented an important attempt to escape from the old man-centred viewpoint. For, instead of assessing the plant's edibility, beauty, usefulness or moral status, all of which came ultimately to be regarded as wholly irrelevant, the naturalists were seeking its intrinsic qualities; structure alone was the source of distinction between species.[86] The process of change was slow, for in the late eighteenth century there were still those who wanted a natural system of classification because they thought it would provide a more reliable key to the uses and virtues of medicinal plants.[87] But the eventual outcome was the emergence of a totally new mode of perception. As a modern commentator remarks, 'it was indeed a revolution in the manner of understanding plants when, instead of describing their usefulness, abundance, size, smell and colour, attention was given exclusively to the disposition and form of the parts of the flower and seed.'[88] By the later seventeenth century botany was ceasing to be a mere branch of medicine; plants were now increasingly studied for their own sake. Of course, the practical uses of plants were still important, but, as the followers of Linnaeus would emphasize, a naturalist was not the same

* 'With these obscene processes and prurient apparitions,' John Ruskin would write, 'the gentle and happy scholar of flowers has nothing whatever to do'; *Proserpina* (1875–86), in *Works*, ed. E. T. Cook and Alexander Wedderburn (1903–12), xxv. 390–91.

thing as a physician, a chemist, a farmer or a gardener, and his principles of classification were different.[89]

In zoology the revolution was less obvious, for Aristotle had already laid the foundation of a classification based on the animal's physical structure. But more attention was now paid to the creature's internal anatomy, as opposed to its external characteristics; the distinction between 'wild' animals and 'tame' ones was abandoned; and utilitarian considerations became distinctly less conspicuous in zoological writing.[90] During the course of the eighteenth century some commentators even came to accept that all systems of classification were artificial, man-made contrivances. Nature herself knew nothing of classes, orders, genera and species: only individuals existed.[91] In the nineteenth century, of course, the timeless mode of classification exemplified by Linnaeus would give way to a more dynamic form of biological science in which descent became the essential criterion and the immutability of species was no longer assumed. In the meantime there was inevitably some tension between the assumption of an unchanging natural order and the increasing awareness of the fact of change in human society.[92]

The scientists thus gradually rejected the man-centred symbolism which had been so central to earlier natural history. Francis Bacon had observed that the emblematic meanings conventionally given the creatures were not inherent in them, but only human inventions. But it was John Ray and his friend Francis Willoughby who were the first English naturalists to emancipate themselves explicitly from the emblematic tradition.

> We have wholly omitted [wrote Ray in 1678] what we find in other authors concerning ... hieroglyphics, emblems, morals, fables, presages or aught else appertaining to divinity, ethics, grammar or any sort of human learning; and present ... only what properly relates to natural history.[93]

This manifesto was easier to issue than to implement. Traces of the old attitude survived in their work, as in that of their successors. But the breach with the allegorical tradition was to prove decisive. The new science was totally hostile to symbolic thinking; and in the eighteenth century the emblem books were gradually relegated into a mere amusement for children.[94] It was in a well-meaning attempt to bring inherited convention into line with reality that Benjamin Franklin objected to the choice of the bald-headed eagle as the American national emblem, because it was 'a bird of bad moral character', 'a rank coward', who lived 'by sharping and robbing' and was therefore ill suited to represent the 'brave and honest Republic of America'.[95]

Long before that, many had come to think it absurd to endow animals with moral qualities at all. The brutes were capable neither of law nor of reason, maintained Jeremy Taylor in 1660; they merely followed their natural instincts. He therefore denounced 'all those discourses concerning the abstinence of beasts, their gratitude, their hospitality, their fidelity, their chastity and marriages'. 'The instinct of the wolf,' agreed William Jones of Nayland in 1771, 'is not cruelty but appetite.'[96] Political analogies also became less fashionable. Thomas Hobbes was not alone in assuming that human nature was so different from animal nature that the behaviour of ants and bees was wholly irrelevant to the political activities of man.[97] There was no justification, explained the Dutch entomologist Swammerdam, for the popular belief that the government of bees was carried on by prudence and judgement, with rewards and punishments and a system of laws; bees merely followed the promptings of nature. In the eighteenth century Réaumur and other French naturalists would say the same.[98] It was in this spirit that the late-eighteenth-century critic Lord Kames condemned the passage in Shakespeare's *Richard II* (Act v, sc. i) where the deposed king is reproached by his wife for not resisting his fate, as would the lion, king of the beasts. 'This comparison,' ruled Kames, 'has scarce any force: a man and a lion are of different species.' It was no longer acceptable to draw social parallels between the human and the natural world. 'The real habits of animals,' wrote Hartley Coleridge in 1835, 'should be carefully observed and they should not be described as performing human actions to which their natural actions have no imaginable analogy or resemblance.'[99]

It also became unfashionable to regard any animal species as intrinsically ugly. There was something beautiful in every creature, Aristotle had said: natural objects of all kinds should be studied without inhibition or distaste.[100] This view was reiterated in the Elizabethan period. 'If a horse be beautiful in his kind, and a dog in his,' asked Thomas Muffett, 'why should not the beetle be so in its kind? Unless we measure the forms of all things by our own, that what is not like us must be held to be ugly.' 'I cannot tell by what logic we call a toad, a bear or an elephant ugly,' agreed Sir Thomas Browne. The scientists thus revived the view which had long ago been proclaimed by Plato: anything which did its work well was beautiful: an ass's ear or a hog's snout were as well-constructed for the practical purposes which they had to serve as were any features of the human body.[101] Augustan confidence in God's design enabled the third Earl of Shaftesbury to assert that serpents, savage beasts and poisonous insects were 'beauteous in themselves', and that even 'a dunghill or heap of any seeming vile and horrid matter' was sufficient to show the beauty of nature.[102]*

Armed with these principles, the naturalists struggled to contemplate the whole animal world with detached curiosity. They did not find it easy. Ray and Willoughby could not prevent themselves from castigating the quail as 'a bird no less salacious than the partridge, infamous also for obscene and unnatural lust'. Even Linnaeus mingled his zoological descriptions with moral and aesthetic judgements; the versions put out by his English editors and adaptors were particularly free with their use of such terms as 'loathsome' and 'disgusting'. But others battled manfully to attain a dispassionate perspective. 'The hog is certainly the most impure and filthy of all quadrupeds,' said Pennant rather doubtfully, 'but we should reflect that filthiness is an idea merely relative to ourselves.' The general appearance of the toad was such as 'to strike us with disgust and horror'. But those with the resolution to view it calmly would observe that at least its eyes were very fine.[103]

In the later eighteenth century we see the general emergence of this romantic *point de vue spectaculaire*, delighting in the world's diversity and reluctant to judge it by human standards. The incomparable Gilbert White, in his letters to Pennant (1767–80) and Daines Barrington (1769–87), shows endless wonder at the ingenuity of animal instinct, immense curiosity towards animate nature in every form, a respect for all living beings and an almost complete lack of repugnance for toads, spiders and other creatures conventionally thought repulsive.† The same outlook is apparent in the art of George Stubbs (1724–1806), who, by contrast with earlier animal painters, is controlled, detached and utterly unanthropomorphic. After him came Thomas Bewick, of whom Edmund Blunden wrote that, 'if Bewick has any meaning, it is that of a world in which the dog, the plover, the farmer's wife, the tramp, the old pollard are all personalities to be watched and interpreted without bias in favour of the human species.' Bewick's near-contemporary John

* The aesthete Uvedale Price later commented on 'a picture of Wovermans, in which the principal objects were a dung-cart just loaded; some carrion lying on the dung; a dirty fellow with a dirty shovel; the dunghill itself, and a dog, that from his attitude seemed likely to add to it'; *Essays on the Picturesque* (1810), ii. xiv.

† Though even he could describe the frog as 'vile' and the buzzard as 'dastardly'; *The Natural History and Antiquities of Selborne* (1789), letters to Pennant, xvii; *Gilbert White's Journals*, ed. Walter Johnson (1931; reprint, Newton Abbot, 1970), 130.

Constable did not specialize in painting animals, but he took the same view: 'I never saw an ugly thing in my life.'[104]

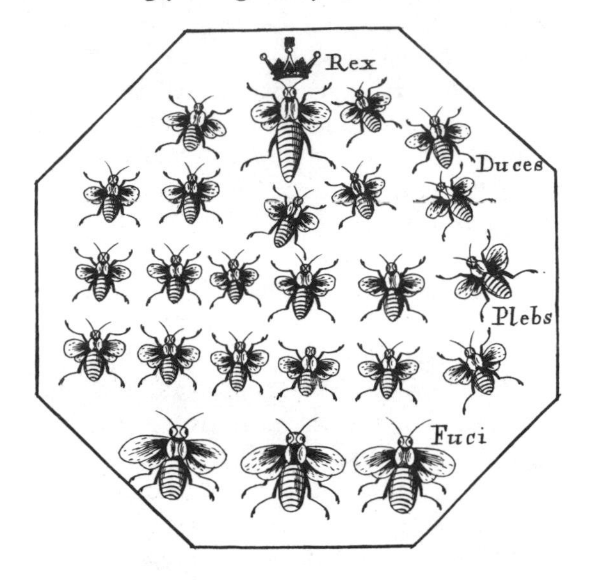

This revolution in perception – for it was no less – did not just affect the professional scientists or that growing body of gentry, clergymen and ladies who made the pursuit of amateur natural history one of the most characteristic middle-class recreations of the eighteenth century; it also had a traumatic effect upon the outlook of ordinary people, by making obsolete many assumptions which had been integral to popular attitudes to nature.

At the beginning of our period the agricultural worker had been justifiably regarded as knowledgeable about the natural world which surrounded him. He had, for example, a very large vocabulary enabling him to draw complicated distinctions between different kinds of domestic animals. All pastoral peoples have a large vocabulary for this purpose: the Nuer of the Sudan have six terms for the shape of their cattle's horns, ten main colour terms and twelve for particular combinations of white and grey; and the Lapland herders have some fifty names for the colour of their reindeer.[1] The English rural vocabulary was equally large. We know this retrospectively from the many local dialect glossaries which were assembled in the early nineteenth century and which constitute a marvellous, and still largely unexploited, source for the mental outlook of the common people on the eve of industrialization.[2] Here we find an infinity of terms categorizing sheep, cattle and other domestic beasts by age, sex, appearance, with great attention to every associated detail. In Northamptonshire, for example, a clot of cowdung was called a bumbal; in Cumberland it was, if in firm condition, 'clap'; if semi-fluid, 'swat'.[3]

A similarly copious terminology existed for wild nature. Hunters had their minutely detailed vocabulary for the beasts of the chase, distinguishing them by their age and sex and naming their tracks, their cries, their anatomy, their droppings and their behaviour. Sir Thomas Urquhart drew on such lore in his translation of Rabelais, when he wrote of

> The barking of curs, bawling of mastiffs, bleating of sheep ... tattling of jackdaws, grunting of swine, girning of boars, yelping of foxes, mewing of cats, cheeping of mice, squeaking of weasels, croaking of frogs, crowing of cocks, kekling of hens ... chanting of swans, chattering of jays, peeping of chickens, singing of larks, creaking of geese, chirping of swallows, bumbling of bees, rammage of hawks, chirming of linnets, croaking of ravens, screeching of owls ... [and so on and so on][4]

There was an equally baroque abundance of collective nouns for each species.[5]

Considerable knowledge of the natural world was thus taken for granted, and he was a poor fellow who could not tell a hawk from a heronshaw. It may be that the origins of this knowledge lay, as the anthropologist Claude Lévi-Strauss has suggested, in the universal desire of all peoples, however 'primitive', to know and classify their biological environment, simply for the sake of knowledge and for the satisfaction of imposing some pattern upon their surroundings. As one observer writes of the cattle-herding Dinka of the Sudan, 'the imaginative satisfactions provided by their herds are scarcely less important than the material benefits.'[6] In early modern England, similarly, the popular taxonomy of plants, birds, beasts and fishes was more elaborate than purely utilitarian considerations required; and much of it had symbolic or emotional value. Nevertheless, it was the practical aspect of this popular knowledge of the natural world which seems to have been uppermost. Interest in wild birds and animals was powerfully stimulated by the desire of hunters and fowlers to catch them. Plenty of countrymen could attract hares or game-birds by simulating their cries;

and one seventeenth-century author tells us that he knew some who could imitate the notes of twenty different kinds of bird.[7]*

The plant world was just as much an object of practical concern. The use of herbs for medicinal purposes was universal at a popular level. It generated a vast lore about the healing properties of plants, to be transmitted orally or written down in the herbals which gained a wider circulation with the birth of printing and continued to be published through the eighteenth century and beyond. Since 1652 there have been over a hundred editions of Culpeper's *Herbal* alone; and it had scores of rivals.[8] *The Grete Herball* offered plant remedies for every complaint, from superfluous facial hair to 'stench of the armholes' or 'swollen ballocks'.[9] Countrywomen planted clary (*salvia*) in the garden as salve for their husbands' backache. They used Solomon's seal for bone fractures, lichen for tuberculosis, bindweed for cancer of the mouth.[10] When cattle had the murrain, the husbandmen cut a hole in the beast's ear and inserted a root of bearfoot.[11] Whether in his family herbal on the cottage shelf or in the medicinal plants growing in the garden outside, the country-dweller demonstrated his practical knowledge of and dependence on the plant world. The future Chartist Thomas Cooper (born 1805) tells us how as a boy in Lincolnshire he learned to recognize agrimony, wood betony, wood sage and other herbs used as medicines by the poor.[12] All country-dwellers knew where to find plants with which to make ointments, laxatives, purgatives, narcotics or cures for warts and ringworm. Of course, their identifications were not always correct: a person died at Havant in 1758 after taking four spoonfuls of juice from the root of hemlock dropwort which had been mistaken for water parsnip.[13] But the very occurrence of such accidents encouraged the formation of an alert attitude to the plants of field, wood and hedgerow.

So did the habit of eating a much wider range of wild plants than we do today. 'The usual manner' of making salads, observed the Jacobean herbalist John Parkinson, was 'to take the young buds and leaves of everything almost that groweth, as well in the gardens as in the fields, and put them altogether.' In 1667 John Worlidge recommended boiled elder shoots, hop buds, nettle tips, pea leaves, nasturtium and larkspur.[14] Country folk cooked chickweed like spinach, boiled ground ivy, made soup from primroses and pickled the samphire which grew on cliffs ('incredibly dangerous to gather', thought the herbalist Robert Turner).[15]

* When the present author was a child in the Vale of Glamorgan, his most frequent companion was a farm labourer's son who knew how to catch hares and rabbits with his bare hands.

Plants were also used for an infinity of other practical purposes. Reeds were cut for thatching and rushes for lights. Thistledown was gathered for pillows, cushions and mattresses.[16] Almond leaves kept moths from clothes.[17] Puffballs were used to smoke out bees and to carry fire to a neighbour's house.[18] Housewives scoured their dishes with horse-tail, made dye out of alder bark and carried their butter to market wrapped in burdock leaves.[19] Beggars used buttercup roots or crowfoot to ulcerate their flesh and excite the compassion of passers-by.[20] In early Tudor Northumberland small boys glued their arrows with slime from the roots of crowtoes (*Scilla nutans*) or harebells. In Elizabethan Lancashire women used asphodel to dye their hair yellow.[21] Other herbs were employed, less plausibly, to encourage conception, to prevent the feet from tiring on long walks and to stop small children from crying.[22]

It is not surprising, therefore, that natural history at first depended for its progress on absorbing much popular lore. The first Tudor ornithologist, William Turner, got much information from fowlers and bird-catchers, who, as Sir Thomas Browne said later, could identify many uncommon species of wild duck unknown to naturalists.[23] The first professional plant-hunter, Thomas Willisel, who acted as paid researcher to Christopher Merrett and John Ray, was an uneducated ex-soldier.[24] The bird-illustrator Eleazar Albin followed Willoughby's *Ornithology*, but for the names of many species he had to fall back on those used by the London bird-catchers. The botanist William Curtis (1746–99) acquired his taste for flowers from an ostler who had studied herbals. Sir Joseph Banks, the future President of the Royal Society, as a schoolboy paid herb-women to teach him the names of flowers.[25] Physicians and apothecaries had long depended for their supplies upon such persons, what William Turner called 'the old wives that gather herbs'.[26] Similarly, the early geologists had much recourse to miners and stonemasons.[27] As late as 1820 the Rev. James Grierson, in his paper to the Wernerian Society (of Edinburgh) on 'the natural history and habits of the mole', acknowledged his debt to Mr Robert Fletcher, 'a most experienced and scientific mole-catcher'.[28] In such areas book learning was notoriously inadequate. A scholar might read Aristotle and Pliny, remarked a Jacobean authority on bee-keeping, 'but when he cometh abroad to put his skill in practice every silly woman is ready to deride his learned ignorance'. In 1752 John Hill noticed that no learned writer on birds had ever listed the summer teal, although it was common at Whittlesea Mere in Cambridgeshire and well known to the local inhabitants.[29]

But popular knowledge was soon eclipsed by the more thoroughgoing inquiries of the scientists, whose viewpoint was not narrowly utilitarian and who rapidly became disillusioned to discover that there were limits to rural curiosity. Husbandmen, remarked Samuel Hartlib, were not naturalists; and their knowledge of local vegetation often left much to be desired. 'As far as I could ever find by conversing with farmers', wrote Benjamin Stillingfleet in the 1750s, 'our common people know scarce any of the grasses by names.'[30] Many plants, particularly small ones, were discovered by the early botanists to have no vernacular name at all. Popular knowledge of so-called 'vermin' was equally unreliable. Slow-worms and grass-snakes were indiscriminately persecuted as venomous.[31] Hedgehogs, badgers and toads were believed to suck milk from cows' udders.[32] Gamekeepers slaughtered innocent jays and woodpeckers. Gardeners destroyed worms, because they allegedly gnawed the roots of plants. The shrew's bite was said to make cattle swell up and die. Not one gardener in twenty, thought William Swainson in 1834, knew the difference between a fly and a beetle.[33] Knowledge of animal diseases and their cures was negligible.

Traditional husbandry thus came to seem riddled with ignorance. To beget male lambs, the wind had to be in the north; hens should always sit on an odd number of eggs; holly seeds wouldn't sprout unless first digested by a thrush; elm trees would grow from old chips and shavings.[34] Bee-keeping was surrounded by erroneous notions about the insect and its habits.[35] Much popular lore about wild nature proved equally spurious. The vulgar were said to believe that the osprey had one webbed foot, so that it could swim like a duck, and one cloven one, so that it could catch prey like a hawk.[36] Toads were regularly killed because they were thought to have a jewel in their heads or because they were believed to be dangerous to cattle.[37] In Staffordshire the mole was thought to have only one drop of blood; in Surrey they said it had only one ear.[38] In eighteenth-century Hampshire it was believed that the nightjar inflicted a fatal disease upon young calves. In Jacobean Cleveland it was well known that any wild goose which flew over Whitby would instantly drop dead; and that to catch a seal it was first necessary to dress as a woman.[39] Elsewhere it was asserted that the bittern used a reed as a pipe; that robins would bury dead persons with moss and leaves; that the hare changed its sex and slept with one eye open; that the badger's legs were shorter on one side than the other; that cuckoo's spit was poisonous; that puffballs would make a man blind; and that snakes could not die until sunset.[40]

The naturalists also discovered that for most people in the early modern period the plant world was alive with symbolic meaning. Certain trees and shrubs – rowan, vervain, mistletoe, angelica – were worn or hung up to give protection against witchcraft.[41] Others – bays, beeches and horseleeks – were planted near the house to save it from lightning.[42] Rosemary, it was said, was a 'holy tree'; it would flourish only if its cultivators were 'rightful and just' and it would not exceed Jesus's height or his age (33 years).[43] A divining-rod had to be made of hazel, while in the Forest of Dean the miners would not take an oath on the Bible unless they also held a holly stick.[44] Many wild plants brought bad luck to those who picked them: in Staffordshire St Bertram's ash grew over a spring bearing his name and it was supposedly dangerous to break a bough from it. In Cheshire it was bad luck to cut down a mountain ash.[45] Motherdee (red campion) was so called because it would kill the parents of the child who picked it. Moonwort was known as 'unshoe the horse' because it loosened the shoes of horses who trod on it; in the Civil War a whole troop of the Earl of Essex's cavalry was said to have suddenly lost their shoes in this way on the downs near Tiverton.[46] If a woman stepped over a cyclamen she would miscarry; John Gerard put sticks over the ones in his garden to prevent this happening.[47] Other plants had special healing powers. Dr Plot was disconcerted to find that countryfolk dealt with swellings in cattle by whipping them with branches from the tree in which they had first incarcerated the shrew mouse held responsible. Curing ruptured children by passing them through the cleft in an ash tree was a widespread remedy.[48] At Daventry the dane-weed (dwarf elder) allegedly grew from the blood of Danish invaders slain centuries ago and would bleed if cut on the anniversary of the battle.[49] Innumerable charms and observances accompanied the gathering of medicinal herbs.[50] Much of this lore was implicit in the continuing use of protective plants on ritual occasions: box, rosemary and bays at weddings and funerals; holly, ivy and mistletoe at Christmas; willow branches on Palm Sunday; green boughs and blossom to bring good luck on May Day.[51]

Crucial to these practices was the ancient assumption that man and nature were locked into one interacting world. There were analogies and correspondences between the species, and human fortunes could be sympathetically expressed, influenced and even foretold by plants, birds and animals. Hedgehogs, swallows, owls, cattle and cats, all gave out signs of a future change in the weather.[52] Sailors watched for the stormy petrel, while the housewife used the cricket on the hearth as her barometer.[53] If the ash was out before the oak, they were bound to have a soak. Worms in oak apples presaged a pestilential year.[54] Human fortunes could also be predicted in this way. Ladybirds, four-leaved clover and black cats were all lucky. So were swallows nesting under the eaves.[55] On the other hand, it was very bad luck, even a sign of death, to meet a hare, or to hear a howling dog, a singing swan, a chirping cricket, a croaking raven, a death-watch beetle, a bittern, or a screech-owl.[56] A dog-rose was unlucky and one should never form a plan when sitting near one; foxgloves similarly had 'knowledge' in them.[57] If a house was about to fall down (a more common event in the seventeenth century than now) the disaster would be preceded by the mass exodus of the rats who lived there; just as lice and other vermin would leave the body of a dying man.[58]

In accordance with such beliefs, some species were especially venerated, for not all the natural world was available for uninhibited exploitation. In parts of England the robin and the swallow were more or less sacred and treated with cautious respect.[59] 'Man superstitiously dares not hurt me,' says the robin in one seventeenth-century poem, 'For if I'm killed or hurt, ill luck shall be.' 'To rob a swallow's nest,' lamented a Jacobean preacher, 'is, from old beldames' catechisms, held a more fearful sacrilege than to steal a chalice out of a church.' 'Dick took a wren's nest from his cottage-side,' runs an eighteenth-century verse, 'And ere a twelve-month past his mother dy'd!'[60] House-martins, cranes, daddy-longlegs, crickets, cuckoos, ladybirds, owls, spotted fly-catchers: all in different parts of the country were species which, as one observer put it, 'superstition protects from wanton injury.'[61] Conversely, others were the victims of superstition. In many places the wren and the squirrel were ritually hunted at Christmas or on St Stephen's or New Year's Day.[62] In Northamptonshire it was good luck to kill the first wasp of the season.[63] On Shrove Tuesday the practice of throwing stones and cudgels at tethered cocks was almost universal.[64]

It is not easy to disentangle the various sources of these diverse practices and beliefs, what Wordsworth called 'the traditionary sympathies of a most rustic ignorance'.[65] As with the different brands of religious heterodoxy found in Tudor and Stuart England, it is often hard to tell how seriously these notions were held and how widely (for many of them were highly local), let alone whether they fitted together in one coherent cosmology or are better regarded as isolated 'superstitions', long severed from their original moorings in an integrated view of the world.[66] Many derived from classical sources, like Pliny's

Natural History, and were learned errors rather than vulgar ones. Sir Thomas Browne in the seventeenth century and William Cobbett in the early nineteenth, both of them acute observers, held the classical writers responsible for the bulk of English rural superstitions. It was because such ideas had been put into general circulation by the compilers of the herbals and the emblem books that the English, like the Romans before them, went on believing that the whitethorn was a fortunate tree and that owls and ravens were ominous.[67] Other practices, like hunting wrens at Christmas or eating geese at Michaelmas, were possibly relics of pre-Christian seasonal rituals.[68] Some survived in a Christianized version: the whitethorn was protective because Christ's crown had been made from it, whereas the elder was the tree on which Judas had hanged himself; wild geese could not fly over Whitby because a medieval abbess had expelled them for eating all the corn.[69]

Other beliefs were remnants of different and, by our standards, erroneous methods of classification: treating slow-worms as snakes, frogs as female toads, wrens as wives of robins or cuckoos as hawks in summer plumage.[70] Sometimes they reflected the special status of certain species: hares, owls and ravens, for example, around whom so many prognosticatory beliefs tended to congregate. But one can only speculate as to whether these creatures owed this special status to their associations in pagan mythology or to some anomaly of appearance or behaviour which made them hard to fit into existing systems of classification or to some more practical consideration. The raven *may* have been fatal because it was associated with carrion.[71] The swallow *may* have been fortunate because the ancients had noticed that it did not frequent malarial districts.[72] The elder tree, the rowan and the whitethorn *may* have had special significance because these were the shrubs which grew up after the primeval forest clearings and thus became associated with human habitation.[73] But we cannot know for certain.

On the other hand, it does seem clear that the reason that many natural events were deemed ominous was that they seemed to blur those crucial categories of 'wild' and 'tame' around which so much popular thinking revolved. The encroachment of wild creatures into the human domain was always alarming: if a town was suddenly infested by jays and owls, for example, or if a wild bee flew into a cottage, or a shoal of sharks accompanied a ship, or a raven nested on a church steeple, or a jackdaw came down the chimney, or a mouse ran over one's foot, or a robin tapped at the window (the latter a notorious form of 'call', which even in Victorian times would make healthy men take to their beds).[74] In 1593 it was feared that the plague in London would get worse because a heron perched on the top of St Peter's, Cornhill, and stayed there all afternoon. In 1604 the House of Commons rejected a bill after the speech of its Puritan sponsor had been interrupted by the flight of a jackdaw through the Chamber – an indisputably bad omen.[75] The attitude resembled that of those African peoples amongst whom misfortune is expected whenever the world of the bush encroaches upon that of human settlement; the same outlook is implicit in the long-surviving notion that it is unlucky to bring certain wild plants into the house.[76]

Another occasion for regarding an event as alarming was when it seemed to defy the regularities of nature. It was a bad sign if an apple-tree bore blossom and fruit simultaneously or if a hen crowed like a cock (whereupon many seventeenth-century housewives instantly wrang the bird's neck).[77]

All these notions reflected an older, and altogether different, way of looking at the natural world from that which the scientists of seventeenth-century England were now seeking to introduce, for to them these beliefs were merely examples of popular 'ignorance', proof that the testimony of the uneducated was unreliable.

Many of these popular ideas had already been condemned by moralists as ungodly and superstitious. The clergy had attacked divination based on the observation of natural phenomena by arguing that God no longer used birds or animals to signify his intentions; it was, therefore, wrong to see owls, hares or cats as ominous.[78] Vehement Protestants, deeply sensitive to any apparent survivals of popery or paganism, were, like some of their medieval predecessors, strongly hostile to the notion that vegetation might have any protective power and were unsympathetic to the symbolic use of plants. To put a garland of flowers on a hearse at a funeral was monstrous idolatry, urged an Essex lecturer in 1617.[79] To hang up evergreens at Christmas was paganism, thought William Prynne; and in December 1647 the Lord Mayor of London went round the City pulling down the holly and ivy with which inhabitants had decorated their houses. Churches should not be adorned with garlands and maypoles were idolatrous.[80] It was even thought wrong to stick a piece of rosemary in the joint of meat when it was brought to the table.[81]

This Protestant attack on older ways of looking at the natural world was now strongly reinforced by the writings of the scientists. From the later seventeenth century, the denunciation of 'vulgar errors' became an increasingly obsessive theme, as the compilers of the many county natural histories written or projected during the later Stuart period picked their way through all the legends of prognosticatory springs, portentous birds and similar marvels. The common people would 'believe any strange thing', lamented John Morton in his *Natural History of Northamptonshire* (1712); 'That which looks like truth is at often times as possible with them as that which really is so.'[82] Even in the sixteenth century, the herbalists had deplored popular credulity in accepting stories of the mandrake root, which supposedly grew under the gallows from the seed of hanged men and bore the shape of a human figure. The little puppets sold by Tudor pedlars as 'mandrakes' were the fabrications of confidence-tricksters, warned William Turner.[83] Popular ignorance of the new astronomy had also long been remarked upon. The 'common sort' supposedly did not know that the earth was round and had no idea of the distance which separated the stars from each other. Some countrymen allegedly believed that the sun was no bigger than a cartwheel.[84] By the later seventeenth century the scientific attitude towards popular errors had become aggressively rationalist. The pack of hounds which the colliers of Wednesbury heard flying through the air, explained Dr Plot, was only a flock of wild geese. The apparently ominous eruption of springs at certain times of the year had a perfectly natural explanation, thought Joshua Childrey. If the oak near which Charles I set up his standard had subsequently turned white, said William Borlase, that was just because an infection of the tree had variegated the leaves. If the Glastonbury Thorn blossomed on Christmas Day, then, said Sir Hugh Platt, someone must have applied 'some philosophical medicine ... to the root thereof'. The will-of-the-wisp, thought John Ray, was merely a glow-worm.[85]

Equally frowned upon by the new naturalists were the monsters and fabulous animals which had been portrayed in the medieval bestiaries and had survived in the writings of Aldrovandi, Rondelet, Gesner and the other sixteenth-century continental zoologists. Not all the common people had believed in the existence of centaurs, basilisks or monk-fish. Indeed Topsell observed in 1607 that 'the vulgar sort' were sceptical of the unicorn, scarcely believing in the existence of 'any beast but such as is in their own flocks'.[86] But classical mythology had its effect upon popular beliefs. Dragons and unicorns lingered on in heraldry and folklore; and in the early seventeenth century it was still possible to spread a report that Sussex had been devastated by a 'monstrous serpent'.[87] The herbalist John Parkinson declared in 1640 that the unicorn lived 'far remote from these parts and in huge, vast wildernesses among other most fierce and wild beasts'; and Robert Lovell in 1661 could divide serpents into (a) 'vulgar and lesser' serpents and (b) 'dragons', subdividing the latter into those with legs and those without. Early members of the Royal Society even conducted controlled experiments on portions of unicorn's horn.[88] But Ray was emphatic about omitting griffins and phoenixes from his edition of Willoughby's *Ornithology* (1678); and the unicorn's horn turned out to be the tusk of either the narwhal or the rhinoceros. Tales of dragons, thought Richard Bradley, related in fact to crocodiles and alligators.[89] As classificatory systems hardened, attitudes towards reports of any exotic creature grew sceptical, sometimes excessively so: 'It be a question whether there *be* any such animal,' says Ray of the hippopotamus (though he later came to accept that there was). As for the pigmies, satyrs and sphinxes of classical antiquity, they were merely apes and monkeys, explained Edward Tyson in 1669.[90] The belief in the existence of mermen and mermaids proved longer-lasting, but it too was explained away by sceptics in naturalistic terms. 'It is probably from an imperfect view of this fish [the sea-cow],' suggested John Hill in 1752, 'that the opinion of mermaids, mermen and sirens first arose. This creature has a way of raising itself upright and standing for some minutes together, half out of water; a person who should look at it from a distance in front of such a posture would see something like hands and breasts and this seems to be all that has given origin to the reports of seeing mermaids, etc.'[91]

The later seventeenth century was thus a decisive period in the separation of popular from learned views of the natural world. The rupture proved a stimulus to the activities of educated collectors like Sir Thomas Browne, John Aubrey, John Ray and Henry Bourne, who conducted semi-anthropological inquiries into popular beliefs, customs, dialects and proverbs.[92] But it made serious eighteenth-century naturalists contemptuous of popular lore. The botanist Peter Collinson rejected 'hearsay stories of ignorant peasants', because there was no depending on the reports of 'credulous people'. The belief that butter was yellow because the cows ate yellow crowfoot, said Benjamin Stillingfleet, 'shows how very incurious the country people are in relation to things they are every day conversant with'. Common labourers, agreed William Smellie, were 'totally unqualified for examining every circumstance with the discerning eye of a philosopher'.[93] 'Ah, sir!' said an eighteenth-century ostler, after failing to answer a long series of questions put him by a gentleman about the animal in his charge, 'considering that I have lived thirteen years in a stable, tis surprising to think how *little* I knows of a horse.'[94]

iii. NOMENCLATURE

The gulf between popular and learned ways of looking at the natural world was further widened by the introduction of a new Latin terminology to supersede the vivid vernacular names with which ordinary people had been in the habit of identifying the plants, birds and animals around them. These vernacular names were seldom peculiar to the English language, but usually had close continental equivalents. Indeed many of today's supposedly 'old-fashioned' flower names are post-medieval in origin; early English botanists, finding that many indigenous plants had no English name at all, did not hesitate either to coin anglicized versions of European names (*Superba Austriaca*, for example, becoming London Pride) or to invent new labels altogether. It is to Turner that we owe 'cow parsnip' and to John Gerard 'traveller's joy'. But many of the vernacular names were ancient and had genuine popular roots.[1]

Picturesque plant-names, with their strong visual, emotional and human connotations, fitted well into popular cosmology. Some were biblical or religious names, usually inherited from a Catholic past: Christ's Ladder, Star of Bethlehem, Solomon's Seal, and all the names which allude to saints, as in St John's wort, or to the Virgin, as in ladies' cushions or ladies' smocks. Conversely, there are over fifty supposedly ugly or unpleasant plants whose names begin with 'Devil –'.[2] Some were based on supposed similarities to parts of animals: hound's tongue, bearfoot, cat's tail, bird's eyes, cranesbill, coltsfoot, goat's beard; some on the plant's smell: hounds' piss (*Cynoglossum*),[3] stinking arrach; some on its edibility: poor man's pepper, sauce alone, hedge mustard, fat hen; some on fancied resemblances to parts of the human body: miller's thumb, old man's beard, maidenhair, dead man's fingers; or an item of clothing: bachelor's buttons, shepherd's purse, fool's cap, ladies' slippers. A large proportion alluded to proposed medicinal properties: navelwort, lungwort, kidney beans (so called 'for they strengthened the kidneys'),[4] feverfew, saxifrage (because it would break up the stone), pissabed (dandelion). Others were frankly poetic, 'idle and foolish names', as Parkinson calls them:[5] patience; honesty; thrift; love in idleness; goodnight at noon; son-before-the-father (because the flowers appear before the leaves); courtship and matrimony (alluding to the deterioration in the scent after the flower was picked); and welcome-home-husband-though-never-so-drunk. The herbalists thought that women were responsible for many of these names: 'our London gentlewomen have named it [swallow wort] Silken Cisley'; 'our women have named [oxlips] jack-an-apes on horseback'.[6]

Vernacular names for birds, insects and other wild creatures were based on similar principles. The hen-harrier was so called, noted William Turner, because it slaughtered countrymen's fowls; and the shrike was the 'butcher bird' because it impaled its victims on thorns.[7] The large hairy caterpillar was 'the devil's finger ring'. Pea-eating birds were 'pudding bags'. Frogs were 'Dutch nightingales' or (in Lincolnshire) 'Boston waits'. The woodlouse when touched turned itself into a ball, like a Dutch cheese, and was consequently known as a 'cheese-bob'.[8] Nearly every kind of wild bird had an accepted Christian name. Some of these have survived into modern times: tom tit, jackdaw, robin redbreast, jenny wren, poll parrot, and, of course, dicky bird. But in the early nineteenth century, when the dialect dictionaries were assembled, the range was much wider; and so it probably was in early modern times. We have Jack Baker (the chaffinch), Bessie Blackers (the reed bunting), Billy Owl (the barn owl) and so on down the alphabet through 'little Matthew Martin' (the bullfinch)[9] to Will Wagtail.

Canon C. E. Raven, a pioneer investigator whose sympathies were firmly with the moderns, considered that these vernacular names revealed 'a folk-lore full of charming fancies and quaint beliefs, but wholly unscientific in its attitude'.[10] The learned naturalists of our period disliked them for similar reasons. Many were unintelligible outside a particular locality. As a Northcountryman who moved south under Edward VI to become Dean of Wells and later travelled extensively abroad, the pioneer botanist William Turner was particularly sensitive to the bewildering variety of English regional usage; and his herbal often draws attention to the differences between northern and southern nomenclature. What was 'plantain' in the south, for example, was called 'waybread' in the north.[11] The herbals show that it was common for plants to have at least half a dozen wholly different names: so that ground ivy was also known as catsfoot, alehoof, Gill go by the ground, Gill creep by the ground, tun hoof or haymaids; and ladies' bedstraw was also cheese rennet, gallion, pettimugget, maid's hair or wild rosemary.[12] Mulleyn (*Candelaria*) was variously known as high-taper, hagtaper, woollens, Jupiter's staff, hare's beard and bullock's lungwort.[13] John Gerard wrote of treacle mustard: 'we call this herb in English penny flower or money flower, silver plate, pricksong-wort; in Norfolk sattin and white sattin and among our women it is called honesty.'[14] No less than fifty different local names have been recorded for *Caltha palustris*, the marsh marigold.[15] Insect names were subject to similar regional variations. Hairy caterpillars were 'hairy worms' in Yorkshire; 'millers' in Herefordshire; 'devil's rings' in the south.[16] Even apples bore different labels in different parts of the country: a Jacobean expert complained that 'John-Apples be in some places called Dewzings or Long-Lasters; and Goodings be called Old Wives.'[17]

Many vernacular names were hopelessly volatile, leaping from plant to plant according to local whim. At least ten different species were known in one place or another as cuckoo flower; over twenty were bachelor's buttons or dead man's fingers. Any flower out of the ordinary, observed Parkinson, would be indiscriminately labelled 'jack-an-apes on horseback'.[18] Bird names were equally volatile: the name 'Peggy' was bestowed alike upon the whitethroat, the blackcap, the garden warbler, the willow warbler and the chiffchaff. Billy Biter could be the blackcap, the great titmouse or the blue titmouse.[19]

It is no wonder that the early naturalists devoted so much energy to compiling dictionaries of Latin names and their vernacular equivalents. In the seventeenth century Sir Thomas Browne declared himself 'much unsatisfied' with the names given to local birds by the Norfolk countrymen; and in the nineteenth century the American Ornithologists Union campaigned for a regular nomenclature for birds in place of the infinity of local and regional names.[20] The bewildering range of local plant nomenclature was unpopular because it gave scope for imposture by nurserymen, who, as Parkinson complained, 'do so change the names of most fruit they sell [so] that they deliver but very few true names to any'.

The growth of a national market in plants and flowers thus generated pressure towards standardization; and it was in order to avoid frauds and confusion through the same plant being sold under different labels that a London Society of Gardeners put out its *Catalogus Plantarum* in 1730.[21] The propaganda of agricultural improvers, irritated by the obscurities of regional terminology, had a similar effect. As one country gentleman urged in 1743, 'all plants, whether grasses or weeds, should have the true botanical name given them, or it is impossible to find them out, as almost every county gives a different name.'[22]

The old names were also disliked by Protestants when they had Popish associations with the Virgin or with saints, or indeed if they had any religious implication at all.[23] Any names which preserved the tradition of a plant's supposedly religious or protective significance were quite unacceptable. Equally contentious were names which perpetuated claims to a non-existent healing power. In the later seventeenth century, scientific opinion became increasingly hostile to the doctrine of signatures, the belief that every plant had some human use, and that its colour, shape and texture were designed to give some external indication of that use, so that, for example, spotted herbs cured spots, yellow ones healed jaundice, and adder's tongue was good for snake-bites. Though upheld by mid-seventeenth-century herbalists,[24] the belief in signatures was rejected by John Ray and Nehemiah Grew as wholly unempirical and rapidly disappeared from official botany.[25]* Meanwhile, the enormous influx, from the early decades of the seventeenth century onwards, of new drugs from America and the East made learned doctors increasingly indifferent to local herbal remedies. As early as 1656 the art of the herbalist was said to have grown 'contemptible'.[26] The use of simples, said a mid-eighteenth-century writer, was now much neglected; native plants had 'made way for a farrago of exotics imported,

* The Cambridge Platonist Henry More thought that the world would have been very dull if *all* plants had been marked with external indicators of their uses; 'The rarity of it is the delight.' See *An Antidote against Atheism* (2nd edn, 1655), 99.

and palmed on us'.[27] It was among country folk and the poor that the use of the old herbal remedies would survive. In the late nineteenth century the rural classes were said still to use the old herbals and, 'in the north of England at any rate', to 'collect vast quantities of medicinal plants'.[28] But, for scientists, plant-names which preserved the memory of fancied potencies and resemblances had long become unacceptable.

Finally, the old vernacular names for plants and animals were disliked because they were thought too coarse. Anyone who wants evidence of the way in which polite sensibilities have changed with the centuries need only consider the briskly anatomical nature of this now suppressed terminology, for in the seventeenth-century countryside there grew black maidenhair, naked ladies, pissabed (or shitabed), mares fart and priest's ballocks. In the herb garden could be found horse pistle and prick madam; while in the orchard the open arse (or medlar) was a popular fruit. Even the black beetle was twitch-ballock and the long-tailed titmouse bum-towel.[29] Many of today's more fanciful flower names – lords and ladies, for example – are deliberate inventions of the nineteenth century, designed to obliterate some unacceptable indecency of the past; and some present-day survivors still conceal a grosser meaning given them by liberal shepherds in bygone days. The mid eighteenth century was a transitional period, when bowdlerization had begun, but not been completed. Even so genteel a figure as Robert Smith, official rat-catcher to George II's daughter, the Princess Amelia, could occasionally drop his guard, as when he refers in the cold print of his book on how to catch vermin (1768) to a bird called 'the large brown, white arse, ring-tailed hawk'.[30] But then he lived in an age when Wittenham Clumps, those pleasant twin hills in Berkshire, were still known (after the local landowner's wife) as Mrs Dunch's buttocks.[31]*

The gulf between popular and learned terminology was an old one and is often mentioned in the Tudor herbals. It was reflected in the description of animals, no less than of plants. Sir William Petty's father observed that a country butcher could be an excellent anatomist, but that he used a different language, calling a tendon a 'buckle', a membrane a 'vilme' and an artery a 'pipe'.[32] The gap widened as learned scientists wrote in Latin for an international audience. Standard botanical handbooks, like John Ray's comprehensive three-volume survey of the world's flora, *Historia Plantarum* (1686–1704), and his more portable

* The bowdlerization of English place-names has been a steady development since the late eighteenth century. In Northamptonshire alone, Buttocks Booth became Boothville, Pisford became Pitsford and Shitlanger was turned into Shutlanger; John Steane, *The Northamptonshire Landscape* (1974), 253n.

Synopsis (3rd edition, 1784), had always been in Latin, but the decisive step was the speedy adoption in the late 1750s and early 1760s of Linnaeus's standardization of binomial nomenclature, as set out in his *Species Plantarum* (1753) and the tenth edition of his *Systema Naturae* (1758). This system has lasted ever since; botanical names used before 1753 have no standing in the modern nomenclature unless adopted by the Swedish naturalist.

For Linnaeus every plant, regardless of local vernacular practice, had to have two Latin names, one indicating the genus, and the other the species; and the rules he set out in his *Critica Botanica* (1737) were strict, permitting no names based on the plant's scent, taste, medical properties, moral character or religious significance, all of which he considered to be highly subjective qualities, varying according to the beholder. 'If a genus known for a long time past and familiar even to ordinary people, should bear an absolutely erroneous name,' he declared, 'it must be expunged.'[33] Linnaeus's actual practice was much less rigid, for he retained a good deal of the old anthropomorphic terminology in his new formulae. But such inconsistency did not matter, for his plant-names were now in botanical Latin, a different language from that of Cicero, but one equally remote from ordinary people. The Linnaean system of nomenclature was introduced to England by James Lee in his *Introduction to the Science of Botany* (1760) and William Hudson in his *Flora Anglica* (1762). From 1768 it was more generally disseminated via the eighth and subsequent editions of Philip Miller's highly popular *Gardener's Dictionary*.[34] In France there was a serious attempt at a word-for-word translation into the vernacular of the new botanical terminology. But in England such efforts met with little success. Instead, the botanists stuck to the Latin.[35] The old vernacular names were either forgotten altogether or declined in status to survive as the makeshift equipment of the rustic and the amateur.

A contemporary protested that to give plants hard Latin names when they already had easy English ones would mean that in future only learned botanists would be able to identify them: 'to class them botanically . . . , so that nobody but a botanist can find them out, appears . . . something like writing an English grammar in Hebrew. You explain a thing by making it unintelligible.'[36] Poets similarly lamented the disappearance of the evocative and vernacular names of the past. As William Whitehead wrote of *The Goat's Beard* (now *Spiraea Ulmaria*)

> The flaunting woodbine revell'd there
> Sacred to the goats; and bore their name

Till botanists of modern fame
New-fangled titles chose to give
To almost all the plants that live.[37]

But the professional attitude was unyielding. Vulgar names were an obstacle to science. 'Those who wish to remain ignorant of the Latin language,' said John Berkenhout in 1789, 'have no business with the study of botany.' A decade or so later, the farmers who still used 'vulgar, provincial names' to identify the pests which attacked their crops found themselves unable to communicate with the naturalists, who did not know which species they were talking about.[38] In the nineteenth century there was a brief, sentimental attempt by John Ruskin and some other gardening writers to revive or invent English names for garden flowers and wild plants.[39] But by that time the learned world had permanently discarded the language of ordinary discourse.

iv. CHANGING PERSPECTIVES

In all sorts of ways the *content* of learned natural history had also changed, as many deeply-rooted popular notions had come to be discarded. The invention of the microscope enabled late-seventeenth-century entomologists on the continent, like Redi, Leeuwenhoek, Swammerdam, and Malpighi, to conduct controlled experiments proving that, contrary to universal belief, worms and flies were not spontaneously generated out of decaying meat, and that maggots only appeared in putrefying matter if flies had access to it. This demonstration came as a great relief to the pious, for whom spontaneous generation had threatened to make a Creator unnecessary. Previously, all intellectuals had accepted spontaneous (or as it was sometimes called 'equivocal') generation. The author of *Oceana*, James Harrington, when imprisoned in the Tower, believed that his perspiration was turning into bees; and the Netherlandish physician J. B. Van Helmont (1577–1644) had a famous remedy for hatching mice from dirty shirts.[1] The doctrine had Aristotle behind it and was difficult to dislodge. 'Artificial brains', said a Jacobean author scathingly, had vainly tried to breed oysters off the Yorkshire coast by transplanting them from elsewhere, being ignorant that oysters bred by 'a seminary virtue in the slime of the sea'.[2] But in the early eighteenth century all the educated knew that the doctrine of spontaneous generation was now 'exploded and rejected'.[3] There was still doubt about the generation of microscopic organisms, which many free-thinkers chose to see as spontaneous until the time of Pasteur. But it was only the vulgar who continued to see it raining frogs and to believe that insects grew out of horse-hair or rotten wood.[4]

Equally repudiated was the closely-associated belief that the barnacle (or brent) goose was hatched from shells which grew on trees or rotten driftwood. This notion was said by William Turner in early Tudor times to be confirmed by the unanimous testimony of all the longshoremen of England, Scotland and Ireland. In fact it had been explicitly rejected by the Emperor Frederick II in the twelfth century and by the scholastic philosopher Albert the Great in the thirteenth.[5] John Gerard accepted it in his *Herball* (1597), but when Thomas Johnson came to revise and expand that work in 1633, he knew that Dutch sailors had observed the geese to be hatched from ordinary eggs like any other bird. The notion died out among the learned during the seventeenth century.[6] Similarly, scientists had by the end of the eighteenth century come to agree that swallows really did migrate in winter to warmer countries instead of going to the moon, as Charles Morton suggested in 1703, or burying themselves under water or hanging upside down in caves, as Aristotle, Izaak Walton, Dr Johnson and even Gilbert White believed.[7] But many countryfolk continued to claim to have fished them up from the bottom of ponds, matted together like a swarm of bees, and to have found cuckoos passing the winter in hollow trees.[8]

On the Bodleian Library's copy of a mid-seventeenth-century work of natural history (John Jonston's *An History of the Wonderful Things of Nature*, English translation 1657), some anonymous eighteenth-century reader has written a note, contrasting 'the amazing ignorance of the seventeenth with the knowledge of the eighteenth centuries'. Knowledge about the natural world, he observes, has increased 'with an inconceivable rapidity'; to compare Jonston with Buffon is to see that 'in one century more light was thrown on this science than had been elicited in the preceding period of near 5,700 years[!]'. The vulgar errors, the indecent expressions and the incredible lies of the past 'must lead a reflecting mind to rejoice in having existence in a more enlightened ... era'.[9]

Enlightened it may have been. But by eroding the old vocabulary, with its rich symbolic overtones, the naturalists had completed their onslaught on the long established notion that nature was responsive to human affairs. This was the most important and most destructive way in which they shattered the assumptions of the past. In place of a natural world redolent with human analogy and symbolic meaning, and sensitive to man's behaviour, they constructed a detached natural scene to be viewed and studied by the observer from the outside, as if by peering through a window, in the secure knowledge that the objects of contemplation inhabited a separate realm, offering no omens or signs, without human meaning or significance. 'One of the nicest things' about studying wild creatures, thought Gavin Maxwell's zoologist aunt, 'is that we are interested in them, while they're not interested in us.'[10] No Tudor zoologist could have said that.

Of course, the new scientific naturalists were far from having totally separated the natural world from the human one. On the contrary. Thomas Hobbes had rejected political arguments based on the practice of ants and bees, but later social theorists proved reluctant to dispense with biological metaphors. Indeed biology itself retained an obvious social component. The natural world was still frequently seen as a projection of human social relationships, actual or ideal. In the eighteenth century the Great Chain of Being offered a buttress for the human social hierarchy, while Adam Smith's contemporaries had no difficulty about perceiving in 'the economy of nature' such fashionable phenomena as the Division of Labour, the abhorrence of waste, and the operation of the Invisible Hand. As a commentator put it in 1745,

> Though there be a numberless variety of creatures, and each individual seems to be acting as for himself, and to have his own

> private ends in view; yet ... all of them together ... do in event conspire to the strength or convenience, to the beauty, harmony or perfection of the whole; and, what is more, contribute in some manner and degree to the advantage and happiness of each other.[11]

Nature, agreed an observer in the 1820s, was 'so remarkably simple in all her actions, economical in her ways, and frugal of her means'.[12] A few decades later, Karl Marx would criticize Charles Darwin for representing the natural state of the animal kingdom as one of free competition and for seeing among the beasts and plants his own English society, 'with its division of labour, competition, opening up of new markets, "inventions" and the Malthusian "struggle for existence"'.[13] By portraying the most ferocious forms of competition as in the natural order of things, Darwin was in the tradition of those many earlier writers who had urged that the lower classes should accept their hardships cheerfully because nature would ensure that all was for the best. Famine and death, he taught, were the means by which the continued production of higher animals was ensured; meanwhile, 'no fear is felt ... death is generally prompt ... and the vigorous, the healthy, and the happy survive and multiply.'[14]

In the later twentieth century, 'sociobiology' would once again attempt to justify the conviction of some Enlightenment theorists that 'the social laws by which ... animals are governed might open views into the social nature of man'.[15] But the modern attempt to expound the biological determinants of human behaviour by scrutinizing that of animals has to encounter the opposition of those who believe that nature and culture are different entities. Sociobiology has accordingly been denounced as 'ideologically prescriptive' and attacked for attempting to prove that the particular arrangements of Western society are biologically inevitable.[16] Appeals to what is supposedly 'natural' continue to be made in support of social and political programmes, but they meet much more resistance than was once the case.

For the seventeenth and eighteenth centuries had seen a fundamental departure from the assumptions of the past. Instead of perceiving nature primarily in terms of its analogies and resemblances to men, the naturalists had begun to try to study it in its own right. They were by no means indifferent to nature's human uses, but they did not make those uses central to their perceptions. A neutral, supposedly objective, taxonomy had replaced more man-centred methods of classification. Scientists rejected the belief that natural phenomena were to be understood in terms of their human meaning, just as they castigated the vulgar error that birds, beasts and plants could respond sympathetically to human behaviour. The conviction that animals and vegetation had religious or symbolic meaning for men remained an article of faith for many Victorian country folk,[17] but it no longer had the support of intellectuals; for the educated were now coming to believe that the natural world had its own independent existence and was to be perceived accordingly. 'Never to see or describe any interesting appearance in nature, without connecting it by dim analogies with the moral world,' observed S. T. Coleridge, 'proves faintness of Impression. Nature has her proper interest; and he will know what it is, who believes and feels, that every thing has a Life of its own.'[18]

Henceforth, the systematic investigation of nature would be conducted on the assumption that plants and animals should be studied for their own sake, independent of their utility or meaning for man. This was a return to that separation of human society from nature which had been pioneered by the ancient Greek atomists Leucippus and Democritus, who had maintained that nature followed its own regularities and was wholly unresponsive to the moral behaviour of human beings.[19] This essentially modern view of causality had been overlaid by centuries of Christian teaching, when nature was portrayed as the creation of an omnipotent deity and her laws were seen not just as impersonal regularities but as moral norms. Now, once again, the scientists returned to the view that nature and human society were fundamentally distinct.

Yet, although the naturalists were coming to discard many of the anthropomorphic assumptions of the past, it was hard for others to stop seeing the natural world as a reflection of themselves. Even as the older view was driven out by the scientists, it began to creep back in the form of pathetic fallacy of the Romantic poets and travellers, for whom nature served as a mirror to their own moods and emotions. To understand that the natural world was autonomous, only to be understood in non-human terms, was still an almost impossible lesson to grasp.

III
MEN AND ANIMALS

I have sent unto you . . . a beast, the creature of God, sometime wild, but now tame, to comfort your heart at such time as you be weary of praying.

A monk of Christ Church, Canterbury, to Lady Lisle, 1536; *The Lisle Letters*, ed. Muriel St Clare Byrne (1981), iii. 350.

'I have,' said a lady who was present, 'been for a long time accustomed to consider animals as mere machines, actuated by the unerring hand of Providence, to do those things which are necessary for the preservation of themselves and their offspring; but the sight of the Learned Pig, which has lately been shewn in London, has deranged these ideas and I know not what to think.'

Sarah Trimmer, *Fabulous Histories designed for the Instruction of Children* (3rd edn, 1788), 71.

Having proved mens & brutes bodies on one type: almost superfluous to consider minds.

Charles Darwin, *Notebooks on Transmutation of Species*, ed. Sir Gavin de Beer (*Bulletin of the British Museum* (*Natural History*), *Historical Series*), 2 (1959–63), 163.

In the first chapter it was suggested that theologians and philosophers in early modern England tended to take a man-centred view of the natural world; and in the second it was argued that the rise of natural history helped to undermine this anthropocentric view, both in its learned and its popular versions. The next task is to examine some of the ways in which people's actual experience of animals, on the farms and in their houses, conflicted with the theological orthodoxies of the time and ultimately encouraged intellectuals to develop an altogether different view of man's relationship to other species.

i. DOMESTIC COMPANIONS

Although the theologians preached a strict separation between man and nature, the actual practice of early modern England was by no means so rigid. Just as popular attitudes to wild nature presupposed that men, plants and wild creatures were inextricably bound up in one great community, so human relations with domestic animals were much closer than official religion implied.

It is perfectly true that cattle, pigs, horses, sheep and poultry were not kept for sentimental reasons. They were there to work or be eaten, or both. A late-seventeenth-century preacher remarked of the beasts kept for food that 'these are only fed for slaughter: we kill and eat them, and regard not their cries and strugglings when the knife is thrust to their very hearts.'[1] The killing indeed could be a protracted business, for, though cattle were normally pole-axed before being slaughtered, pigs, calves, sheep and poultry died more slowly. In order to make their meat white, calves, and sometimes lambs, were stuck in the neck so that the blood would run out; then the wound was stopped and the animal allowed to linger on for another day. As Thomas Hardy's Arabella would explain to Jude, pigs should not be slaughtered quickly:

> the meat must be well bled and to do that he must die slow ... I was brought up to it and I know. Every good butcher keeps un bleeding long. He ought to be up till eight or ten minutes dying, at least.[2]

Earlier in their lives, moreover, male animals intended for food would normally have been castrated. The triple justification for this age-old practice was that it made the beast easier to handle, prevented it from dissipating its energies on sexual activity and helped to produce fatter, healthier and better-tasting meat. In addition, the stones of gelded lambs were, as one seventeenth-century Yorkshire farmer put it, 'a very dainty dish' when fried with parsley.[3] Not just lambs, but pigs, calves, cocks and rabbits were castrated accordingly. Ungelded bulls were deemed unfit for eating unless they were first baited by dogs; the violent exercise involved would, it was thought, help to thin the animal's blood and make its flesh tender.[4] In the late medieval and early modern period, accordingly, most towns had a rule making it compulsory to have a bull baited before it was slaughtered by the butcher.[5] The same practice was recommended in the case of he-goats.[6] Battery farming, moreover, is not a twentieth-century invention. In

Elizabethan times the usual way of 'brawning' pigs was to keep them 'in so close a room that they cannot turn themselves round about ... whereby they are forced always to lie on their bellies'. ('After he is brawned for your turn,' the formula continued, 'thrust a knife into one of his flanks and let him run with it till he die; [or] gently bait him with muzzled dogs.') 'They feed in pain,' said a contemporary, 'lie in pain and sleep in pain.'[7] Poultry and game-birds were often fattened in darkness and confinement, sometimes being blinded as well. 'The cock being gelded,' it was explained, 'he is called a capon and is crammed in a coop.'[8] Geese were thought to put on weight if the webs of their feet were nailed to the floor; and it was the custom of some seventeenth-century housewives to cut the legs off living fowl in the belief that it made their flesh more tender.[9] In 1686 Sir Robert Southwell announced a 'new invention of a ox-house, where the cattle are ... to eat and drink in the same crib and not to stir until they be fitted for the slaughter'. Dorset lambs were specially reared for the Christmas tables of the nobility and gentry by being imprisoned in little dark cabins.[10]

Nevertheless, relations with domestic animals were more intimate than such bald facts might suggest. Beasts, after all, were relatively more numerous than they are today; and they lived much closer to their owners. In modern England there are three people for every one sheep. At the beginning of the sixteenth century the ratio was the other way round.[11] The animals were also less sharply segregated. By the sixteenth century it had become customary for the English to boast that they kept their domestic stock at a distance; they despised the Irish, the Welsh and the Scots because many of them ate and slept under the same roof as their cattle, 'very beastly and rudely in respect of civility', as a contemporary put it.[12] In Wales, where it was a tradition that cows gave better milk if they could see the fire, it was said in 1682 with some exaggeration that 'every edifice' was 'a Noah's Ark', where cows, pigs, chickens and the human family all lay together promiscuously.[13] But an Elizabethan recalled that 'till of late years', the inhabitants of Cheshire had also lived like the Anglo-Saxons, with a fire in the middle of the house and the oxen under the same roof:[14] and similar arrangements had existed in many parts of medieval England, particularly in the Highland zone. In the 1590s Joseph Hall wrote of the Northern husbandman in his smoky cottage:

> At his bed's feet feeden his stalled team,
> His swine beneath, his pullen o'er the beam.
> A starved tenement, such as I guess
> Stands straggling in the wastes of Holderness.[15]

In the sixteenth and seventeenth centuries the so-called 'long-house', or combination of house and byre, in which men and cattle slept under the same roof, usually separated by a low wall or cross passage, but entering by the same door and enjoying internal access from one part to the other, was evolving into an exclusively human residence, with the erection of a party wall or a separate entrance for the cattle or the conversion of the byre to other purposes and the moving out of the animals to separate buildings in the farmyard.[16] In the mansion houses of southern England, noted William Harrison in 1577, it was not customary to keep animals under the same roof 'as in some of the Northern parts of our country'.[17] This change possibly reflected the growth of larger farms; it certainly revealed an increasing distaste for living in close proximity to animals. But houses of the older type or of a style closely similar to it, with men and animals under the same roof, were still being built in the northern and western regions of England.[18] In all parts of the country it remained common for unmarried farm labourers to sleep above the stable or cowshed, while in eighteenth-century Lincolnshire it was noticed that the owners of geese treated their birds 'with great kindness, lodging them very often in the same room with themselves'.[19]

In the towns of the early modern period animals were everywhere, and the efforts of municipal authorities to prevent the inhabitants from keeping pigs or milking their cows in the street proved largely ineffective.[20] The London poulterers kept thousands of live birds in their cellars and attics, while one Jacobean starchmaker is known to have had two hundred pigs in his backyard.[21] In 1842 Edwin Chadwick found that fowls were still being reared in town bedrooms and that not just dogs but even horses lived inside town houses.[22] For centuries, wandering pigs were a notorious hazard of urban life: they sometimes started fires by brushing straw into dying embers and they often bit or even killed small children; Sir Hugh Cholmley (born 1600) was attacked by a sow at the age of eight, and reports of child deaths from this cause continued into the nineteenth century. As Oliver Goldsmith remarked, few, even in cities, were unacquainted with the hog and its way of living.[23]

Dwelling in such proximity to men, these animals were often thought of as individuals, particularly since, by modern standards, herds were usually small. Shepherds knew the faces of their sheep as well as those of their neighbours, and some farmers could trace stolen cattle by distinguishing their hoof prints; in Hanoverian Bury St Edmunds a man was hanged for sheep-stealing on the oath of an accuser who swore to the countenance of the sheep in question.[24] Sheep or pigs were not usually given individual names, but cows always were; not human ones,* for distance had to be preserved, but flower names, like Marigold or Lily, or descriptive epithets, often suggestive of an affectionate attitude on the owner's part. In Tudor Essex there were cattle called Gentle, Brown Snout, Old White Lock, Button and Lovely. In Yorkshire, Lovely again and Motherlike, Goldlocks, Bride, Winsome, and Welcome Home. Bullocks had flatter, less emotive labels, though some were called Dearlove or Proudlook. Working oxen, yoked together, bore stereotyped pairs of names, designed to sound distinct from each other when the ploughman called them out: Crisp and Curly or Hawk and Pheasant (a combination popular over at least four centuries).[25]

These domestic beasts were often dressed up with bells, ribbons and other finery.[26] They were also frequently spoken to, for their owners, unlike Cartesian intellectuals, never thought them incapable of understanding. 'Hillo, ho, ho, boy! Come, bird, come,' says Hamlet; and the dialect dictionaries give us a fine range of such modes of address. Geese and chickens were called to their food: 'Yuly, yuly!', 'Coom biddy' (come, I bid thee); sent away: 'Shoo, shoo!' 'Shough, shough!' Pigs were called: 'sic, sic, sic' in the north; 'chuck, chuck' in Hampshire; 'sug, sug' in Norfolk; 'sook, sook' in Devon. 'Bawk up,' said the Suffolk milkmaid, as she tied up her cows. 'Rynt thee,' said her Cheshire counterpart, meaning 'Move over, I've finished.' 'How up, how up,' shouted the men as they drove the cattle.[27]

Even bees could be communicated with, for, when they swarmed, their owners would whistle, clap their hands, ring bells and tinkle basins and kettles. This was an ancient practice, going back to Roman times, but still universally observed in eighteenth-century England.[28] Its original purpose seems to have been to warn neighbours of the approaching swarm and to prevent disputes by establishing the owner's right in advance. 'The tinkling,' as one expert put it, 'secures a legal right to follow your swarm upon another person's grounds in order to hive them.'[29] But by early modern times the noise was widely regarded by country people as a means of addressing the bees themselves. It was thought to prevent them from flying very far; it made them 'knit' and encouraged them to settle sooner.[30]

For working animals there was a much wider vocabulary. 'Horses and mules,' said a seventeenth-century author, 'understand carters' language,

* In 1698, however, a Dorset squire referred to 'my old bursten-belly'd cow called Matthew'; *The Retrospective Review*, i (1853), 411.

who with their terms of art, as "Gee" and "Ree" and the like, will make them go or stop, turn on the right hand or the left, as they please.'[31] 'Hayt Scot, Hayt Brock!' says the carter in Chaucer's *Friar's Tale*; and in nineteenth-century Suffolk 'Scot' and 'Brock' were still names for carthorses, and 'heit' was still a call, meaning 'turn to the left'.[32] When eighteenth-century travellers heard the bass voices of the ploughmen and the counter-tenor of their boys driving ox teams they were reminded of the chanting of a cathedral choir.[33] As for injunctions to horses: 'Or', 'Whor', 'Woot', 'Hoot', 'Ree', 'Heeck', 'Wo', 'Wey', 'Prut', 'Put': the list is endless. It was an ancient language, much of it Celtic or Anglo-Saxon.[34] Aristocratic horsemen used a more lordly vocabulary. The riding masters recommended 'Ha, Villain!', 'Diablo!', and 'such-like threatenings'; or, if the horse was to be praised, 'Holla! So boy, there boy, there.'[35] Professional horse-breakers acquired the reputation of being able to communicate by whistling or by mysterious whispering in the animal's ear; in fact, they probably exploited the horse's sense of smell.[36] As an eighteenth-century writer observed, the conventional methods of training dogs and horses would have been absurd if the animals had been machines, lacking understanding.[37] Dogs and horses, Friedrich Engels would say a century later, had so learned to understand human beings that 'anyone who has much to do with such animals will hardly be able to escape the conviction that there are plenty of cases where they now *feel* their inability to speak is a defect, although unfortunately it can no longer be remedied, owing to the vocal chords being too specialized in a definite direction.'[38]

Domestic beasts were often treated as morally responsible. Dogs and horses were trained by an elaborate system of rewards and punishments, developing an individual 'character' in the process. 'Their affections, passions, appetites and antipathies,' wrote the third Earl of Shaftesbury, were 'as duly regarded as those in human kind, under the strictest discipline of education.'[39] Some were stubborn or stupid, others bright and willing. If idle or malicious, they were punished like humans. 'The Captain's bitch killed a lamb yesterday,' wrote the Virginian gentleman William Byrd in 1710, 'for which we put her into a house with a ram that beat her violently to break her of that bad custom.'[40] England has no real counterpart to that curiosity of continental legal history, the trial and execution of homicidal animals. Animals were not capable of guilt, said the moralists. The Old Testament prescribed death for beasts involved in homicide or bestiality, not as a punishment, but as a symbolic way of expressing abhorrence of the crime and respect for human life.[41] Yet the early Christian Church in England had prescribed the killing of animals polluted by sexual intercourse with humans and of bees who had stung a man to death.[42] When Englishmen moved to Massachusetts in the seventeenth century, they followed biblical precedent by ordering (and carrying out) the execution of animals involved in cases of bestiality; at Tyburn in 1679 a woman and a dog were hanged together for the same reason.[43] There were also plenty of informal trials of animals. Elizabethan sailors revenged the injuries they had suffered from sharks by catching a shoal of the fish and torturing them. When a bear killed a child in the reign of James I, the king ordered that it should be baited to death; the same fate befell a savage horse in 1682. In the countryside dogs caught poaching or killing sheep were frequently hanged in a grotesque imitation of Tyburn.[44]

In many ways, therefore, domestic beasts were subsidiary members of the human community, bound by mutual self-interest to their owners, who were dependent on their fertility and wellbeing. As Sir Kenelm Digby remarked in 1658, 'there's not the meanest cottager but hath a cow to furnish his family with milk; 'tis the principal sustenance of the poorer sort of people, ... which makes them very careful of the good keeping and health of their cows.'[45] On a ship, dogs and cats were so much a recognized part of the crew that the first Statute of Westminster (1275) ruled that a vessel was technically not abandoned so long as either animal remained aboard. When the fishing ship *Anne* set out from Hull in 1532 it naturally carried 'a dog and a cat, with all other necessaries'.[46]

Bees were also within the human community, and it was believed that they would not thrive unless treated accordingly. Bees would hate you, said an authority, if you did not love them.[47] They would not make honey if their owners were dirty, quarrelsome, or unchaste.[48] They were not to be ignominiously bought for money, but only exchanged for food or some useful commodity.[49] When there was a death in the family they had to be told at once and given a share of the funeral repast, else they would die themselves or leave in umbrage.[50] If what was said of bees was right, thought Sir William Petty, 'their souls seem ... like the souls of men.'[51]

It is therefore perfectly true that, as one late-seventeenth-century observer contemptuously put it, 'farmers and poor people' made 'very little difference between themselves and their beasts'.[52] They went out with them in the fields in the morning, toiled with them all day and returned home with them in the evening. Their very language expressed their sense of affinity between them and their animals, for many descriptive terms applied equally to either. Children were 'kids', 'cubs' or 'urchins'; a boy apprentice was a 'colt'; and the same term was indiscriminately used for a weakly child or the runt of a litter. A woman expecting a baby was said to have 'got upon the nest'. Her husband would address her affectionately as 'duck' or 'hen', less affectionately as 'cow', 'shrew', 'bitch' or 'vixen'. When she grew old she would become a 'crone', that is a ewe who has lost her teeth. In rural Northamptonshire a foolish person was said to be 'as wise as Walton's calf, who ran nine miles to suck a bull'.[53] The same sense of affinity between men and animals was exhibited at higher social levels. Queen Elizabeth gave all her courtiers animal nicknames; and in 1579 her future Lord Chancellor, Sir Christopher Hatton, even signed a letter to her, 'your majesty's sheep'.[54]

This continuing use of animal analogy and metaphor in daily speech reinforced the feeling that men and beasts inhabited the same moral universe and that terms of praise or reproach could be applied interchangeably to either. Of course, such analogies are still used today, but they lack the immediacy conveyed in the early modern period by the sheer proximity of animal life. Some men are still as bald as a coot. But how many of us have ever seen the bird in question? Already in the early modern period the growth of towns and industry was eroding this familiarity. Addison remarked in 1711 that the London streets were 'quite full of Blue Boars, Black Swans and Red Lions'. But, in fact, animal devices were being less commonly used as street signs in the eighteenth century than they had been in earlier periods; and it is revealing that, even in the late Middle Ages, merchants' marks rarely employed animal emblems in contrast to aristocratic heraldry with its almost totemic devices.[55]

ii. PRIVILEGED SPECIES

Certain favoured animals, however, remained close to human society and may even have been growing closer. The first of these was the horse. It is true that England was proverbially a hell for horses[1] and that many were literally ridden to death. 'Pursuivants that ride from post to post,' says an early Tudor phrasebook, 'destroy many horses.'[2] When used for draught and burden they could be harshly treated. 'How often have I seen them fainting under their loads,' exclaimed a preacher in 1669, 'wrought off their legs and turned out, with galled backs, into the fields or highways to shift for a little grass. Many times have I heard and pitied them, groaning under unreasonable burdens and beaten on by merciless drivers till, at last, by such cruel usage, they have been destroyed and cast into a ditch for dogs' meat.'[3] Seventeenth-century illustrations of carts and carriages invariably show their drivers carrying enormous whips; and in the 1720s John Gay described a London street scene:

> The lashing Whip resounds, the Horses strain,
> And Blood in Anguish bursts the swelling Vein.[4]

Horses employed for socially more pretentious purposes might also be severely used. After a day in the field, wrote a riding-master in 1655, it 'would pity the heart of him who loveth a horse to see them so bemired, blooded, spurred, lamentably spent, tired out'.[5] When worn out they were quickly discarded. Of a horse that is no use for work, remarked a preacher, 'every man will say, better knock him on the head than keep him ... His skin, though not worth much, is yet better worth than the whole beast besides.' The suffering of old hunters, thought one eighteenth-century traveller, was one of the most disagreeable spectacles to be seen on the English roads.[6] As Gulliver sheepishly explained to the Houyhnhnms:

> horses were the most generous and comely animals we had ... and [when] they belonged to persons of quality, employed in travelling, racing or drawing chariots, they were treated with much kindness and care, till they fell into diseases, or became foundered in the feet and then they were sold and used to all kind of drudgery till they died; after which their skins were stripped and sold for what they were worth and their bodies left to be devoured by dogs and birds of prey.

One morning in 1581 Sir Thomas Wroth counted 2,100 horses travelling between Shoreditch and Enfield; but another observer added that within the next seven years 2,000 of them would be dead in some ditch through overwork.[7]

Nevertheless, so long as the horse contributed to its owner's self-esteem, it was highly valued. Often, its upkeep was a greater burden on its owner than were the wages of a human servant.[8] The horse, said Edward Topsell, was 'the most noble and necessary' of quadrupeds. It was praised for its aristocratic qualities of courage and 'generosity' and credited with many semi-human attributes. Gervase Markham had no doubt that horses felt all the passions of love, hatred, joy and sorrow.[9] The Duke of Newcastle's horses were so highly trained 'that they wanted nothing of reasonable creatures, but speaking'.[10] In Elizabethan times the riding-manuals prescribed some ruthless techniques for dealing with recalcitrant mounts, including putting hot irons on their buttocks, blazing straw about their ears and a hedgehog or 'a shrewd cat' under their tail.[11] By the seventeenth century techniques had softened and it became normal for the books to advise humane methods, urging that there should be 'a sincere and incorporated friendship' between horse and rider. The horse, said one riding-master, was 'the most loving creature to man of all other brute creatures'; and there were many owners like Cowper's Jack, who

> Lived in the saddle, loved the chase, the course
> And always, ere he mounted, kissed his horse.[12]

The hawk was another creature much cherished by its master. It too was praised for its 'greatness of spirit'; and a close relationship was usually established between the bird and its trainer. As one Jacobean falconer remarked, 'there cannot be too much familiarity between the man and the hawk'; the owner's aim should be to have his hawk 'in love' with him.[13]

But the most favoured of all animals was the dog. Dogs were ubiquitous in early modern England: Fynes Moryson thought England had proportionately more of them than any other country.[14] As a means of protecting private property, the household mastiff was a great deal more important than the village constable or Justice of the Peace; and even the forest law, which was hostile to privately-owned dogs, conceded the need for 'curs ... to bark about men's houses in the night'. The mastiff, ruled Sir Edward Coke, was a necessity to defend the house and to give warning of the approach of thieves and robbers.[15] Municipalities struggled to ensure that such creatures were locked up or muzzled during the daytime, but they remained a notorious hazard.[16] In the mid-seventeenth-century diary of Ralph Josselin we meet a mad dog which bites a pig, which subsequently dies ('blessed be God it was not a child'); a great mastiff bitch which runs mad and snaps at Josselin's son; 'Coleman's dog', which attacks Josselin himself; Mr Clark's dog, which 'flew on me and rent my coat very much'; and the diarist's own dogs, who kill a lamb and a sheep. It is not surprising that in one of his dreams Josselin 'had much ado to keep a fierce mastiff off'. The biographer of Seth Ward, the great bishop of Salisbury, tells how, when the bishop once went to stay 'at a gentleman's house ... if I mistake not in Cambridgeshire', he went out at night to the 'necessary house'; it was at the end of a long garden, where he was unlucky enough to meet a large mastiff, which had been chained up in the day but was now running loose. We have a graphic account of the ensuing contest, which ended with Dr Ward pinning the dog to the ground and lying on top of him, uncertain what to do next.[17]

In David Loggan's late-seventeenth-century engravings of Cambridge there are dogs everywhere. St John's has a huge creature at the entrance; Trinity has two outside and three in the front court, one of which is being egged on to fight the college's captive eagle. King's has a dog on the lawn and two fighting inside the chapel (Christ's and Trinity, by contrast, employed a special servant to keep dogs out of the chapel[18]). The overall total comes to thirty-five, while in the companion pictures of Oxford there are no fewer than fifty-six dogs, many of them fighting each other or running after passers-by.[19]

Many of these seventeenth-century dogs had practical functions. They pulled carts, sleds, even ploughs.[20] They were indispensable to shepherds, drovers, farmers and butchers. In big houses they served as turnspits.[21] A few were even used to track criminals.[22] Often there was a close attachment between dog and owner, particularly in the case of sheepdogs, whose marvellous skills were understandably admired.[23] But usually these working dogs seem to have been regarded unsentimentally; and they were generally hanged or drowned when they had outlived their usefulness. 'My old dog Quon was killed,' wrote a Dorset farmer in 1698, 'and baked for his grease, of which he yielded 11 lbs.'[24] It was not these necessary animals, but the unnecessary ones, hounds and lap-dogs in particular, which received the real affection and the highest status.

Then, as now, the passion for unnecessary dogs started with the royal family. The Stuarts were obsessed by them. James I had his favourite hounds, Jowler and Jewell, the latter unfortunately shot by his wife, Anne, in mistake for a deer. Anne herself was painted by Paul van Somer along with a horse and five dogs; their son Henry had his favourite horse done lifesize in 1611 by a Florentine artist and himself portrayed in the company of a red mastiff; while their daughter Elizabeth, the Winter Queen, lived surrounded by dogs, birds and horses and was notorious for preferring her pets to her children. To James, even Robert Cecil was his 'little beagle', Buckingham his 'dog Steenie'.[25] When presented with Dr Caius's monograph on the antiquity of Cambridge University, the king is said to have remarked rather ungraciously, 'What shall I do with this book? Give me rather Dr Caius's *De Canibus*.' It is not surprising that James was accused in 1617 of loving his dogs more than his subjects.[26]

His successors continued the tradition. Charles I's wife gave birth prematurely in 1628 after being involved in a fight between large dogs in the gallery at Greenwich; later in life she visited the diarist John Evelyn and 'recounted ... many observable stories of the sagacity of some dogs that she had formerly had'. Charles's nephew Rupert had a white poodle, Boy, who became a celebrated figure in the 1640s, attracting much satirical comment; and the king himself only parted with his own dog after receiving sentence of death in 1649.[27] Charles II was notorious for playing with his dog at the Council table, and everyone knows someone who knew someone who knew someone who saw him walking with his spaniels.[28] His brother James took his dogs to sea with him when he was Admiral and in 1682 in a bad shipwreck, when many sailors were drowned, was alleged to have disgraced himself by crying out, 'Save the dogs and Col. Churchill!' This was certainly a gross libel, for some at least of the dogs had to fend for themselves; it is known that Sir Charles Scarborough, the Duke's physician, engaged with Mumper, the Duke's dog, in a humiliating tussle for the last remaining plank.[29]

The aristocracy had similar tastes. As the proverb said, 'he cannot be a gentleman who loveth not a dog.'[30] Greyhounds and spaniels were always acceptable gifts between aristocrats;[31] and a gentleman's hounds were treated with much indulgence. When masters returned from the hunt, observed an early Stuart commentator, they would often show 'more care for their dogs than of their servants and make them lie down by them, and often the servant is beaten for the dog; you may see in some men's houses fair and fat dogs to run up and down and men pale and wan to walk feebly.' Later in the century the godly Ambrose Barnes related with disgust how, in his father's household, when the dinner was keeping warm in front of the fire for the huntsmen's return, the hounds would often run before them into the kitchen and snatch the meat away, while no one dared complain.[32] Hounds were often better fed than the servants, and they were sometimes better housed; John Byng noted in 1794 that the Duke of Bedford's foxhound kennels towered proudly over the hovels of his labourers.[33] As was repeatedly pointed out, hounds consumed food which could have been used to relieve the poor;[34] and the gentry's practice of compelling their tenants to rear young puppies for them was also attacked. 'How oft may we see greedy landlords force their tenants to feed their dogs with what should feed their own children,' exclaimed a late-seventeenth-century preacher, 'A barbarous custom!'[35]

At the royal court and in great houses dogs were everywhere. Late medieval books of courtesy reminded the page that before his lord went to bed he should drive the dogs and cats out of the bedroom; and they cautioned guests at banquets against stroking dogs or cats while sitting at the table.[36] Lord Savage's house at Long Melford occasioned special comment in 1619 because it was so neatly kept, without a dog or a cat 'to cause any nastiness within the body of the house'.[37] More typical was Woodlands, the Dorset home of Henry Hastings, second son of the fourth Earl of Huntingdon, where in 1638, it was said, the great hall was strewn with marrow bones and swarmed with hawks, hounds, spaniels and terriers. The walls were hung with the skins of recently-killed foxes and polecats, while in the parlour favoured dogs lay around the hearth. There were litters of cats on the chairs and on the tables stood hawks' hoods, bells and hats full of pheasants' eggs.[38] By the later seventeenth century polite society was coming to despise this old way of housekeeping 'with dogs' turds and marrow-bones as ornaments in the hall'. The forecourt in front of great houses was kept free from dogs; and in due course the new fastidiousness would lead to the supersession of the dog turnspit by a mechanical jack.[39]* But illustrations suggest that dogs still ran round the tables at banquets; and when the Duchess of Marlborough went to Scarborough in 1732 she was unable to sleep at night because of the 'barking and howling of dogs and hounds, which is kept all around me for the entertainment of fine gentlemen in this place'.[40]

Lower down the social scale the story was the same. An Englishman, thought John Bunyan, would rather go for a walk with a dog than with

* At Park House, Gateshead, in 1723 William Cotesworth ordered the dog-wheel to be moved 'on purpose to keep the dog from the fire, the wheel out of the way and the dog prevented [from] shitting upon anything it could'; Edward Hughes, *North Country Life in the Eighteenth Century* (1952–65), i. 30–31.

a Christian: 'Some men cannot go half a mile from home, but they must have dogs at their heels.' Indeed, dog-ownership was widespread; as Adam Smith later remarked, the poorest family could usually manage to keep one without extra expense.[41] In Elizabethan New Romney all dog-owners had to register their property; as a result we have a fine list of the town's dogs starting with the Mayor's 'great, bald, branded mastiff', down through umpteen spaniels, shocks, and turnspits to Mr Downton's red cur without a tail.[42] The dog population of England was periodically reduced at times of plague, when municipal authorities endeavoured (not without strenuous resistance by the dog-owners) to have the animals destroyed as a sanitary measure.[43] But their numbers do not seem to have been permanently affected. There had been complaints about the excessive number of dogs since at least the 1530s, yet it was one of the few matters which even Thomas Cromwell did nothing about.[44] A tax on dogs was periodically suggested, and in the eighteenth century there were several unsuccessful Bills on the subject. Only in 1796 did the dog-tax at last come in.[45] By then it was said that there was 'scarce a villager who has not his dog'; and the total dog population was thought to approach a million, of which the majority were kept more for pleasure than for practical need.[46]

This state of affairs reflected a steady rehabilitation of the animal in question. The Eastern view of dogs as filthy scavengers had been transmitted via the Bible to medieval England and was still widely current in the sixteenth century. The Book of Revelation suggested that at the Resurrection dogs, like other unclean beings, would be excluded from the New Jerusalem. Some commentators thought this meant 'men of dogged impudency and maliciousness',[47] but most took it literally. Chaucer has nothing good to say about the dog and neither has Shakespeare. In popular proverbs there was no suggestion that the dog might be faithful and affectionate; instead, we have 'as greedy as a dog', 'a surly as a butcher's dog', and 'a dog's life'.[48] Fine ladies, observed the Elizabethan Thomas Muffett, hated lice, even 'more than dogs and vipers'.[49] The dog, said a Jacobean preacher, was an emblem of greed and shamelessness: 'a most unclean and filthy creature which goeth publicly and promiscuously to generation'. Dogs were filthy, beastly, quarrelsome creatures, agreed the sectary George Foster.[50] In 1662 the preacher Thomas Brooks classified dogs with 'vermin', and in eighteenth-century painting the dog frequently remained a symbol of man's baser parts: he represented gluttony, lust, coarse bodily functions and general disruptiveness.[51] 'In all countries and languages,' declared a mid-eighteenth-century author, '"Dog" is a name of contempt.'[52] As

Freud would observe, dogs were reprehensible because they had no horror of excrement and no shame about their sexual functions.[53]

Yet there were dogs and dogs.

> Hounds and greyhounds, mongrels, spaniels, curs,
> Shoughs, water-rugs and demi-wolves, are clept
> All by the name of dogs.

The mastiffs and mongrels were lecherous, incestuous, filthy and truculent, and the butcher's cur snarling, angry, peevish and sullen. But the hound, by contrast, was noble, sagacious, generous, intelligent, faithful and obedient.[54]

The reason for this distinction was essentially social. Dogs differed in status because their owners did. As an early-eighteenth-century author observed, people tended to have dogs appropriate to their social position. The squire owned hounds and the aristocratic sportsman possessed greyhounds and setters. But the tinker would be followed by a mongrel, and 'yelping curs' were the property of 'alley scoundrels'.

> For e'ery mortal that is prone to
> Keep a dog, will pick out one
> Whose qualities are like his own.[55]

One feature of the game laws was that, since the late fourteenth century, they had confined the ownership of hunting dogs to persons above a specified social level. This was the distinction recognized by the Mayor of Liverpool in 1567, when he ordered that mastiffs and bandogs (i.e. watchdogs) should be kept tied up and not allowed to roam the streets, 'for avoidance of sundry unconvenience as for hurting of greyhounds, hounds and spaniels, *that is gentlemen's dogs*'.[56]

The higher status of the hound thus reflected the hunting tastes of the medieval nobility; and it was no accident that it was its fidelity which was its most celebrated quality, for that was also the essential virtue of the chivalric knight. By the twelfth century there were many famous exemplars of canine fidelity;[57] and they nearly all were hounds. The only dog to become a saint was a French greyhound, unjustly killed after saving a child from a snake in the diocese of Lyons; in the thirteenth century the common people knew him as St Guinefort, and healing miracles on sickly children were performed at his tomb until the Dominicans suppressed the cult.[58] On medieval tombs the hound lay at his master's feet as an emblem of fidelity.

In Tudor times this tradition of the dog's loyalty proved capable of wider extension. It was not a hound but a mastiff which saved the life of Elizabeth I's master of the Armoury, Sir Henry Lee; he had it painted with himself in a large portrait, to which he attached verses celebrating the animal's love and devotion.[59] Praise of dogs in Elizabethan and Jacobean literature was not confined to a particular breed, but was often expressed in general terms. The dog was the best of all animals because it was the creature which came nearest to man; it was 'a natural, kind and loving thing'.[60] To associate only vices with dogs was absurd, observed Sir John Davies in his epigram *In Cineam* (written by 1594):

> Thou sayest thou art as weary as a dog,
> As angry, sick, and hungry as a dog,
> As dull and melancholy as a dog,
> As lazy, sleepy, idle as a dog.
> But why dost thou compare thee to a dog?
> In that for which all men despise a dog,
> I will compare thee better to a dog.
> Thou art as fair and comely as a dog,
> Thou art as true and honest as a dog,
> Thou art as kind and liberal as a dog,
> Thou art as wise and valiant as a dog.

Dogs, agreed John Worlidge in 1669, were 'the most observant and affectionate of all beasts whatever to mankind. They love even to the loss of their lives in defence of their master, his cattle, goods, etc.'[61]

By this time the taste for dogs had progressed so far as to be an issue on which contemporaries were prepared to overrule even the Bible. In Scripture, said the preacher Joseph Caryl, 'dog' was 'a term, not only of some diminution, but of utmost disgrace'; whereas, in fact, 'some dogs have very many good qualities in them'. The dog had been regarded as vile and ignominious, agreed Timothy Nourse in 1686, but, 'upon a just consideration', it should be seen as 'the greatest emblem of heroic virtue', eminent for 'fidelity, gratitude and courage'.[62]

Courage, however, was scarcely a feature of the other dog cherished in these years, the lady's lap-dog: usually a toy spaniel in the early sixteenth century and a pug in the seventeenth (Chinese dogs did not become popular until the nineteenth century). The fashion had started in the later Middle Ages and the essential requirement was that the pet dog should be very small. Various recipes even taught how to keep it artificially so.[63] 'The smaller they be,' wrote Abraham Fleming in 1576,

> the more pleasure they provoke, as more meet playfellows for mincing mistresses to bear in their bosoms, to keep company withal in their

> chambers, to succour with sleep in bed, and nourish with meat at board, to lay in their laps and lick their lips as they ride in wagons.[64]

No well-to-do woman was complete without a pet of this kind; and preachers lamented that fashionable ladies neglected their children, preferring to 'embrace a whelp or a puppy'. Anne Boleyn was so attached to her pet dog that no one dared inform her when it died and it was left to the King to tell her.[65] Pepys has a memorable vignette of Mrs Penington's little dog, which she rashly took to bed with her, with predictably unfortunate results, and over which she grieved so much when it died.[66]

The publication of sentimental books about dogs did not get under way until the nineteenth century, with Joseph Taylor's *The General Character of the Dog* (1804). That was the period which saw the rise of the dog show (1859), the foundation of the Kennel Club (1873) and the writing of innumerable poems about dogs with human eyes. By the mid nineteenth century it was a commonplace that there was 'no civilized land where the canine race is more the companion of man than in Great Britain or any nation which has so many valuable varieties of it'.[67]

But it was in early modern times that the foundations of this canine obsession had been laid. By the eighteenth century the dog was generally recognized as 'the most intelligent of all known quadrupeds' and praised as the 'trusty servant and humble companion of man'.[68] A long anecdotal tradition of dumb mastiffs dying of grief for the death of their masters has preceded the sentimental poems of the Romantic era about loyal old dogs. There was also a pronounced tendency to regard the dog as a symbol of the nation. English dogs had been in demand since Roman times and in the Elizabethan age it was customary to claim that they were better than those of any other country.[69] As early as 1619 the Virginia General Assembly passed a law prohibiting the selling or giving to Indians of dogs 'of the English race'; and Peter Beckford would later assert that English hounds were the best in the world.[70] But it was above all the bulldog, clinging tenaciously to its much larger opponent, which was singled out for special commendation. The British bulldog was described as 'probably the most courageous creature in the world'. 'The courage of bull-dogs,' agreed David Hume, 'seems peculiar to England.'[71] In the eighteenth century the 'ancient genuine race of true bred English bulldogs' became an accepted national emblem: 'excelling in fight, victorious over their enemies, undaunted in death'.[72] The animal neatly united the twin preoccupations of the eighteenth-century ruling class: a concern with pedigree and breeding, and a taste for aggressive war.

Cats were slower to rise in status. In the Middle Ages they were kept in houses for protection against rats and mice. Only occasionally do they appear as companions and objects of affection, as in the ninth-century poem by an Irish monk about his cat, Pangur Ban, or the fifteenth-century tomb at Old Cleeve, Somerset, which shows a man with his feet resting on a cat, which in turn has its paws resting on a mouse.[73] Many householders deliberately refrained from feeding them, so as to ensure that they had an incentive to hunt. The cat, ruled Topsell in 1607, was 'an unclean and impure beast that liveth only upon vermin and by ravening'. An allergy to cats was common and the dangers of their breath much discussed in the medical books.[74] An early Tudor textbook contains for translation into Latin the simple sentence, 'I hate cats (*horreo aluros sive feles sive cattos*).' It was only in the sixteenth century that Dick Whittington seems to have acquired his cat.[75]

Yet by the early Stuart period there were plenty of authentic cat-lovers, like the third Earl of Southampton, whose portrait commemorating his sojourn in the Tower after Essex's rebellion shows an extremely sleek and alert cat to have been his companion in imprisonment. Archbishop Laud was particularly fond of cats and in the late 1630s was given one of the earliest imported tabbies, then valued at £5 each, but soon to become so common as to supersede the old English cat, which was blue and white.[76]* In the same decade the house of the prominent Leeds merchant John Harrison had holes cut in the doors so as to allow cats free passage 'even into the best room of the house', as a later antiquary noted with evident surprise. By the reign of Charles II, according to Defoe, few London families were without them, 'some having several, sometimes five or six in a house'.[77]

It is true that many people still regarded cats as fair game for any sport. On New Year's Day 1638 in Ely Cathedral there was 'a great noise and disturbance near the choir' occasioned by the roasting of a live cat tied to a spit by one William Smyth in the presence of a large and boisterous crowd. A few years later Parliamentary troopers used hounds to hunt cats up and down Lichfield Cathedral.[78] During the Pope-burning processions of the reign of Charles II it was the practice to

* It was in Laudian Oxford that the future Ranter Abiezer Coppe, then a post-master of Merton College, kept 'a wanton huswife in his chamber ... to whom carrying several times meat, at the hour of refection, he would make answer, when being asked by the way what he would do with it, that "it was a bit for his cat" '; Anthony Wood, *Athenae Oxonienses*, ed. Philip Bliss (Oxford, 1813–20), iii. 959–60.

stuff the burning effigies with live cats so that their screams might add dramatic effect. At country fairs a popular sport was that of shooting at a cat suspended in a basket. As Alexander Pope remarked in 1713, 'the conceit that a cat has nine lives has cost at least nine lives in ten of the whole race of them.'[79]

But attitudes were changing. When Delia, the cat belonging to Walter Stonehouse, Rector of Darfield, Yorkshire, died in the mid seventeenth century he buried her in his garden and wrote a Latin verse epitaph.[80] In the eighteenth century the domestic cat established itself as a creature to be cosseted and cherished for its companionship. The antiquary William Stukeley was deeply affected by the death of his cat, Tit, 'an uncommon creature and of all I ever knew the most sensible, most loving and indeed with many other engaging qualities'. He grieved for her 'exceedingly', believing that she had 'sense so far superior to her kind' and 'such incontestable ways of testifying her love to her master and mistress'. English cats, like English dogs, thought Christopher Smart, were the best in Europe.[81] It is likely that the cat gained in popularity as standards of domestic cleanliness rose. In 1809 William Bingley thought that it was because of the animals' cleanliness and elegance that some people were 'passionately fond of cats'. But he also noted that they exhibited 'many pleasing traits of character' and were 'susceptible of considerable educational attainments'. In the mid nineteenth century the cats'-meat man interviewed by Henry Mayhew told him that in London there was at least one cat for every ten people; and that they were twice as numerous as dogs.[82] The first cat show was in 1871.

Pet-keeping had been fashionable among the well-to-do in the Middle Ages, and monks and nuns were repeatedly (and vainly) forbidden to keep them. Pet monkeys were imported in the thirteenth century.[83] But it was in the sixteenth and seventeenth centuries that pets seemed to have really established themselves as a normal feature of the middle-class household, especially in the towns, where animals were less likely to be functional necessities and where an increasing number of people could afford to support creatures lacking any productive value. There were pet monkeys, tortoises, otters, rabbits and squirrels (of whom Topsell remarked that, apart from their tendency to devour woollen garments, they were 'sweet, sportful beasts and ... very pleasant playfellows in a house').[84] On the farms there were 'pet' lambs, reared by hand and sentimentally cherished.[85] In the eighteenth century, as sympathies widened, there would be pet hares,[86] pet mice,[87] pet hedgehogs, even pet bats[88] and pet toads. Thomas Pennant reports that in 1768 one Devonshire gentry household had been keeping a tame toad as the family pet for the past thirty-six years. He commended the 'good sense of a family which soared above all vulgar prejudices', and noted that 'even ladies so far conquered the horrors instilled into them by nurses as to desire to see it'.[89]

Finally, there were cage-birds, kept either for their song (as with canaries, nightingales, goldfinches, larks and linnets) or for their imitation of the human voice (as with parrots, magpies and jackdaws). 'We think it no great matter for a man to cause a pie or popinjay to utter certain distinct words and speeches,' remarked an Elizabethan preacher. James I had a pet kingfisher and Charles II a pet starling.[90] Commercial bird-dealers had made their appearance by Tudor times, and in the late seventeenth century there was a large London market in singing-birds, some caught at home by professional bird-catchers, others exotics imported from the tropics. Canaries, which had been imported annually by the thousand since the mid sixteenth century, were by this time bred domestically and said to be so plentiful that 'even mean persons' could afford to buy them.[91] In the eighteenth century jays, thrushes, bullfinches, starlings, wrens, cuckoos and wild birds of every kind were captured and sold in the London bird markets.[92] 'The chaffinch,' as a Victorian commentator observed, was always 'a favourite cage-bird with the lower classes.'[93] These birds could be the object of close affection. Samuel Pepys was 'much troubled' when his canary died, while in her old age the nineteenth-century scientist Mary Somerville was still upset by the recollection of the death of her pet goldfinch when she was a child.[94]

Certain wild birds became honorary pets without being captured. This was particularly the case with the robin, whose distinctive appearance and readiness to frequent human habitation in search of winter food had endeared him to generations of householders. In Jacobean times the robin was already hailed as the bird 'that best of all loves men'; and in the ensuing centuries innumerable poets would celebrate 'the household bird',

> the bird whom man loves best,
> The pious bird with the scarlet breast.[95]

It was entirely appropriate that Joseph Taylor, having treated *The General Character of the Dog* in 1804, should subsequently turn to *Tales of the Robin* (1808), an anthology of poetic tributes to this supposedly 'harmless' and 'innocent' 'little friend' of man.

In response to the emergence of the pet as an animal kept for private emotional gratification, the law itself was gradually adapted to accommodate the new notion that a pet could be a piece of property, even when not used for draught or food. In a case of 1521 a judge denied that property could exist in tamed animals whose only use was to give pleasure; they might be possessed, but they could not be owned. But the other judges disagreed. 'If I have a popinjay or thrush which sings and refreshes my spirit,' said one, 'it is a great comfort to me and if any man takes it from me he does me a great wrong.' Even if not subject to larceny, such birds could be owned, and there could be a private action to get them back.[96] In 1588 it was conceded that a dog, 'being a thing that is tame by industry of man', could indeed be an object of property and that the law recognized four kinds of dog, namely mastiffs, hounds (including greyhounds), spaniels and tumblers. Another case in 1611 confirmed that monkeys and parrots could be commercial property.[97] Even so, the law was reluctant to prosecute those who stole pets. Michael Dalton's *Country Justice* explained in 1655 that it was not larceny to steal dogs, apes, squirrels, parrots and singing birds, if kept only for pleasure, even 'though they be in the house and tame'.[98]

Three particular features distinguished the pet from other animals. First, it was allowed into the house. Abraham Fleming defined a dog as 'a creature domestical or household servant, brought up at home with offals of the trencher and fragments of victuals'. But household servants did not necessarily eat with the family. Henry Carey (d. 1743) has a poem about the farmer whose mastiffs were never let indoors. His wife kept the house

> too nice and neat
> For dogs to traipse with dirty feet.

His hound by contrast was

> Caress'd and lov'd by every soul
> He ranged the house without control.[99]

Even today it is the farmer's working dog who is unlikely to be allowed into the house, whereas pets are more favourably treated; just as the officers who excluded dogs from the court of Henry VIII made an exception for the ladies' spaniels.[100]

Pets also went to church, despite the constant efforts of Tudor bishops to make congregations leave their dogs, hawks and monkeys behind. In the reign of Edward VI a proclamation even complained of 'the common bringing of horses and mules' into churches.[101] There was an official dog-whipper in almost every parish church[102] and one of the main purposes of the Laudian communion rails was to keep dogs away from the altar.[103] Yet the church court records abound in cases like that of the Yorkshirewoman who, in 1632, while the minister was delivering communion, 'did dangle a dog on her knee and kiss him with her lips', or the Cambridgeshire man who in 1593 brought to church a dog which disturbed the congregation because of all the bells it was wearing.[104] When lightning struck Widecombe Church in 1638 the casualties included a dog running through the chancel door; when it struck again at the Cornish church of St Anthony in Meneage on Whit Sunday 1640, it killed 'one dog in the belfry and another at the feet of one kneeling to receive the cup'.[105] In the mid eighteenth century a contemporary observed that 'we may often see a footman following his lady to church with a large common-prayer book under one arm and a snarling cur under the other'.[106]

The second feature of the pet was that it was given an individual personal name. This distinguished it from all other creatures. Of course, wild birds might, as we have seen,* be called Jack or Bessie, but such names were attached to the whole species, not to an individual. Birds were only metaphorically human. Dogs, horses and other domestic animals who were adjuncts to human society (metonymical humans, as Claude Lévi-Strauss calls them) had also long been given names.[107] But their names were only semi-human and emphasized their social distance. Since classical times it has been customary to give dogs short names which were easy to call out. Hounds usually bore descriptive labels. When Peter Beckford came to write his *Thoughts on Hunting* in 1781, it was customary to give all the foxhounds in a single litter a name beginning with the same initial. Beckford proposed a long list of suitable names, mostly epithets or occupations; under 'E', for example, he suggested Eager, Earnest, Effort, Elegant, Envoy. Many of these hound-names were aggressively masculine in their connotations: Arrogant, Active, Angry, Bachelor, Barbarous, Boisterous, Ranter. Ringwood, Bellman, Jowler and Merryman had been popular for centuries.[108]

Horses also normally lacked genuine human names. In Tudor Yorkshire, for example, they might be called Bayard, Rivers, Sharlocke, Greywood, Burrill, Galloway, Greenwood or Throstle.[109] In modern times racehorses are rigorously individualized, no two animals having the same name; their names, however, are not human and they have no

* Above, p. 82.

descriptive value. This was to some extent true of the names given to English thoroughbred stock from the start, though in the seventeenth and early eighteenth centuries there was a discernible tendency to give mares distinctly erotic names: Sweet Lips, Miss Hip, Lady Thigh, Venus, Brown Betty, Creeping Kate and Darling are all recorded.[110]

In none of these cases was there any risk of confusion between animal and human. But certain special beasts had a more ambiguous nomenclature. The bears used by the Elizabethans for baiting, for example, often had names quite indistinguishable from human ones, like 'Harry Hunks', 'George Stone', 'Little Bess of Bromley' and 'Ned of Canterbury'.[111] Apparently they were named after their owners. Certain horses were similarly associated with their masters. Among the Tudor gentry it was even customary to give the animal the surname of the family from whom it had been acquired. Thus Lord Petre's horses included Bay Wadham, Pied White and Sorrel Greville, while in Sir Francis Walsingham's stable in 1589 there were Grey Bingham, given him by Sir Richard Bingham, Bay Sidney, given by one of the Sidney family, and Pied Markham, given by Mr Robert Markham, father of Gervase Markham.[112] When Lord Williams of Thame died in 1559 he left a mare called Maude Mullford and a great horse called Sorrell Williams.[113] In 1846 the Two Thousand Guineas was won by a horse owned by Sir Tatton Sykes and also called Sir Tatton Sykes.[114]

The more the animal was doted on by its owner, the more likely was it to bear a human name. Henry, first Lord Berkeley, had two particularly choice falcons, whom he named Stella and Kate; and in 1626 the Essex gentleman Sir Gawen Harvey bequeathed his kennel of beagles to Bishop Harsnet, 'all but Nancy'.[115] The dog at the foot of the brass of Sir Bryan de Stapleton (dated 1438) bore the label 'Jakke', while in Deerhurst Church, Gloucestershire, a brass of *c.* 1400 records that the pet dog of the wife of Sir John Cassy was called 'Terri'.[116] Of course, pet animals often had names which were non-human or mock-human, like Shock, Bouncer or Towser. But there was a recurring tendency, which in the eighteenth century became very pronounced, to give pets human names; and the shift was indicative of a closer bond between pet and owner. When we find Christopher Smart's cat called Jeoffry, and Shenstone's Lucy, and Gilbert White's tortoise Timothy,* and Southey's old spaniel Phillis, we know that we are confronted by a relationship of altogether greater intimacy. In earlier times moralists had insisted

* A post-mortem revealed that 'Timothy' was in fact female; Cecil S. Emden, *Gilbert White in his Village* (1956), 101.

that animals should never be given Christian names.[117] But in the late eighteenth century Priscilla Wakefield described two linnets, Robert and Henry, named after the owner's favourite schoolfriends; a friend of the antiquary Joseph Ritson had a dog called Ritson; and Jeremy Bentham owned a tom-cat called Sir John Langborn, though as he grew older and more sedate, his name was changed, first to the Reverend John Langborn and then to the Reverend Doctor John Langborn. In France, where different attitudes prevail, dogs have never been given Christian names.[118]

Thirdly, the pet was never eaten. This was not for gastronomic reasons. As an early Stuart author remarked, 'Cats *have* sometimes been eaten, by some of purpose, and by others unawares, who never found any offence by this food.'[119] It was not because of their taste, but because of their close relationship to human society that such animals were not consumed. Of course, cats and dogs were unacceptable as food anyway because they were carnivorous; Topsell explained that the flesh of cats was dangerous 'by reason of their daily food, eating rats and mice, wrens and other birds which feed on poison'; and the poet Cowper referred succinctly to 'flesh obscene of dog'. But it was the social position of the animal as much as its diet which created the prohibition. As Bernard Mandeville put it, some people would not eat 'any creatures they have daily seen and been acquainted with'.[120] Hence such traumas as that described by Mary Howitt in her poem on *The Sale of the Pet Lamb*; or the embarrassment of an English traveller when invited to share a meal with the natives of the Sandwich Islands. 'The idea of eating so faithful an animal as a dog,' he reported, 'prevented any of us in joining in this part of the feast; although, to do the meat justice, it really looked very well when roasted.'[121] The dog was one of man's best friends; it could not be food as well.

The same was true of the horse, although here the taboo was reinforced by the association of horse-flesh with Northern paganism. In the mid fifteenth century the Prior of Hatfield had deftly managed to bring an end to the burdensome annual custom of providing a harvest dinner for the tenants of his lordship of Canfield by deciding to serve them horse-meat. It was only the wild Irish who ate horses, 'contrary to nature'.[122] Francis Bacon records that in England 'some gluttons' would eat colts' flesh baked, but the popular attitude was expressed by the breathless title of one seventeenth-century ballad: *News from More-Lane, or a mad, knavish and uncivil frolick of a tapster dwelling there, who buying a fat coult for eighteen pence, the mare being dead and he not knowing how to bring the coult up by hand, killed it, and had it baked in a pastie and invited many of his neighbours to the feast and telling of them what it was, the conceit thereof made them all sick, as by the following ditty you shall hear ...*[123] In the late eighteenth century the notion lingered that horse-meat was bitter and unpalatable, and there was little support for occasional suggestions that it might become a regular feature of the English diet.[124] The sale of horse-flesh for human consumption was legalized in many European countries during the first half of the nineteenth century. But in the 1860s an organized attempt to convert the English to horse-meat proved a total failure, despite the holding of a special dinner at the Langham Hotel, where the menu began with *consommé de cheval*, moved through *mayonnaises de homard à l'huile de Rosinante* and ended with 'boiled withers'.[125]

Some of the objections to eating horses were practical: the animals were too expensive to rear for food alone, and if only old ones were eaten the meat would be tough and suitable only for the working classes. But the latter in turn strongly objected to being fobbed off with food which their superiors spurned. Underlying the argument, however, it is not difficult to detect the feeling that the horse was too noble a creature and too near to man to undergo such a fate. It was only in extreme famine that the English would break such tacit prohibitions. In the hard conditions of the 1620s dogs' flesh was thought 'a dainty dish' in many houses, and cats' meat was turned into 'good pottage'. Horses were eaten during the Civil War, but only as a last resort.[126]

In the same way the extension of an anthropomorphic attitude to wild singing-birds meant that larks, linnets and thrushes, once so popular a feature of the English diet, would slowly disappear from the middle-class menu. This disappearance was only a gradual process, for at the end of the nineteenth century thousands of larks were still being annually offered for sale in the London poultry markets.[127] Even so, the consumption of small birds was less common in Victorian times than it had been in the Stuart age. It has been said that 'there is no rational explanation why in England it is considered unkind to shoot any small bird except the snipe, whereas in Italy small birds are fair game'.[128] But it was surely growing prosperity which first led the English to regard small birds as too much trouble to prepare for the table, 'scarce worth the dressing, much less powder and shot', as John Ray put it. The relative poverty of the past explained why, as Thomas Pennant noticed in the late eighteenth century, 'our ancestors were as general devourers of small birds as the Italians are at present'.[129] The change was also assisted by the growth of new sensibilities. When John Ray visited Italy in 1665 he was disconcerted to discover that the Italians 'spare not the least and most innocent birds ... robin redbreasts, finches of all kinds, titmice, wagtails, wrens ... One would think that in a short time they should destroy all the birds of these kinds in the country.' A century later Tobias Smollett noted that one could travel through the South of France

> without hearing the song of blackbird, thrush, linnet, gold-finch or any other bird whatsoever. All is silent and solitary. The poor birds are destroyed, or driven for refuge, into other countries, by the savage persecution of the people, who spare no pains to kill, and catch them for their own subsistence.

When Mountstuart Elphinstone, the ex-Governor of Bombay, was travelling in Italy in the 1840s, he reacted with horror to the local habit of cooking nightingales, goldfinches and, worst of all, robins: 'What! Robins! Our household birds! I would as soon eat a child.' Yet in the Elizabethan age 'robin red-breasts' had been 'esteemed a light and good meat'.[130]

By 1700 all the symptoms of obsessive pet-keeping were in evidence. Pets were often fed better than the servants. They were adorned with rings, ribbons, feathers and bells; and they became an increasingly regular feature of painted family groups, usually as a symbol of fidelity, domesticity and completeness, though sometimes (as in the case of dogs) as an emblem of mischievous irreverence.[131] Meanwhile the aristocracy showed an increasing desire to be surrounded by individual portraits of their favourite dogs, birds and horses. The fashion had started by Jacobean times, when Sir John Harington owned 'an excellent picture, curiously limned', of his marvellous dog, Bungey, whose life and exploits he retailed at length to the young Prince Henry.[132] At Welbeck after the Restoration there were over a dozen life-sized portraits of the Duke of Newcastle's horses, while in 1681 in the hall of the Earl of Bridgewater's house at Ashridge, 'some good horses which my lord hath been owner of are drawn in full proportion'.[133] Artists like John Wootton (1688(?)–1765) and James Seymour (1702–52) helped to popularize this kind of portraiture. The purpose of painting, thought a late-eighteenth-century gentleman, was not to make indecent representations of naked Venuses and dying saints, but to survey one's ancestry, and restore the memory of former horses and faithful dogs.[134] As Benjamin Marshall, a racehorse painter (d. 1835), remarked, 'I discover many a man who will pay me fifty guineas for painting his horse who thinks ten guineas too much to pay for painting his wife.'[135]

When cherished pets died, the bereaved owners might be deeply upset. They would commemorate their passing in epitaphs and elegies, a classical convention which was now revived, sometimes facetiously, but often with real sincerity.[136] In the eighteenth century the remains of pets might be covered with an obelisk or sculptured tomb.[137] If the owner preceded them to the grave they might attend the funeral; and, from the late seventeenth century onwards, they could even hope to receive a legacy for their maintenance. The fourth Earl of Chesterfield (d. 1773) provided for his horse. The first Earl of Eldon (d. 1838) left money both for his coach-horses and his dog Pinter. Frances Stuart, Duchess of Richmond (d. 1702), bequeathed a legacy for the upkeep of her cats.[138]

A characteristic pet-owner was the bachelor clergyman-poet Robert Herrick, who in the 1630s had a cat, a spaniel, a pet lamb, a sparrow and a pig (the last was taught to drink beer out of a tankard).[139] Equally typical was Isabella, daughter of Sir Allen Apsley and widow of Sir William Wentworth (d. 1692). In the reign of Queen Anne she had her bitch Fubs, her monkey Pug (whose portrait she had painted when he died), her parrot, and five other dogs. Fubs would be taken to church, where she 'sat very orderly', and her mistress could never relax when she was out of her sight, 'especially if a dog happens to be shot, then I am out of my wits till I see her'. Another pet dog, Pearl, fell ill and was dressed in 'little nightgowns made fit for it and its legs ... put into sleeves'.[140]

Pets were company for the lonely, relaxation for the tired, a compensation for the childless. They manifested those virtues which humans so often proved to lack ('faithful dog, faithless man', said the epitaphs); and they held out a model to domestic servants:

> Here lies a pattern for the human race,
> A Dog that did his work, and knew his place.[141]

It was no accident that dogs were so cherished for their unswerving loyalty, what the Victorians called 'Love of Master', later institutionalized at Crufts in the 'Obedience' championships. They were valued because they were either idealized servants who never complained or model children who never grew up. As both Burke and Dr Johnson observed, affection for pets was seldom untinged by a touch of adult superiority and contempt.[142]

Only very occasionally did pet-keeping encourage the desire to fuse the different orders of creation. The frequency with which bestiality was denounced by contemporary moralists suggests that the temptation could be a real one. 'It is a pit out of which the few that do fall into it do hardly recover,' warned Richard Capel in 1633.[143] We have Samuel Pepys's confession of the jealousy he felt ('God forgive me') when a dog was brought to line his little bitch. But, as a later-seventeenth-century preacher remarked, 'Such crimes as these are rarely heard of among us.'[144] The assize records do not suggest that bestiality was a common event; in the reign of Elizabeth there were ten cases in Kent, eight in Essex, five in Sussex, four in Hertfordshire and three in Surrey. Kent, Hertfordshire and Sussex had none at all under James; and there were only nine cases in Essex between 1620 and 1680.[145] It was a rural crime, most often involving cows and horses, and it seems seldom to have arisen as an extension of the emotional feelings between owner and pet.

Today the scale of Western European pet-keeping is undoubtedly unique in human history. It reflects the tendency of modern men and women to withdraw into their own small family unit for their greatest emotional satisfactions. It has grown rapidly with urbanization; the irony is that constricted, garden-less flats actually encourage pet-ownership. Sterilized, isolated, and usually deprived of contact with other animals, the pet is a creature of its owner's way of life; and the fact that so many people feel it necessary to maintain a dependent animal for the sake of emotional completeness tells us something about the atomistic society in which we live.[146] The spread of pet-keeping among the urban middle classes in the early modern period is thus a development of genuine social, psychological, and indeed commercial importance.

But it also had intellectual implications. It encouraged the middle classes to form optimistic conclusions about animal intelligence; it gave rise to innumerable anecdotes about animal sagacity;* it stimulated the notion that animals could have character and individual personality; and it created the psychological foundation for the view that some animals at least were entitled to moral consideration. It is no coincidence that many, if not indeed the majority, of those who wrote on behalf of animals in the eighteenth century were, like Pope or Cowper or Bentham, persons who had themselves formed close relationships with cats, dogs or other pets. (Jeremy Bentham indeed was one of those unfortunates who shared an equal liking for cats and for mice.)[147] Lord Erskine was famous for having several favourite dogs, a favourite goose, a favourite macaw, even favourite leeches, whom he named after two surgeons of

* Cf. below, pp. 121, 126–7.

the day. It is no surprise to learn that it was he who in 1809 proposed a parliamentary motion against cruelty to animals.[148]

I cannot be
Unkind, t'a beast that loveth me.

Andrew Marvell's *Nymph Complaining for the Death of her Faun* is a reminder of the emotional bonds which could be forged between pet and owner; while Chaucer's Prioress, who wept not just when someone beat her dogs, but also when a mouse was caught in a trap, shows how sympathies could be extended outwards from pets to other animals.[149] As a hostile commentator wrote of Frances Power Cobbe, leader of the anti-vivisectionists in the 1870s:

> Her dog and her cat are a great deal to her; and it is the idea of their suffering which excites her ... She is not defending a right inherent in sentient things as such; she is doing special pleading for some of them for which she has a special liking.[150]

iii. THE NARROWING GAP

There is no doubt that it was the observation of household pets which buttressed the claims for animal intelligence and character. We can see this in the sixteenth century, when Dr Caius declared that 'some dogs there be which will not suffer fiery coals to lie scattered about the hearth, but with their paws will rake up the burning coals, musing and studying first with themselves how it might conveniently be done'. We can see it in the Stuart age, when pigeon-fanciers like William Ramesey spent hours observing their birds, delighting in their 'mutual love, chastity and constancy, obedience of the hen to the cock pigeon, care of the cock to the young'; and when John Aubrey saw 'sparks of justice and detestation of oppression' in 'a brave, brinded-mastiff dog in Jermyn Street, that, for the sake of righting wrong, fell severely on a great, cruel mastiff that ... had near killed ... a poor, little, harmless cur'.[1] We see it everywhere in the eighteenth century, whether in Thomas Gainsborough describing his bull terrier, Bumper ('a most remarkable, sagacious cur'); or in Richard Dean claiming that he knew pets who would 'sooner be hanged than pilfer or steal, even under the greatest temptation'; or in Henry Needler recalling the horse in Portsmouth dockyard which always stopped work when it heard the 12 o'clock bell; or in Erasmus Darwin refuting the old notion that animals had no rights because they could not make contracts:

> Does not daily observation convince us that they form contracts of friendship with each other, [and] with mankind? When puppies and kittens play together is there not a tacit contract that they will not hurt each other? And does not your favourite dog expect you should give him his daily food, for his services and attention to you?[2]

By this time, anecdotes testifying to the 'extraordinary sagacity' of dogs, cats and horses were becoming a well-established literary genre.[3] Like most literary genres, its models were classical. Aelian, Pliny and Plutarch had all told tales of intelligent animals like the conscientious performing elephant who got up in the middle of the night to practise his dancing when he thought his performance was slipping. But it was direct experience rather than classical tradition which did most to stimulate the mounting belief in animal intelligence. Observation of pets, coupled with experience of domestic animals, provided support for the view that pets could be rational, sensitive and responsive. As George Eliot would remark, 'Everyone with a large acquaintance with decent and "gentleman-like" dogs ... must admit their share in the highest humanities: and what is true of them is true, to a greater or lesser extent, of animals generally.'[4]

It is against this pet-keeping background that we should view the growing tendency in the early modern period for scientists and intellectuals to break down the rigid boundaries between animals and men which earlier theorists had tried to raise.

The attack on conventional orthodoxy came from two separate directions. There were those who said that men were morally no better than animals, possibly even worse; and there were those who said that animals were intellectually almost as good as men. In the former category were the sceptics and libertines, who, like the Cynic philosophers of antiquity, denigrated the claims of humanity, urging that men were beastlike in their inclinations and capable of vices of which animals never dreamed. It was a humanist commonplace that man's very possession of reason and free choice enabled him to descend to infinitely greater moral depths than could the brute; so-called animal instinct was much less fallible than reason.[5] Scores of commentators pointed out that beasts did not get drunk or tell lies, were not sadistic and did not make war on their own species. 'No one beast, be it never so bad, can be matched unto man,' said a Jacobean bishop; and John Locke agreed: 'the busy mind of man' could 'carry him to a brutality below the level of beasts when he quits his reason'.[6] It was certain, wrote Thomas Tryon in 1683, that lions and tigers were 'not more savage and cruel, geese and asses not half so stupid, foxes and donkeys less knavish and ridiculous, wolves not more ravenous, nor goats more lascivious than abundance of those grave, bearded animals that pride themselves with the empty title of rational souls'.[7]

Some of these sceptics went so far as to maintain that men, like animals, were doomed to extinction when they died. This was the outlook of many seventeenth-century French *libertins* and it had its adherents in England, and not just at a high intellectual level. We shall never know just how many village materialists there were who, like the Wiltshire gentleman John Derpier in 1622, believed 'there was no God, and no Resurrection, and that men died a death like beasts'. But scores of such doubters have been uncovered in the records of the church courts and they continue to be turned up.[8] Owners of domestic animals, well aware that the corpses of men and beasts decomposed in externally indistinguishable fashion, did not always find it easy to believe that the souls of the one survived while those of the other were obliterated. In many parts of medieval Europe there were peasant sceptics who believed that men and oxen died a similar death. In the words of Ecclesiastes (iii. 19), 'that which befalleth the sons of men befalleth beasts; even one thing befalleth them: as the one dieth so dieth the other; yea they have all one breath; so that a man hath no preeminence above a beast.'[9]

A less extreme position was that of the mortalists who taught that the soul slept with the body until the General Resurrection, when both would rise together. This belief was widely held in early modern England.[10] What united its adherents, whether individuals like Thomas Hobbes or John Milton, or whole sects like the Muggletonians and many of the Ranters and Familists, was the belief that men were not essentially distinct from the rest of nature. 'If man be fallen,' wrote Richard Overton in his *Man's Mortality* (1644), 'and the beasts be cursed for his sake, man must [be] equally mortal with them.'[11] It was this which made the heresy so offensive. By rejecting the Christian dualism of body and soul, and denying that the spirit could exist independent of the body, the mortalists were not just weakening the belief in rewards and punishments on which the good behaviour of the lower classes was thought to depend: they were also removing the essential prop by which man's right to rule the lower species was usually supported. When Overton argued that men and beasts were equally mortal he was accused of treason to the human race. 'This dangerous traitor,' said an opponent, was 'trying to rob man of his superiority.'[12] To predicate mortality in the soul, agreed Sir Kenelm Digby, 'taketh away all morality and changeth men into beasts'. 'Atheists' and 'epicures' were thus associated with the view that animals and humans were equal to the extent that they shared a common mortality.[13] For mortalists, man's pre-eminence over the beasts was something which only became evident at the Resurrection.[14]

In the eighteenth century this heretical onslaught on man's supposed uniqueness was powerfully reinforced by the materialism of French thinkers like La Mettrie, who repudiated the ancient distinction between mind and matter, body and soul, on which the whole separation of man from nature had been founded. Though recognizing that man's capacities were greatly superior, they saw a similar organic explanation for animal and human intelligence alike. 'From animals to men,' wrote La Mettrie, 'the transition is not violent.'[15] Men, thought the sceptic Viscount Bolingbroke, were essentially animals themselves. Humanity's superior attainments were considerable, he wrote, 'but though they are great they do not take us out of the class of animality ... the metaphysician, who fancies himself wrapped up pure intellect, ... will feel hunger and thirst, and roar out in a fit of the stone.' Bolingbroke, commented one of his readers, had employed the highest degree of human understanding to prove himself a brute. Materialism of this kind made a number of converts in England. 'The ancients were fond of making brutes to be men,' remarked Edward Tyson in 1699, but 'now, most unphilosophically, the humour is to make men but mere brutes and matter.'[16] The antiquary Martin Folkes, President of the Royal Society (1741–53), was a prominent infidel and scoffer, notorious for greeting with loud laughter any references at meetings of the Royal Society to Moses or the Deluge or other religious matters. 'He thinks there is no difference between us and animals,' noted a contemporary, and 'professes himself a godfather to all monkeys.' It was persons like this who were denounced by Burke in Parliament in 1773 as 'the sceptics, who labour to degrade us below the brutes; the sceptics who would fain persuade us that we are inferior to those animals which wallow in a sty'. Yet an increasing number of scientists would, like Sir Everard Home, F.R.S. (1756–1832), maintain that the human mind was but a nice arrangement of matter and that men excelled other animals only by the superiority of their physical organization.[17]

Heretics and materialists who argued that man was nothing more than an animal had a following, but until the nineteenth century they remained a minority, untypical of educated opinion in general. On the other hand, far more people were prepared to concede that animals did not fall very short of men. After all, the intellectual difference between man and beast had long been represented by many as one of degree, not of kind. As a commentator remarked in 1679, 'That there are some footsteps of reason, some strictures and emissions of ratiocination in the actions of some brutes, is too vulgarly known and too commonly granted to be doubted.'[18] Medieval Christianity had portrayed man as *sui generis*, but in classical antiquity it had been a commonplace that there were resemblances between men and animals and that reason was not a uniquely human quality.[19] Even in Christian times the theologians represented man as closer to the animals than to God.[20]

The widely-held notion of the great chain of being was equally ambiguous. On the one hand it postulated a clear hierarchy of creation with man well above the beasts and well below the angels. On the other, it suggested that there were no breaks in the chain, but that each species imperceptibly moved into the next so that the line dividing men from animals was highly indistinct.[21] So when in 1615 John Preston and Matthew Wren conducted a public disputation at Cambridge before James

I on the subject of whether or not dogs could reason they were not just playing a donnish game: they were grappling with a topic of notorious philosophical perplexity.[22] And the harder the philosophers struggled to clarify it, the more elusive did the old distinction between reason and instinct become. In the seventeenth century the most common view held by intellectuals was that beasts had a kind of reason, but an inferior one. They possessed sensibility, imagination and memory, but no powers of reflection; as John Locke put it, they could not compare ideas or reason abstractly.[23] Yet Locke's own philosophy seemed to reduce all thought to sensation, of which beasts were certainly capable; and many of his contemporaries saw nothing strange in God's remark to Milton's Adam that beasts 'reason not contemptibly'.[24] Sir Matthew Hale thought foxes, dogs, apes, horses and elephants displayed 'sagacity, providence, disciplinableness and a something like unto discursive ratiocination'. The Dean of Winchester, preaching at Whitehall in 1683, stressed the similarities between animals and men. 'Even in that which we pretend our peculiar prerogative, ratiocination, they seem to have a share ... Their knowledge extendeth not only to simple objects, but it is evident by the subtlety and docibility which is so wonderful in many of them, even to propositions, assumptions and deductions.' Brutes were not mere automata, said the mathematician Humphry Ditton. 'Their actions plainly show thought and design.'[25]

In the mid eighteenth century David Hume conceded animals the power of 'experimental reasoning', adding that if they were not guided by their reason in their ordinary actions, then 'neither are children; neither are the generality of mankind in their ordinary actions and conclusions'. David Hartley similarly thought that brutes had more reason than man's ignorance of their language permitted him to appreciate. The example of dogs and horses suggested that, if pains were taken, then animal docility and sagacity could reach a surprisingly high level; sagacious quadrupeds could be compared to dumb persons who had arrived at adult age and were possessed of much knowledge, but were unable to express their thoughts save by gestures. In the later eighteenth century the most common view was that animals could indeed think and reason, though in an inferior way.[26]

As the anecdotes testifying to the 'extraordinary sagacity of the brute creation' began to accumulate, it thus became increasingly difficult to maintain that man's mental superiority was more than one of degree or to be certain that *all* men were more rational than *all* animals. Impressive achievements of animal instinct had always been admired, like the skill with which birds and bees created their nests or cared for their young. 'In all the days of my life,' related an Elizabethan seaman of the wild ducks of South America,

> I have not seen greater art and curiosity in creatures devoid of reason than in the placing and making of their nests – all the hill being so full of them that the greatest mathematician of the world could not devise how to place one more than there was upon the hill, leaving only one pathway for a fowl to pass betwixt.

A country clergyman in the reign of Queen Anne even claimed to have seen a group of crows planting a grove of oaks, which, twenty-five years later, were high enough to offer a safe site for their nests.[27] 'Instinct' on this scale seemed hard to distinguish from reason. Of course, many philosophers continued to maintain that instinct, unlike reason, was incapable of improvement; the difference was of kind, not degree, and the barrier between man and animals recognizably distinct.[28] But others regarded 'animal instinct' and 'human reason' as merely different degrees of the same quality. Man was only superior because he had acquired a greater number of such 'instincts'. 'Experimental reasoning,' said David Hume, was 'nothing but a species of instinct.'[29]

In urging the rationality of animals, the intellectuals were merely restating what many uneducated persons had always thought. Farm labourers knew that animals could be taught to perform many complicated operations. Shepherds had never doubted the sagacity of their sheepdogs. Horse-trainers had always regarded it as axiomatic that their charges had memory, imagination and judgement.[30] The bee, thought an agricultural writer in 1616, had 'a kind of wisdom coming near unto the understanding of man'. The sagacity of hogs in providing against cold at night was 'quite wonderful', wrote William Cobbett.[31] Popular fairgrounds had regularly featured a so-called 'learned' dog or pig or other performing animal, some of them frauds, no doubt, but others very impressive, like Morocco 'the wonderful white horse', who could count, dance, and, on a famous occasion in 1600, was said to have climbed the steeple of St Paul's Cathedral.[32] When Edward Fenton and his crew captured a pirate ship in 1582 they found on board a dog that could dance and sing pricksong. Some baboons, thought Topsell, could write and pick out letters.[33] Even untrained animals could display much shrewdness. 'Dogs,' said Bacon, 'know the dog-killer. When, as in times of infection, some petty fellow is sent out to kill the dogs ... though they have never seen him before, yet they will all come forth and bark and fly at him.' 'We daily see,' agreed Sir Kenelm Digby, 'that dogs will have an aversion of glovers that make their ware of dogs' skins.'[34]

Country folk also believed in the intelligence of wild creatures. Rooks, it was said, could spot a gun and knew what it meant. The hedgehog was regarded by some as 'the very emblem of craft and cunning'. Hares could dance in time to music. Seals, when hunted on the Cornish coast, were said to turn round and hurl stones at their pursuers. Even vermin were respected by the eighteenth-century royal rat-catcher, Robert Smith, for their 'wonderful wiliness and sagacity'.[35]

Those acquainted with animals, moreover, did not necessarily believe that language was unique to man. Jacobean lawyers maintained that anyone who thought that birds and beasts could converse like characters in Aesop should be legally written off as an idiot; and in the eighteenth century educated writers became increasingly hostile to anthropomorphic stories in which animals behaved like human beings, urging that 'all fables which ascribe reason and speech to animals should be withheld from children, as being only vehicles of deception'.[36] But the inhabitants of rural England had never doubted that wild creatures could communicate or that domestic ones could express their feelings. Folk tales abounded in wise birds and beasts who gave advice or rescued men from some predicament. Little birds told tales and ladybirds flew home after children had recited the appropriate rhyme.[37] The song of wild birds was often interpreted anthropomorphically by country people. The great titmouse said 'sit ye down'; the quail cried 'wet my lips! wet my lips!'; and the chaffinch sang 'pay your rent'.[38] The bird-fanciers even believed that birds sang in local dialect.[39]

By the mid seventeenth century there were also plenty of educated persons prepared to credit animals with a form of language, declaring that birds and beasts, through movement, sound and gesture, could convey their thoughts as well as men. 'In animals,' wrote Nathanael Homes in 1661, 'every man may see that through their sensitive knowledge they have a suitable will to take or refuse an object; to express their desires with sounds or notes of voice; to express their affections of love and hatred by sociableness and conflicts, of joy and sorrow by other notes and noises.'[40] That humans did not usually understand animal language proved nothing; after all, how many Englishmen understood Japanese? 'We understand them no more than they us,' observed Montaigne, 'By the same reason may they as well esteem us beasts as we them.'[41] Sir Kenelm Digby had several acquaintances who held 'that beasts use discourse upon occasions'; and in the next century the aesthete William Gilpin was merely following the fashion when he said that animals could be seen 'conversing with each other in short pithy, uninterrupted sentences, which are no doubt expressive of their own enjoyments and of their social feelings'.[42] John Locke was much impressed by the reports of a talking parrot which had belonged to Prince Maurice of Nassau. There was an even more famous story of Henry VIII's parrot, which fell into the Thames and squawked out, 'A boat! a boat! twenty pounds for a boat!' When a passing waterman picked him up and took him to the King for his reward, the bird changed its tune and said, 'Give the knave a groat.'[43]

In the seventeenth century no one had greater faith in animal capacity than Margaret Cavendish, Duchess of Newcastle, an eccentric and ill-disciplined but highly individual writer. In her poems and essays (of the 1650s and 1660s) she rejected the whole anthropocentric tradition, applying a sort of cultural relativism to the differences between the species and arguing that men had no monopoly of sense or reason. Beasts, she maintained, could experience the whole range of human passions and had their own type of reason and language, which was very probably as deep and expressive as anything available to humans. Man's advantage stemmed solely from his shape; and it was mere arrogance which made him think himself intellectually superior. His 'pride', 'self-conceit' and 'presumption' had misled him into judging other creatures by human standards, not realizing that language and reason could take non-human form.

> For what man knows whether fish do not know more of the nature of water, and ebbing and flowing and the saltness of the sea? Or whether birds do not know more of the nature and degrees of air, or the cause of tempests? Or whether worms do not know more of the nature of the earth and how plants are produced? Or bees of the several sorts of juices of flowers than men? ... Man may have one way of knowledge ... and other creatures another way, and yet other creatures' manner or way may be [as] intelligible and instructive to each other as Man's ...

Even vegetables and minerals had their own brand of reason.

> But because they do not act in such manner or way as Man, Man judgeth them to be without sense and reason; and because they do not prate and talk as Man, Man believes they have not so much wit as he has.

It was, therefore, 'the ignorance of men concerning the creatures' which was the 'cause of despising other creatures, imagining themselves as petty gods in nature'.[44]

These ideas reflected the influence of Montaigne and of French *libertin* writing;[45] and most contemporary readers would have thought them extravagant nonsense. But the sentiments are remarkably close to modern attacks on human 'species-centredness', what a psychologist, writing a few years ago, termed 'the innate selfishness and arrogance that make us ascribe consciousness and self-awareness to ourselves and withhold them from other species'.[46] In fact, Margaret Cavendish ended with the disappointingly conventional conclusion that man was still unique:

> All other creatures only in sense join,
> But man has something more which is divine;
> He hath a mind and doth to Heav'n aspire.[47]

In the early modern period, therefore, there was an increasing tendency to credit animals with reason, intelligence, language and almost every other human quality. Domestic beasts, thought Archbishop Abbot, were sensitive to their owners' state of mind. 'I think that I do not abuse the word to say that some of them in some things have a kind of fellow feeling with us.' 'The elephant', wrote Oliver Goldsmith, 'gathers flowers with great pleasure and attention; it picks them up one by one, unites them into a nosegay, and seems charmed with the perfume.'[48] No one thought that either wild or domestic animals were completely human, but by the eighteenth century they were generally credited with a great number of human characteristics. The boundary between man and beast thus became a great deal more indistinct than the theologians would have liked. 'The sense of the separation of man from the rest of animal creation,' writes a modern historian, 'was beginning to break down.'[49]

Perhaps most decisive of all was the revelation by comparative anatomy of the similarity between the structure of human and animal bodies. Everyone knows of the turmoil into which European thought was thrown by the discovery of the great apes of Africa and South-East Asia.[50] Since Vesalius, anatomists had been embarrassed by their inability to find some respect in which the human brain differed in structure from that of higher animals. 'In the brain, which we term the seat of reason,' observed Sir Thomas Browne, 'there is not anything of moment more than I can discover in the crany of a beast.'[51] With the growth of comparative anatomy in the later seventeenth century matters became even more alarming. In 1698 Edward Tyson, anatomy reader at Surgeons' Hall, dissected an infant chimpanzee (which he called an orang-outang), showing, as he thought, its essential resemblance to the human frame. 'Anatomists discourse how like brutes' organs are to ours,' commented Matthew Prior.[52] In his *Systema Naturae* Linnaeus rejected earlier distinctions between 'rational' animals (i.e. men) and 'irrational' (non-humans). He firmly classified man as part of the animal creation, placing him in the same order (*primates*) as that which included not just apes but even bats, and in the same genus (*homo*) as the orang-outang (*homo sylvestris*).[53]

Finally, in 1774, the Scottish sage Lord Monboddo, in pursuit of his thesis that speech was not universal among human beings, asserted that the orang-outangs were not animals at all, but a race of men who had not yet learned to speak but had been left behind in an arrested state of development; in the woods of Angola, he thought, there were still herds of wild men living in a brutish condition without use of speech.[54]* For this notion Monboddo was indebted to Jean-Jacques Rousseau, who had a few years earlier suggested that the orang-outangs were men who had no occasion to develop their faculties. For Rousseau, language was an invention of human society, not an innate human attribute.[55] In the same spirit, Monboddo maintained that human superiority was only the result of centuries of social evolution. It reflected not intrinsic merit, but the advantage of living in society, where language and knowledge were artificially generated, conserved and handed down. In his 'natural' state man had lived without clothes, without houses and without language.

The implications of Linnaeus's inclusion of man among the animals were resisted by many contemporary scientists, who, like Buffon, continued to accept the doctrine of absolute discontinuity between humans and non-humans;[56] and Monboddo's orang-outangs were the object of much satirical comment. But there were many who inclined to the Scotsman's view of human development. After all, the doctrine of man's social evolution was an ancient one. In classical antiquity, Protagoras, Diodorus Siculus, Lucretius, Horace, Cicero and Vitruvius had all suggested that man had made only a gradual ascent from a bestial condition, developing language and civilization over a long period of time. The Epicureans, as an eighteenth-century preacher explained, held that 'man was at first but one degree removed from the beasts of the field, and was not till after many ages refined and improved into the being he now is'.[57] Christianity had usually been hostile to such

* In the East Indies, however, there was a native tradition that the ape was a human being who deliberately chose not to speak, so as to avoid being put to work by white men; Anthony Le Grand, *An Entire Body of Philosophy*, trans. Richard Blome (1694), ii. 237

evolutionary doctrines, but even within the Christian tradition there were those who believed that, after the Fall, men had ranged over the earth like animals, only slowly developing the social virtues. 'Stories many and diverse,' noted Thomas Starkey in the 1530s, told how men had once wandered abroad in the fields and woods, 'none otherwise than you see now brute beasts do.' According to one mid-Tudor commentator, human beings after the loss of Eden had, for a time, 'like brute beasts grazed upon the ground'.[58] The discovery in the sixteenth and seventeenth centuries of the naked 'savages' of the New World was a dramatic reminder of the condition in which all men had once lived. It was a reasonable inference that the ancient Britons had been, as John Aubrey put it, 'almost as savage as the beasts whose skins were their only raiment'. An early-eighteenth-century poet recalled that, before the Romans arrived, the inhabitants

> Savage and wild, by commerce unrefin'd,
> Differ'd but little from the brutal kind;
> Uncultivated, ignorant and rude,
> A painted herd, they rang'd the Plains and Wood,
> And prey'd upon their fellow brutes for food.[59]

It was because of their interest in these evolutionary doctrines that eighteenth-century philosophers pounced on the 'wild boys', periodically discovered in the woods, naked and inarticulate, as visible proof of the condition to which any human would revert, were he excluded from society.[60*]

It was not only wild boys whom it was believed social life could make capable of improvement. If humans could evolve socially, then why should not animals do the same? Already contemporary naturalists had suggested that many so-called animal 'instincts' had been learned by experience. Fear of humans, for example, was an impulse which wild creatures had gradually acquired and then transmitted to their descendants.[61] The arts of insects, the migration of birds, even the cleanly habits of the cat, had been built up by trial and error, claimed Erasmus Darwin.[62] Bird-song was also an acquired skill.[63] It followed that, given the right conditions, animals might progress yet further. As Bishop Butler remarked, it was impossible to ascertain what latent powers and capacities the brutes might have; like a 'great part' of humans, who died in infancy or childhood, they went to their deaths with their potentialities

* At least fifty-three such feral children are known to have been reported since the fourteenth century; see Lucien Malson, *Wolf Children* (Eng. trans., 1972), 80–82.

unrealized. Lord Monboddo knew an experienced animal-trainer who thought that, if only beasts lived longer and had sufficient care devoted to them, there was no telling what heights they might not reach. Tyson found no physical reason why chimpanzees should not be able to speak, and his investigations bolstered the hope that one day they might be taught to do so.[64] In 1661 Pepys saw a strange creature (probably a chimpanzee or gorilla) from Guinea; it was 'so much like a man in most things ... I cannot believe but that it is a monster got of a man and a she-baboon. I do believe it already understands much English; and I am of the mind it might be taught to speak and make signs.'[65] Sir Ashton Lever (1729–88) indeed possessed an orang-outang that was said to have learned to articulate several sounds.[66] Many believed with Erasmus Darwin that pigs would have progressed much further if it were not for their confinement and the short lives men allowed them. John Stuart Mill was in this tradition when he wrote that education had the power to conquer the animal instincts, not just in men, but 'to no inconsiderable extent' in domesticated beasts as well.[67]

The growing belief in the social evolution of mankind thus encouraged the view that men were only beasts who had managed to better themselves. It thereby dealt a serious blow to the doctrine of human uniqueness. An even greater challenge was presented by the discovery of man's biological evolution. Since the early Greeks, there had been a tradition that man had descended from the animals.[68] Strongly repudiated by Christianity, the notion had nevertheless enjoyed a subterranean life in intellectual circles, periodically surfacing in the writings of those acquainted with the works of Diodorus Siculus, Lucretius and the other sceptical thinkers of antiquity. In 1653, for example, John Bulwer reported that a contemporary philosopher had said to him in conversation 'that man was a mere artificial creature and was at first but a kind of ape or baboon who, through his industry (by degrees) in time, had improved his figure and his reason up to the perfection of man'. Bulwer himself believed that some men had failed to make the transition, remaining on all fours like beasts.[69] In the following year John Hall of Richmond seriously discussed the possibility that man's upright posture was an unnatural development. As an experiment, he proposed that selected children should be excluded from the sight of anyone in a vertical position and allowed to remain on all fours, and that those who fed and taught the children should do so on their hands and knees; this would soon establish whether the upright position was innate or merely learned. Thirty years later, one of John Locke's correspondents remarked that Germans and Scandinavians could not have always lived on two legs only, for otherwise they would not have called gloves 'hand-shoes' (*Handschuhe*).[70]

In the eighteenth century many scientists discussed the possibility that man had evolved from lower forms of life. In Lord Monboddo's view, man was formed

> Not ... at once, but by degrees, and in succession; for he appears at first to be little more than a mere vegetable, hardly deserving the name of a zoophyte; then he gets sense, but sense only so that he is yet little better than a muscle [*i.e.* mussel]; then he becomes an animal of a more complete kind; then a rational creature; and finally a man of intellect and science, which is the summit and completion of our nature.[71]

So long as the time-scale of human history was officially limited to the six thousand years suggested by biblical chronology, it was impossible to accommodate the idea of organic evolution within conventional thinking. Only at the very end of the eighteenth century did pre-historic archaeologists begin to realize that the story of human development might be infinitely longer than had previously been appreciated. Between 1820 and 1840 the geologists vastly extended the supposed age of the earth, while the study of fossils and cave bones established that man had lived as far back as quaternary times.[72] This new temporal framework made it much easier to accept the evolutionary theories of Lamarck and Darwin.

These developments are well known. But historians do not emphasize that some of the ground for the belief in man's animal ancestry had already been prepared at a less intellectual level. As Marx would observe, the aristocratic concern for breeding and ancestry had long encouraged 'the natural, zoological way of thinking'.[73] Centuries of experiment with the selective breeding of domestic animals had generated an awareness that animal stock was malleable; it had also provoked many ironic comments on man's failure to apply the same doctrines of selective breeding to his own species. A gentleman, thought Timothy Nourse in 1686, 'ought at least to be as careful of his race as he is of that of his horses, where the fairest and most beautiful are made choice of for breed'. Implicit in the many subsequent suggestions for the improvement of the human species by eugenic means was the notion that mankind was also a malleable stock and that care was needed to avoid its reverting to 'lower' forms.[74]

In popular mythology, moreover, the line between beasts and humans was constantly crossed and recrossed. Whether in classical tales of Leda and Europa or in nursery stories of frog-princes and animal paramours, the possibility of cohabitation between the different orders of creation was kept alive;[75] and there was an enormous bestiary of supposed hybrid creatures, like the centaurs, satyrs, and minotaurs of classical antiquity or the dog-headed men, pygmies and troglodytes who were believed to be still living in remote parts of the globe. Images of such creatures decorated the margins of early maps and reports of them were a regular ingredient in travellers' tales. In Cuba there were said to be men with tails, as indeed there were rumoured to be nearer home, and not just in Ireland. In 1653 'an ingenious and honest gentleman' reported in confidence that he knew one Kentish family which still had them.[76]* A century later Lord Monboddo said that he could produce legal evidence of a mathematics teacher in Inverness who had had a tail six inches long which he had concealed in his lifetime but was discovered when he died. In 1670 Samuel Gott declared that 'Some credulous admirers of brutes to the disparagement of ourselves and of all of human nature adopt them into the family of mankind, and claim kindred of apes, baboons, marmosets, drills and I know not what bestial forms and satyrs, as but one degree removed from ourselves.'[77]

In England there had always been families who traced their descent from wild animals, like Siward, Edward the Confessor's Earl of Northumberland, whose grandmother had been ravished by a bear; or the Devonshire family of Sucpitches, who in the eighteenth century maintained that their ancestor had been found in the Prussian woods sucking a bitch.[78] The early modern period swarmed with missing links, half-man, half-animal. The wild men who supposedly stalked the forests of medieval Europe, leading a life of bestial self-fulfilment, had not been forgotten, but survived in sculpture, carving, iconography, pageantry and popular fiction.[79] In his classificatory system Linnaeus found room for the wild man (*homo ferus*), 'four-footed, mute and hairy', and cited ten examples encountered over the previous two centuries. In Jacobean Cleveland men told how a Sea Man had been taken up by local fishermen and maintained on a diet of raw fish until he managed to escape. The correspondence of Sir Joseph Banks reveals that between 1797 and 1811 at least three mermaids were sighted off the Scottish coast.[80]

It was also widely believed that offspring could be engendered by

* The men of Kent were said to have tails because one of their ancestors had cut off the tail of Thomas Becket's horses and been cursed accordingly. For this and other traditional explanations see Samuel Pegge, *An Alphabet of Kenticisms*, ed. Walter W. Skeat (English Dialect Society, 1876), 64–5, and Thomas Fuller, *The Worthies of England*, ed. John Freeman (1952), 258–9. On Irishmen with tails see above, pp. 42–3.

sexual unions between man and beast. Some thought the product would be either a man or an animal,[81] but most believed that the result would be a hybrid. Bestiality, declared William Gouge, was 'a cause of abominable monsters'. This kind of unnatural copulation, said William Ramesey, would produce 'a monster, partly having the members of the body according to the man, and partly according to the beast'.[82] Such productions were always given the greatest publicity; as a preacher remarked, 'When a monster is born, the country rings of him.' At Shrewsbury in 1580 an eight-year-old-boy was exhibited with 'both his feet cloven and his right hand also cloven like a sheep's feet'.[83] At Birdham near Chichester about 1674 the dead body of a monster, supposedly begotten on a sheep by a young man, was nailed up in the church porch so that none should miss it. A few years earlier the prurient Anthony Wood went to see the deformed child of an Irishwoman, 'originally begot by a man, but a mastiff dog or monkey gave the semen some sprinkling'.[84]

These hideous stories, and there were many like them, show that, in popular estimation at least, man was not so distinct a species that he could not breed with beasts. It was because the separateness of the human race was thought so precarious, so easily lost, that the boundary had been so tightly guarded. At the end of the seventeenth century, Edward Tyson endeavoured to disprove the belief that deformed births were the result of mixed conceptions,[85] but the tradition lingered.

In the eighteenth century, therefore, popular and learned notions about animals combined to weaken the orthodox doctrine of man's uniqueness. To say that there was no firm line between man and beast was to strike a blow at human pride. That pride, however, was salvaged, at least for Europeans, by the emergence in the late eighteenth century of doctrines which would nowadays be called racialist. Hitherto such doctrines had been notably absent. Orthodox teaching was that all men were biologically distinct from animals and descended from a common stock, differences of colour being merely the result of physical environment – climate, food, soil, or, on some interpretations, excessive body-painting. Polygenism, the concept of different human species, was heretical and 'atheistic'; it was embraced by only the most isolated and heterodox thinkers.[86] It is true that monogenism did not prevent the emergence of notions of racial inferiority, for blackness was usually regarded as a deformity and it was common to explain the different varieties of men in the world by saying that the blacks had degenerated from their common ancestor, Adam, while the whites had stayed constant or even improved. Linnaeus, for example, was a monogenist, but he was free with his bestowal of unfavourable epithets on non-European peoples: the Asiatics were 'severe, haughty, covetous' and the Africans 'crafty, indolent, negligent'. Only the Europeans were 'gentle, acute, inventive'.[87] Nevertheless, the belief that all men had a common ancestor made it much harder to maintain that some were permanently nearer the animal condition than others.

But as the difference between men and animals ceased to appear an absolute one, polygenism became increasingly attractive. It preserved the superiority of the European by showing that it was 'the lowest rank of men' which, as Tyson put it, was closest to 'the higher kind of animals'.[88] The idea that the human race was made up of different species was propounded in 1677 by Sir William Petty, who had been influenced by sailors' accounts of the primitive peoples they had encountered overseas. In the eighteenth century it was embraced by David Hume, Lord Kames, and other philosophers of the Enlightenment; and in the mid nineteenth century it became anthropological orthodoxy.[89] As the notion spread, the effect was to pull down the negro until he was very near the level which the new belief in animal capacity had created for beasts. Already in his *History of Jamaica* (1774) Edward Long could declare that the orang-outang was closer to the negro than was the negro to the white man. By 1799 English racialism was well developed, for that was the year in which the Manchester surgeon Charles White, drawing on the researches of the Dutch anatomist Peter Camper, analysed the 'regular gradation from the white European down through the human species to the brute creation, from which it appears that in those particulars wherein mankind excel brutes, the European excels the African'.[90]

iv. ANIMAL SOULS

Yet though some white men could now comfort themselves that they were at least the 'furthest removed from the brute creation', they were no longer as removed as they would have liked. Neither anatomy nor language nor even the possession of reason could any longer provide an indisputable barrier between them and the beasts. All that was left was the claim that man was the only religious animal, the sole possessor of an immortal soul. This was the only certain distinction between men and brutes, thought a late-seventeenth-century writer: 'other creatures seem in some measure to partake of reason, but not at all of religion.' A hundred and thirty years later, the physician William Lambe agreed. To deny that man belonged physiologically with the monkeys, apes and baboons was to display 'misplaced pride and an ignorant apprehension'. But in his noble part, his rational soul, man was 'distinguished from the whole tribe of animals by a boundary which cannot be passed'.[1] Unfortunately, this allegedly uncrossable boundary was also the one whose existence was hardest to prove.

For at a popular level religion had never been regarded as inaccessible to animals. Protestant theologians were contemptuous of medieval legends about St Francis preaching to the birds or St Anthony of Padua's horse kneeling to receive the host.[2] But many early modern farmers continued to regard their domestic animals, in the way the Jews had done before them, as essentially within the covenant. After all, the animals were supposed to rest on the Sabbath; in the nineteenth century it was even a question among some High Churchmen as to whether they should not also be made to starve on fast days.[3] In the Victorian countryside on Christmas Eve the horses and oxen were rumoured to kneel in their stables and even bees gave out a special buzz.[4]

All animals were thought to have religious instincts. Classical authors taught that fowls had 'a certain ceremonious religion' and that elephants adored the moon.[5] Such traditions were easily Christianized. Psalm 148 declared that all creatures praised the Lord, even 'beasts and all cattle; creeping things and flying fowl'. 'Let man and beast appear before him, and magnify his name together,' sang Christopher Smart. Some theologians, and many poets, regarded bird-song as a kind of hymn-singing. 'The wren and the robin ... sing a treble,' wrote Bishop Goodman in 1624, 'the goldfinch, the nightingale, they join in the mean; the blackbirds, they bear the tenor, while the four-footed beasts with their bleating and bellowing they sing a bass.'[6]

Animals and birds accordingly played a part in the religious prodigies of the period. In 1694 a pamphlet told of the raven in Herefordshire who 'thrice spoke these words distinctly: look into *Colossians* the third and fifteenth'; and in the same decade a young man was converted by reading a few pages of Richard Baxter's *Call to the Unconverted*, which had been accidentally dropped by a lady and 'very strangely' brought into the house by a little dog. When the Puritan martyr John Bastwick was banished to the Scilly Islands in 1637, 'many thousands of robin redbreasts' assembled to greet him '(none of which birds were ever seen in those islands before)'.[7]

There are also hints of popular belief in something very close to the transmigration of souls. The souls of unbaptized children were vulgarly assigned a great number of animal resting-places: they became headless dogs in Devon, wild geese in Lincolnshire, ants in Cornwall, night-jars in Shropshire and Nidderdale. Fishermen sometimes regarded seagulls as the spirits of dead seamen.[8]

In such notions one can see a debased popular version of the doctrine of metempsychosis which had been taught by Plato and Pythagoras. Though condemned by all orthodox theologians, the notion had been intermittently espoused by medieval heretics, and it was revived by some of the Neo-Platonists of the Renaissance. By postulating the movement of the universal soul of the world into every kind of animate creation, it suggested that even beasts had the divine spark within them. It can be traced in some English seventeenth-century Platonist writing; the 'best of philosophers', thought Henry More, were not averse to conceding animals immortal souls.[9] It is also clearly recognizable in the doctrine attributed to the Ranters, who allegedly held that 'when we die we shall be swallowed up into the infinite spirit, as a drop into the ocean, and so be as we were; and if ever we be raised again, we shall rise a horse, a cow, a root, a flower and such like.'[10]

Yet the roots of the idea that even animals might have an after-life were propagated by the theologians themselves, for the mortality of beasts was part of the curse with which Christ had come to do away. In Chapter 8, verse 21, of his Epistle to the Romans St Paul promised that 'the creature itself also should be delivered from the bondage of corruption into the glorious liberty of the children of God'. Medieval schoolmen had said that this meant only men, together with creatures without life, such as the heavens and the elements.[11] Some early Protestant writers, however, put forward the novel view, previously only held by a few isolated commentators, that by 'creatures' was meant all living animals, birds and plants;[12] and in the century after the Reformation the text was subjected to a fascinating mixture of contradictory interpretation. Many said that the meaning of the passage was impossible to know.[13] Others maintained that deliverance for the animals would merely take the form of annihilation.[14] But a substantial proportion of commentators took the view that animals, like the rest of nature, would be restored to the perfection they had enjoyed before the Fall. Thorns, thistles and creatures engendered by putrefaction would disappear, leaving birds, beasts and useful plants to flourish in renewed perfection.[15] John Evelyn, for example, in 1677 heard a sermon 'showing how even the creatures should enjoy a manumission and as much felicity as their nature is capable of, when at the last day they shall no longer groan for their servitude to sinful man'.[16] This did not necessarily mean that *every* animal who had ever lived would be restored, merely that each species would be represented in heaven.[17] But the notion that *all* animals were resurrected was not easily set aside. It was maintained by the early Protestant reformer John Bradford;[18] and it was reiterated during the Civil War period by several radical believers in universal salvation. 'Christ shed his blood for kine and horses . . . as well as for men,' said the Kentish sectary William Bowling in 1646. 'If in man be an immortal spirit,' declared the Leveller Richard Overton, 'then divers other creatures have the like, though not in the same degree.'[19] Overton suggested that even gnats, fleas and toads would share in the General Resurrection. ('Those who are to live [in heaven] amongst all these are likely to have a gallant time of it,' sniggered one of his opponents).[20]

Many contemporaries found it highly offensive to use the Bible to prove animal salvation. In his commentary on Romans Thomas Horton in 1674 found it necessary to refute 'such kind of persons who, from this present place of scripture, would very fondly and absurdly infer a resurrection of beasts. This,' he emphasized, 'does *not* follow from the text, neither has any other good foundation for it.' 'An hereafter for a brute . . . has a strange sound in the ears of a man,' observed Humphry Primatt a century later, 'we cannot bear the thought of it.'[21]

Yet in the later seventeenth century many otherwise orthodox clergy regarded the issue of animal immortality as entirely open. Samuel Clarke told an acquaintance that he thought it possible that the souls of brutes would eventually be resurrected and lodged in Mars, Saturn or some other planet, while the physician Dr Charles Leigh thought there was 'a spiritual immaterial being' in all living creatures. Ralph Josselin dreamed in 1655 that Christ was born in a stable because he was 'the redeemer of man *and beast* out of their bondage by the Fall'.[22] In the eighteenth century those who felt that animal salvation was at the very least a possibility included Bishop Butler, the clergyman William Whiston, the philosopher David Hartley and the writer Robert Wallace.[23] Those who thought it highly probable or even a certainty included animal-lovers like John Hildrop and John Lawrence, Dissenters like Matthew Henry, and Methodists like Adam Clarke and John Wesley.[24] The physician George Cheyne declared in 1740 that 'it seems utterly incredible that any creature . . . should come into this state of being and suffering for no other purpose than we see them attain here . . . There must be some infinitely beautiful, wise and good scene remaining for all sentient and intelligent beings, the discovery of which will ravish and astonish us one day.' When in 1767 Richard Dean wrote a book to prove the future existence of brutes he was correct to preface it by observing that the idea was 'not quite so novel as some folks perhaps have been inclined to imagine'.[25] In the 1770s the Calvinist divine Augustus Toplady declared that beasts had souls in the true sense, adding that he had never heard an argument against the immortality of animals which could not be equally urged against the immortality of man. 'I firmly believe that beasts have souls; souls truly and properly so-called.'[26]

The idea of animal immortality seems to have made more headway in England than anywhere else at this period; and it was undoubtedly to pet-lovers that it made its greatest appeal. It was buttressed by arguments from scripture and by observation of the mental capacities of the animal in question. There were now many who felt that one had only to look into a dog's eyes (always his eyes) to settle the issue. They knew that cats and dogs could dream, which surely showed a spiritual quality, and they thought them capable of good and bad actions. Canine virtue, said the Quaker writer Priscilla Wakefield, was not very different from moral virtue. No wonder that great admirer of canine virtue Alexander Pope shared the belief of his 'poor Indian' that

> admitted to that equal sky
> His faithful dog shall bear him company.[27]

In Victorian times there were many whose religious faith in a just God was sorely tried by the official doctrine that pet animals were doomed to oblivion and that domestic animals had to suffer without hope of posthumous reward. 'The whole subject of the brute creation is to me one of such painful mystery that I dare not approach it,' thought Dr Arnold.[28] John Dunlop, the temperance reformer, noted in his diary for 1 January 1842: 'Tonight Bill brought the landlady's black cat into the room, and played with it. I always get morbid on coming into close quarters with one of the lower animals – it is inexpressibly mournful to me to think that its soul dies and enjoys existence no more.'[29]

The acceptance of evolution posed the dilemma more sharply, for if men had evolved from animals then either animals also had immortal souls or men did not.[30] No wonder that many bereaved Victorian pet-owners, from the philanthropist Lord Shaftesbury to the scientist Mary Somerville, turned away from this 'narrow creed' (as it was called by Southey, grieving for his childhood companion, the spaniel Phillis) to the more optimistic hope that

> There is another world
> For all that live and move – a better one.[31]

When the Rev. J. G. Wood, the popular nature writer, supported the idea of animal immortality, he was inundated by letters welcoming the idea, many from persons with favourite animals now deceased.[32]

In 1816 the future Archbishop Sumner denounced all those writers 'who have taken an extraordinary pleasure in levelling the broad distinction which separates man from the brute creation'. A decade later, in 1827, the young Charles Darwin attended a meeting of the Plinian Society of the University of Edinburgh, where he heard a Mr Grey read a paper 'in which he attempted to prove that the lower animals possess every faculty and propensity of the human mind'.[33] When in 1871 Darwin published his *The Descent of Man*, he would himself argue not only that man and animals were descended from a common ancestor, but also that the mental difference between humans and the existing higher animals was only one of degree. Today, some of his arguments in defence of the latter proposition seem naively anthropomorphic. Intelligence is not unique to man, he says. Neither is language nor even religion. For do not dogs have a conscience? And does not their deep love for their master approach religious devotion?[34] It is not too much to see behind these passages the influence of generations of middle-class stories about animal sagacity and character. Without the long history of pet-keeping in England and without the knowledge accumulated through centuries of experience of domestic animals, it is hard to believe that the author of *The Descent of Man* could have made his case in quite the way he did.

IV
COMPASSION FOR THE BRUTE CREATION

Kill not the Moth nor Butterfly
For the Last Judgment draweth nigh.

William Blake,
'Auguries of Innocence'

i. CRUELTY

When English travellers went abroad in the late eighteenth century they were frequently shocked to see how foreigners treated animals. The Spanish bull-fight had long been notorious for what the first Earl of Clarendon called its 'rudeness and barbarity'. English tourists always went to see it, but usually only once. 'Fifteen or sixteen wretched bulls were massacred,' wrote the fastidious William Beckford after a Portuguese fight in 1787, adding on another occasion, 'I was highly disgusted with the spectacle. It set my nerves on edge and I seemed to feel cuts and slashes the rest of the evening.' It was 'a damnable sport', agreed Robert Southey.[1]

Continental methods of hunting were equally distasteful. When Sir Richard Colt Hoare went after wild boar with the King of Naples in 1786, he was appalled to discover that the boar, so far from being wild, came when whistled for, and that the hunters stuck it with spears when it was held fast by dogs. 'I was ... thoroughly disgusted with this scene of slaughter and butchery ... yet the King and his court seem[ed] to receive great pleasure from the acts of cruelty and to vie with each other in the expertness of doing them.'[2]

The treatment of domestic animals was also lamented. Tobias Smollett felt compassion for the wretched mules and donkeys in the south of France; Beau Brummell in exile was much upset by the way the Normans treated their horses; while Mrs Hervey, wife of the future Earl of Bristol *cum* Bishop of Derry, expressed the feelings of many subsequent Englishwomen when her coach stuck on a journey to Monte Cassino in 1766: 'What hurt me most was their barbarous treatment of the poor mules, whom they beat most unmercifully with their fists, feet, sticks and even stones. I walked about and begged them to be more gentle to them; they laughed at me.'[3]

These reactions reflect that growing concern about the treatment of animals which was one of the most distinctive features of late-eighteenth-century English middle-class culture. They also show the emergence of a belief which by Victorian times had become an entrenched conviction: that the unhappiest animals were those of the Latin countries of southern Europe, because it was there that the old Catholic doctrine that animals had no souls was still maintained.[4]

Yet previously it had been the English themselves who had been notorious among travellers for their cruelty to brutes. The staging of contests between animals was one of their most common forms of recreation. Bulls and bears were 'baited' by being tethered to a stake and then attacked by dogs, usually in succession, but sometimes all together. The dog would make for the bull's nose, often tearing off its ears or skin, while the bull would endeavour to toss the dog into the spectators. If the tethered animal broke loose, scenes of considerable violence ensued. Baiting of this kind was customarily regarded as an appropriate entertainment for royalty or foreign ambassadors. It also took place at country wakes and fairs and in the yards of ale-houses, where local dogs would be invited to challenge an itinerant bull or bear travelling round the country with its keeper. At Stamford and Tutbury there occurred an annual 'bull-running', when the animal, with its ears cropped, tail cut to a stump, body smeared with soap and nose blown full of pepper, was turned loose to see who could catch him in a general free-for-all. Badgers, apes, mules, and even horses might all be baited in similar fashion. Bull-baiting, wrote John Houghton in 1694, 'is a sport the English much delight in; and not only the baser sort, but the greatest ladies'.[5]

Cock-fighting had been equally popular since at least the twelfth century.[6] In the Stuart age it was a normal feature of fairs and race-meetings. The cock was brought up on a carefully chosen diet and specially trained for the fight. Its wings were clipped, its wattle and comb shorn off and its feet equipped with artificial spurs. Cock-fights were usually 'mains', that is contests between two rival teams paired off into a succession of individual combats, as in modern golf matches; a rougher version was the so-called 'Welsh main', which was a knockout competition. Most spectacular of all was the 'battle royal', when a large number of cocks were put into the same pit together, as at Lincoln in 1617, when James I was made 'very merry' by the spectacle.[7] The contests usually expressed regional rivalries, with different teams of cocks representing different villages or the 'gentlemen' of different counties. Meetings often lasted several days and were accompanied by heavy betting, with fresh wagers being laid at every stage of the fight. They involved the mingling of all social ranks, though only of men, for it was emphatically not a woman's sport. The refined Tudor humanist Roger Ascham was a passionate devotee; and when Pepys went to a cockpit in 1663 he saw everyone from 'parliament men' down to 'the poorest prentices, bakers, brewers, butchers, draymen and what not ... all fellows one with another in swearing, cursing and betting'. The cocks themselves had a short life, even the best being unlikely to survive more than a dozen contests.[8]

In the countryside the pursuit and killing of wild animals for sport had been practised since time immemorial. 'No nation,' remarked Fynes Moryson of hunting and hawking, 'so frequently useth these sports as the English.'[9] In Tudor times it was proverbial that 'he cannot be a gentleman which loveth not hawking and hunting';[10] and, though the spread of enclosures and the wider use of the gun would lead to the decline of hawking, the pursuit of wild birds and animals remained an obsessive preoccupation of the English aristocracy until modern times. In the early modern period the prey was hunted either because it could be eaten, like the red and fallow deer, or because it was a pest, like the fox, or because its speed and agility made it an entertaining object of pursuit, as with the hare. Henry VIII's manner of hunting did not differ very much from that of the eighteenth-century King of Naples: he had two or three hundred deer rounded up and then loosed his greyhounds upon them.[11] Frequently, however, the methods of pursuit, capture and kill were highly stylized, and contemporary literature celebrated the majesty of the hunters, the nobility of the hounds and the music of the chase. Hunting, thought Gervase Markham, was 'compounded ... of all the best parts of most refined pleasure'. No music, it was said, could be more 'ravishingly delightful' than the sound of a pack of dogs in full cry.[12]* At the Inner Temple on St Stephen's Day it was customary to bring a fox and a cat into the hall and set hounds upon them. At Sheffield Park in the 1620s the Earl of Shrewsbury allowed his tenants to keep any buck they could kill, provided they used only their bare hands. At

*There was a Jacobean story of a gentleman who was so delighted by the sound of his hounds in full pursuit that he cried out, 'Oh, what a heavenly noise is this!' 'Whereat one gull of the company, who, as it should seem, never heard any dog but a mastiff, holding up his ear as it were towards the sky to hear some noise from the heavens, broke out into these words: "Oh Lord, where is this heavenly noise?" "Why, hark (quoth the gentleman), list awhile, dost thou not hear?" "No (quoth the gull); the curs keep such a bawling I can hear nothing for them."' See *Pasquils Jests* (1604), in *Shakespeare Jest-Books*, ed. W. Carew Hazlitt (1864), iii(2). 83–4.

Smyrna in the late seventeenth century English merchants procured hounds and conducted a hunt, with the dogs following the hare by the scent, which the Turks regarded as 'a prodigious mystery'.[13] At Aleppo in 1716 they went in pursuit of antelopes with hawks, while later in the eighteenth century the Duke of Grafton planned to go to France with his horses and hounds to hunt wolves.[14] In all cases the climax of the hunt was the death of the hunted animal, for, as Montaigne observed, to hunt without killing was like having sexual intercourse without orgasm. 'When he is caught,' wrote the eighteenth-century huntsman Peter Beckford of the fox, 'I like to see hounds eat him eagerly.' In 1788 the poet Cowper witnessed the end of a fox-hunt, when the huntsman threw the fox's body to the pack of hounds, 'screaming like a fiend, "Tear him to pieces!"'[15]

Equally popular was the pursuit of wild birds, either with hawks or, increasingly, with guns. It was a splendid sight, thought a visitor to the fen country in 1635,

> to see a fleet of a hundred or two hundred sail of shell boats and ... punts sailing ... in the pursuit of a rout of fowl, driving them like sheep to their nets ... sometimes they take a pretty feathered army prisoners, two or three thousand at one draught and give no quarter.[16]

Those who engaged in these sports were seldom inhibited by concern for the possible feelings of the animals themselves. Fishing involved the use of live bait, not just small fish, but also frogs.[17] Hawks were nourished on pigeons, hens and other birds. 'I once saw a gentleman,' recalled William Hinde in 1641, 'being about to feed his hawk, pull a live pigeon out of his falconer's bag, and taking her first by both wings, rent them with great violence from her body, and then taking hold of both legs, plucked them asunder in like manner, the body of the poor creature trembling in his hand, while his hawk was tiring upon the other parts, to his great contentment and delight upon his fist.' The *Gentleman's Recreation* (1674) recommended catching a hart in nets, cutting off one of his feet and letting him go to be pursued by young bloodhounds.[18]

The absence of any apparent moral consideration for the hunted animal is well revealed in the famous description of the entertainment provided for Queen Elizabeth when she visited Kenilworth in 1575. First she hunted the hart until it was killed by the hounds after it had taken to water. This, wrote a contemporary, Robert Laneham, was 'pastime delectable in so high a degree as for any person to take pleasure by most senses at once in mine opinion there can be none any way comparable to this'. A few days later a collection of mastiffs was let loose on to a group of thirteen bears. It was 'a sport very pleasant,' says Laneham, 'to see the bear ... shake his ears twice or thrice with the blood; and the slaver about his physiognomy was a matter of a goodly relief.'[19] Later in the reign, when the Queen was older and less energetic, she had to content herself with shooting captive animals; as at Cowdray in 1591, when

> her Highness took a horse with all her train and rode into the park where was a delicate bower prepared, under which were her Highness's musicians placed, and a crossbow by a nymph with a sweet song delivered to her hands to shoot at the deer (about some thirty in number) put into a paddock; of which number she killed three or four and the Countess of Kildare one.[20]

In addition to these stylized and highly formal methods of tormenting animals, there was an infinity of informal ones. Small boys were notorious for amusing themselves in the pursuit and torture of living creatures. In the grammar schools cock-throwing was a widely observed calendar ritual. On Shrove Tuesday the bird was tethered to a stake or buried in the ground up to its neck, while the pupils let fly at it until it was dead. ''Tis the bravest game,' wrote a seventeenth-century poet.[21] Outside school, children robbed birds' nests, hunted squirrels 'with drums, shouts and noises',[22] caught birds and put their eyes out, tied bottles or tin cans to the tails of dogs, killed toads by putting them on one end of a lever and hurling them into the air by striking the other end, dropped cats from great heights to see whether they would land on their feet, cut off pigs' tails as trophies and inflated the bodies of live frogs by blowing into them with a straw.[23] It was 'a familiar experiment among boys,' reported Thomas Willis in 1664, 'to thrust a needle through the head of a hen' to see how long it would survive the experience.[24] The pious agricultural projector John Beale told Robert Boyle that when he was a child he would skin live frogs 'in sport to see what shift they would make when flayed'. When a boy at Eton, he 'threw many frogs into the Thames to see how far they could swim'.[25] There is a wealth of inference to be drawn from the laconic report of a news-writer in 1697 that an eight-year-old boy had accidentally hit and killed another small child with a carelessly aimed brickbat, when they were both 'casting stones at a dog that was to be drowned in a ditch near my house'.[26] No wonder that traditional nursery rhymes portray blind mice having their tails cut off with a carving-knife, blackbirds in a pie and pussy in the well. 'How full of mischief and cruelty are the sports of boys!' lamented the eighteenth-century Evangelical John Fletcher; and the refrain was echoed by scores of observers.[27]

Yet children merely reflected the standards of the adult world. The seventeenth century was an age when country gentlemen would entertain their visitors by putting their dogs to chase tame ducks or by throwing a goose or chicken into a pike-infested pond to watch its struggles.[28] At country fairs there were contests at biting off the heads of live chickens or sparrows.[29] Even a highly cultivated figure like the economist Dudley North could casually remark that, when he and other young gentlemen were with the English traders at Smyrna, they, for diversion's sake, 'tied a dog they had no great respect for to a bush; and fell on him with their scimitars till they had hewed him to pieces to show what heroes they could be upon occasion'.[30] It is not surprising that when Gulliver found himself among the giants of Brobdingnag his fear was that they would 'dash me against the ground, as we usually do any hateful little animal that we have a mind to destroy'. As the historian W. E. H. Lecky remarks, there were two kinds of cruelty: the cruelty which comes from carelessness or indifference; and the cruelty which comes from vindictiveness.[31] In the case of animals what was normally displayed in the early modern period was the cruelty of indifference. For most persons, the beasts were outside the terms of moral reference. Contemporaries resembled those 'primitive' peoples of whom a modern anthropologist writes that they neither seek to inflict pain on animals nor to avoid doing so: 'pain in human beings outside the social circle or in animals tends to be a matter of minimal interest.'[32] It was a world in which much of what would later be regarded as 'cruelty' had not yet been defined as such. A good example of how people were inured to the taking of animal life is provided by the diary kept by the schoolboy Thomas Isham, who grew up in Northamptonshire in the early 1670s. His little journal records much killing of cocks, slaughtering of oxen, drowning of puppies. It tells of coursing for hares, catching martens in traps, killing sparrows with stones and castrating bulls. None of these events evokes any special comment, and it is clear that the child was left emotionally unruffled.[33]

The same indifference is reflected at a more sophisticated level in a simile used by the poet Edmund Waller:

> As a broad bream, to please some curious taste,
> While yet alive, in boiling water cast,
> Vex'd with unwonted heat, he flings about
> The scorching brass, and hurls the liquor out;

The image is purely visual and there is no interest in the feelings of the fish. In the same way Matthew Prior compares the versifiers of his day to a pet squirrel making futile efforts to escape from his captivity:

> didst thou never see
> ('Tis but by way of simile)
> A squirrel spend his little rage,
> In jumping round a rowling cage,
> The cage, as either side turn'd up,
> Striking a ring of bells a'top –?
> Mov'd in the Orb; pleas'd with the chimes,
> The foolish creature thinks he climbs:
> But here or there, turn wood or wire,
> He never gets two inches higher.[34]

What is revealing about this passage is that it is the squirrel, not Prior, who gets into a rage.

Yet, a hundred years later, William Blake's Robin Redbreast[35] would evoke a very different reaction, for by that time the feelings of animals had become a matter of very great concern indeed. Throughout the eighteenth century, and particularly from the 1740s onwards, there was a growing stream of writing on the subject: philosophical essays on the moral treatment of the lower creatures, protests about particular forms of animal cruelty and (from the 1780s) edifying tracts designed to excite in children 'a benevolent conduct to the brute creation'.[36] There were scores of books and innumerable contributions to periodicals and newspapers. There was also a great deal of poetry. This was one of the periods in English history when poets, Shelley's 'unacknowledged legislators', had a powerful influence on educated opinion. The poets were regularly cited by the pamphleteers and quoted by speakers in Parliament; and it is impossible to understand the vehemence of the movement unless one takes into account the works of Pope, Thomson, Gay, Cowper, Smart, Dodsley, Blake, Burns, Wordsworth, Coleridge, Shelley, Byron, Southey, Crabbe and Clare, to name no more.[37] In the early nineteenth century the agitation culminated in the foundation in 1824 of the Society (later the Royal Society) for the Prevention of Cruelty to Animals and the passing (after unsuccessful bills from 1800 onwards) of a series of Acts of Parliament: against cruelty to horses and cattle (1822), against cruelty to dogs (1839 and 1854) and against baiting and cock-fighting (1835 and 1849).[38] In her Jubilee address of 1887 Queen Victoria would comment that 'among other marks of the spread of enlightenment amongst my subjects' she had noticed in particular, 'with real pleasure, the growth of more humane feelings towards the lower animals'.[39]

How did this change come about? As early as 1795, a writer could attribute it to 'the superior humanity of the present over any former period'; and in the mid nineteenth century the historian Lecky declared that the change had been effected 'not by any increase in knowledge or by any process of definite reasoning, but simply by the gradual elevation of the moral standard'.[40] Yet contemporaries were surely wrong to think of people as being more or less humane at one period in history than at another. What had changed was not the sentiment of humanity as such, but the definition of the area within which it was allowed to operate. The historian's task is to explain why the boundary encircling the area of moral concern should have been enlarged so as to embrace other species along with mankind.

ii. NEW ARGUMENTS

There was, of course, nothing new about the idea that unnecessary cruelty to animals was a bad thing. Such a view had been held by many classical moralists; it was put forward by the medieval scholastics; and it was repeatedly urged in the early modern period. But this view did not originally reflect any particular concern for animals; on the contrary, moralists normally condemned the ill-treatment of beasts because they thought it had a brutalizing effect on human character and made men cruel to each other. The ancient Athenians were said to have condemned a child who blinded crows because they thought that one day he would be cruel to men.[1] In the same spirit, William Hogarth's *Four Stages of*

Cruelty (1750–51) suggested that those who began by torturing cats and dogs would end by murdering their fellows. 'If cruelty be allowable ... towards brutes,' wrote John Lawrence in 1798, 'it also involves human creatures; the gradation is much easier than may be imagined, and the example is contagious.'[2] Samuel Richardson's fictional rake Lovelace had been cruel to animals from infancy; and in the nineteenth century it did not go unnoticed that William Palmer, the poisoner of Rugeley (executed 1856), had been notorious as a boy for his cruel experiments on animals.[3] This kind of argument was extensively employed throughout the period. Indeed when Lord Erskine introduced a bill against animal cruelty in 1809, he too urged that cruelty to animals would lead to cruelty to man; the bill, said one contemporary, was really meant to prevent the murder of humans.[4]

It was from this strictly man-centred point of view that the many Old Testament injunctions against cruelty had been conventionally interpreted. 'If any passage in holy scripture seems to forbid us to be cruel to brute animals,' explained Aquinas, 'that is either ... lest through being cruel to animals one becomes cruel to human beings or because injury to an animal leads to the temporal hurt of man.'[5] The Bible contained passages about helping the ass of one's enemy when it lay under its burden (Exodus xxiii. 5; Deuteronomy xxii. 4); allowing animals to rest on the Sabbath (Exodus xxiii. 12); not muzzling the ox when it trod out the corn (Deuteronomy xxv. 4); and urging that 'a righteous man regardeth the life of his beast' (Proverbs xii. 10).[6] Some Tudor and early Stuart commentators ignored these passages altogether. Others treated them allegorically, suggesting, for example, that the muzzled ox stood for inadequately-paid clergy.[7] But many explained them the way Aquinas had done, saying that God restrained the Jews from cruelty to animals 'lest they might learn to practise it upon man' or lest they damage the property of others.[8] Animals had been permitted to rest on the Sabbath so as to release men from the burden of looking after them on that day.[9] These rules about oxen were not made 'for the sakes of these creatures in themselves,' said one, 'God's laws are not aimed to them.'[10] '[God] is mindful of beasts indeed,' said another, 'but it is for our sakes that he is so mindful of them.'[11] As St Paul had explained, God did not take care for oxen.*

Where, then, do we look for the origins of the much more radical view according to which cruelty to animals is wrong regardless of whether or not it has any human consequences?

* Cf. above, p. 24.

The few scholars who have considered this subject recognize that several classical authors, Plutarch and Porphyry in particular, had shown great concern for animals, sometimes even urging vegetarianism.[12] But they tend to regard the period between them and the eighteenth century as very nearly a total blank. The scarce and half-forgotten book by Dix Harwood, *Love for Animals and How it Developed in Great Britain* (1928), is still the best compilation on the subject, but even its author tells us that 'the evidence of sympathetic interest in animals before 1700' is 'very slight'. Professor Peter Singer in his recent work on *Animal Liberation* declares that no one between Porphyry in the third century and Montaigne in the sixteenth ever condemned cruelty to animals in itself; while another notable recent authority, Professor John Passmore, referring to a fourteenth-century Japanese essayist who opposed the caging of wild birds, remarks that such sensibility 'could certainly not be matched in European writers of the same period'.[13]

Yet in fact it was in the same fourteenth century that Chaucer wrote that, however well a cage-bird was looked after, he would

> Leifer in a forest, that is rude and cold
> Go eat worms and such wretchedness

and there are some striking poems of the same period which express keen sympathy with hunted hares and ill-treated beasts.[14] The truth is that such supposedly 'modern' sensibilities were far from unknown in medieval England. It is possible to set aside as unrepresentative the numerous lives of Celtic saints which show them living on terms of equality and affection with wild creatures.[15] But it is harder to disregard all those legends of medieval holy men who, like St Neot, saved stags and hares from the hunters or, like the twelfth-century Northumbrian saint, Godric of Finchale, went out barefoot to rescue shivering creatures from the cold and release birds from snares.[16] King Henry VI could not bear to see animals slaughtered by hunters; and whenever the fifteenth-century mystic Margery Kempe saw a man strike a horse she had a vision of Christ being beaten.[17]

Of course, these individuals were eccentrically tender-minded by the standards of the age, and sometimes the stories about them are meant to be understood allegorically. But what about the decisive case of *Dives and Pauper*? This important, though much neglected, moral treatise on the ten commandments, written not later than 1410 and probably Franciscan in origin, explains that the fifth commandment (against murder) does not forbid the slaughter of animals 'when it is profitable ... for meat or for clothing', or necessary 'to avoid nuisance of the beasts which be noxious to man'. But it does prohibit killing animals for cruelty's sake or out of vanity, and God will take vengeance on those who misuse his creatures. 'And therefore men,' it continues, 'should have ruth of beasts and birds and not harm them without cause ... and therefore they that for cruelty and vanity ... torment beasts or fowl more than ... is speedful [i.e. expedient] to man's living, they sin ... full grievously.'[18]

This is a notable passage and a very embarrassing one to anybody trying to trace some development in English thinking about animal cruelty. For here at the very beginning of the fifteenth century we have a clear statement of a position which differs in no respect whatsoever from that of most eighteenth-century writers on the subject. It says about animals precisely what William Cowper was to write in 1784:

> The sum is this: if man's convenience, health,
> Or safety interfere, *his* rights and claims
> Are paramount, and must extinguish theirs.
> Else they are all – the meanest things that are –
> As free to live and enjoy that life,
> As God was free to form them at the first.[19]

The truth is that one single, coherent and remarkably constant attitude underlay the great bulk of the preaching and pamphleteering against animal cruelty between the fifteenth and nineteenth centuries. This attitude can be easily summarized. Man, it was said, was fully entitled to domesticate animals and to kill them for food and clothing. But he was not to tyrannize or to cause unnecessary suffering. Domestic animals should be allowed food and rest and their deaths should be as painless as possible. Wild animals could be killed if they were needed for food or thought to be harmful. But, although game could be shot and vermin hunted, it was wrong to kill for mere pleasure. It followed that throwing sticks at tethered cocks on Shrove Tuesday or tormenting animals for entertainment's sake was completely unacceptable. So was the staging of contests between animals. On the other hand, bull-baiting was permissible, because it was required by the civil authorities in order to improve the quality of the meat. It was in keeping with this attitude that the great Elizabethan Puritan William Perkins allowed bull-baiting, but strongly condemned cock-fighting and bear-baiting.[20]

Of course, this position left a good deal of room for argument on the question of what cruelty was 'necessary' and what avoidable, as the later debate on the ethics of vivisection would show. But in its essentials it was shared by all the main groups of those who worried about the treatment of animals, whether the strong Protestants or Puritans of the Elizabethan and early Stuart period, the Quakers, Dissenters and Latitudinarians of the later seventeenth century, or the Evangelicals, Methodists, sentimentalists and humanitarians of the eighteenth. So far as their main arguments were concerned there was a notable lack of historical development.

Yet, though the position was constant, the particular preoccupations of these successive agitators changed a good deal. Before the Civil War the attack was concentrated on bear-baiting, cock-fighting and the ill-treatment of domestic animals. Then, in the later seventeenth century, it widened out to embrace hare-hunting, vivisection, the caging of wild birds, brutal methods of slaughter, and the cruelties involved in gastronomic refinements. The type of argument used also changed. The Puritans conducted their agitation in theological terms, as a debate about how God intended men to behave towards the lower creatures. But from the later seventeenth century the tone would become increasingly secular and other considerations would be advanced.

Initially, however, the opponents of animal cruelty drew primarily on the doctrine, which they found to be latent in the Old Testament, of man's stewardship over creation. According to this view, brute creatures had been created to serve man, but they were to be treated with respect and used only for the purposes which the creator had envisaged. God 'will not have us abuse the beasts beyond measure,' wrote John Calvin, 'but to nourish them and to have care of them.' 'If a man spare neither his horse nor his ox nor his ass, therein he betrayeth the wickedness of his nature. And if he say, "Tush, I care not, for it is but a brute beast," I answer again, "Yea, but it is a creature of God."' Animals, like men, were part of God's creation and, within the limits set by human needs, were entitled to life and happiness. Calvin's views were firmly anthropocentric ('True it is that God hath given us the birds for our food, as we know he hath made the whole world for us'). But he nevertheless drew the line at what he called 'extreme' or 'barbarous' cruelty: when God placed the beasts 'in subjection unto us,' he explained, 'he did it with the condition that we should handle them gently.'[21] It was in the same spirit that Sir Philip Sidney urged mankind not to abuse its trust:

> But yet, O man, rage not beyond thy need:
> Deeme it no glory to swell in tyranny.
> Thou art of bloud, joy not to make things bleed;

> Thou fearest death; thinke they are loath to die
> ...
> And you, poor beasts, in patience bide your hell.[22]

Man's rule, said a mid-seventeenth-century divine, was 'subordinate and stewardly, not absolutely to do what he list to do with God's creatures'. Cruelty to beasts, agreed Sir Matthew Hale, was 'tyranny', 'breach of trust' and 'injustice'.[23] It was true that man was 'viceroy of creation', wrote Thomas Tryon. But this rule was

> not absolute or tyrannical, but qualified so as it may most conduce, in the first place to the glory of God; secondly to the real use and benefit of man himself, and not to gratify his fierce and wrathful, or foolish and wanton humour; and thirdly as it best tends to the helping, aiding and assisting those beasts to the obtaining of all the advantages their natures are by the great, beautiful and always beneficent creator made capable of.[24]

This view of man's relationship to animals would have a long life. 'We seem to be in the place of God to them,' reflected the philosopher David Hartley in 1748, 'and we are obliged by the same tenure to be their guardians and benefactors.' God, warned the pamphleteer Humphry Primatt in 1776, would require a strict account from man of the creatures entrusted to his care. It was upon this religious position that he based his *Dissertation on the Duty of Mercy and Sin of Cruelty to Brute Animals*.[25]

Of course, Elizabethan and Jacobean commentators often qualified their views by conceding that men's duties to animals were not as great as their duties to each other. 'If we must have care of beasts,' Calvin had written, 'much more must we have it of human creatures.' The meaning of Proverbs, xii. 10, explained Thomas Wilcox in 1589, was that a good man should be merciful 'to beasts, much more to men ... He is so gentle and courteous that he neglecteth not his own cattle, but giveth them their meat, attendance, and all other things necessary in due time. How much more then doth he care for his household and needy persons!' Three years later Wilcox's interpretation was reiterated by Peter Muffett: 'A just man will not over-toil the poor dumb creature, nor suffer it to want food or looking to. But if he be so pitiful to his beast, much more is he merciful to his servants, his children and his wife.'[26]

Subject to this qualification, many early Stuart commentators repeated the injunction of the Old Testament that a good man should be merciful to his beast, explicitly repudiating the apparent callousness of St Paul. To the Apostle's question, 'Doth God take care for oxen?' said a preacher at Northampton in 1607, 'We may answer affirmatively: "Yes, sure; even for oxen, and for sheep".'[27] The good husbandman, therefore, would ensure that his animals were well fed and watered and that they were allowed to rest on Sunday. He would also be careful not to lose his temper with them or strike them in anger if they failed to perform according to his expectations.[28] Animals, indeed, were to be treated in much the same way as other labourers. Farmers who did not observe the Sabbath in harvest time were dealing 'injuriously with their servants and cattle', thought Edward Elton; 'mercy and compassion is to be extended to the dumb creature that it may sometimes be spared and have some rest from labour.' The Sunday rest, agreed Bishop Babington, had been ordained 'for the relief of servants and brute beasts, which by pitiless worldlings might else be abused'. It was 'barbarous cruelty,' thought John Dod and Robert Cleaver, 'for one to ride his horse hard all the day, and at night to tie him up to the bare rack, without meat to repair and sustain his strength.'[29]

Considerate treatment of animals thus became a religious obligation: 'Love God, love his creatures.'[30] Cruelty, later writers would urge, was an insult to God, a kind of blasphemy against his creation.[31] Man's responsibility was expressed by George Wither in a poem about his horse:

> And though I know this creature lent
> As well for pleasure as for need;
> That I the wrong thereof prevent,
> Let me still carefully take heed.
> For he that wilfully shall dare
> That creature to oppress or grieve,
> Which God to serve him doth prepare
> Himself of mercy doth deprive.
> And he, or his, unless in time
> They do repent of that abuse,
> Shall one day suffer for his crime;
> And want such creatures for their use.[32]

The paradox, therefore, was that it was out of the very contradictions of the old anthropocentric tradition that a new attitude would emerge. That, after all, is how most new ideas appear. Just as modern atheism is probably best understood as a conviction growing out of Christianity, rather than something encroaching upon it from an external source, so consideration for other species has its intellectual roots within the old man-centred doctrine itself. For theologians had always taught that the defects of animals were the direct consequence of Man's fall. Since the beasts were merely innocent victims of Adam's sin, it followed that men should be merciful and forgiving to them. 'Let us in no wise ... misuse any of the poor creatures,' wrote Thomas Draxe in 1613, 'knowing that if there be any defect or untowardness in their nature or any want of duty or observance in them towards us, our sin hath been and is the cause and occasion of it.' 'Seeing all creatures partake with us in our punishment,' agreed Thomas Wilson in the following year, 'it should cause us to be merciful unto them.'[33]

Similarly, because the mutual ferocity of wild animals was a response to Man's sin, all animals having been tame until the Fall, it followed that it was wrong for men to take pleasure from watching fights between them. The Protestant attack on cruel recreations goes back to at least 1550, when Robert Crowley denounced bear-baiting as 'a full ugly sight'.[34] It has been much misunderstood. Macaulay declared in a famous gibe that the Puritans disliked bear-baiting not because of the pain it gave the bear, but because of the pleasure it gave the spectators.[35] There is a fragment of truth in that remark, but not in the way it is usually understood. Puritans lamented the readiness of dogs to fight with bears because they saw it as the result of the Fall and therefore a reminder of Man's sin. 'The antipathy and cruelty which one beast showeth to another is the fruit of our rebellion against God,' wrote William Perkins, 'and should rather move us to mourn than to rejoice.' It was in this spirit that John Spencer, an early-seventeenth-century gentleman, reproached his brother Nicholas for his inordinate delight in cock-fighting: 'You make that a cause of your jollity and merriment which should be a cause of your grief and godly sorrow, for you take delight in the enmity and cruelty of the creatures, which was laid upon them for the sin of man.'

But the Puritans also felt for the animals. 'What Christian heart can take pleasure to see one poor beast to rent, tear and kill another?' asked Philip Stubbes, for 'although they be bloody beasts to mankind and seek his destruction, yet we are not to abuse them for his sake who made them and whose creatures they are. For notwithstanding that they be evil to us and thirst after our blood, yet are they are good creatures in their own nature and kind, and made to set forth the glory and magnificence of the great God ... and therefore for his sake not to be abused.' Bear-baiting, therefore, was 'a filthy, stinking and loathsome game'. 'I think it utterly unlawful,' declared William Hinde, 'for any man to take pleasure in the pain and torture of any creature, or delight himself in the tyranny which the creatures exercise, one over another, or to make a recreation of their brutish cruelty which they practice one upon another.' The same sentiments were expressed by Henry Bedel, John Dod, Robert Cleaver, Thomas Beard, Edward Elton and many other strict Protestant clergy.[36]

It is true that the Puritans disliked animal sports because of their association with noise, gambling and disorder; and it was on those grounds that cock-fighting and cock-throwing were prohibited in the Protectorate ordinance of 1654.[37] But they also expressed strong sympathy with animal sufferings and thought it barbarous to take pleasure in them. It was sinful to 'take delight in the cruel tormenting of a dumb creature,' urged Robert Bolton, or to revel in 'the bleeding miseries of that poor harmless thing which in its kind is much more and far better serviceable to the Creator than thyself'.[38] In the mid seventeenth century the attack was extended to include horse-racing, which was condemned by the Quakers and others because it was 'destructive to the creatures' and involved 'overstraining ... and over-forcing creatures ... beyond their strength'.[39] In the 1650s an anonymous opponent of animal sports urged that 'there ought to be a law to restrain such barbarous cruelty and to preserve the poor, innocent, sensitive creatures which Almighty God hath made and given for better use'.[40] In fact, some municipalities had already begun to act. Maidstone banned cock-throwing in 1653 as 'cruel and un-Christianlike', while Chester had prohibited bear-baiting on similar grounds as early as 1596. Other local authorities followed suit.[41]

Yet the opinion of the all-decisive gentry was slower to change. The Protectorate government succeeded in closing only the public contests. Colonel Pride put an end to bear-baiting at the Hope Garden in London by shooting the bears, but private bear-baiting continued; and with the Restoration both it and cock-fighting and cock-throwing came out into the open once more.[42] When in 1660 the future non-juror, Edmund Ellis, republished the opinions of various Puritan divines on cock-fighting, a friend commented that he could not expect sympathy from what he called 'the generality of gentlemen'; and Ellis himself admitted that in attacking the sport as a recreation unfit for Christians he would be 'accounted a foolhardy and impudent fellow'.[43] Animal sports retained their popularity in the modish world for the rest of the century; just as horse-racing sustained its position as the sport of kings. As late as 1699 the supporters of the old East India Company lost a crucial parliamentary division by ten votes, many M.P.s being absent, 'going to see a tiger baited with dogs'.[44] Nevertheless, the Puritan opposition to cruel sports was sustained after the Restoration by many Quakers and Dissenters and gained fresh support in the eighteenth century from the Methodists and Evangelicals.

But it was not only those brought up in the Puritan tradition who had begun to turn against these 'filthy sports', as Sir John Davies called them in the 1590s.[45] Many shared the view expressed by Montaigne (whose *Essais* were twice translated into English during the seventeenth century) that there was 'a kind of respect and a general duty of humanity which tieth us ... unto brute beasts that have life and sense ... Unto men we owe justice, and to all other creatures that are capable of it, grace and benignity.' For Pepys the animal sports provided 'a very rude and nasty pleasure' and for Evelyn 'butcherly sports or rather barbarous cruelties'. Bull-running, thought Richard Butcher, the mid-seventeenth-century historian of Stamford, could afford entertainment only 'to such as take a pleasure in beastliness and mischief'.[46] By the eighteenth century this outlook had become the orthodoxy of the educated middle classes and all those who, like Steele and Addison, upheld an ideal of cultivated refinement.[47]

A combination of religious piety and bourgeois sensibility thus led to a new and effective campaign against these time-honoured recreations. Cock-throwing was widely attacked in the early Hanoverian provincial press.[48] It was a predominantly plebeian pastime and it seemed exceptionally unsporting. 'What noble entertainment is it for a rational soul,' asked a schoolmaster in 1739, 'to fasten an innocent, weak, defenceless animal to the ground and then dash his bones to pieces with a club?'[49] From 1720 onwards the ritual was prohibited in an increasing number of schools and towns; it was well in decline by the 1750s, though it lingered on in some rural areas into the nineteenth century.[50] Bull-baiting was prohibited at Birmingham in 1773 and was in retreat elsewhere before the century was over. The first attempt (in 1800) to prohibit it by statute was vehemently opposed, but in 1822 it was made illegal on the public highway and in 1835 it was banned altogether. Badger-baiting, dog-fighting, and similar animal contests were suppressed at the same time.[51]

Cock-fighting proved more resilient because it could be defended as the spontaneous expression of the birds' natural instincts; even the cock's artificial spurs could be justified as ensuring a speedier death for the unsuccessful combatant.[52] When James Boswell went to the London cockpit in 1762, he, like Samuel Pepys, John Dunton and many others before him, felt 'sorry for the poor cocks'. But when 'he looked around to see if any of the spectators pitied them when mangled and torn in a most cruel manner,' he 'could not observe the smallest relenting sign in any countenance'. Nevertheless, an increasing number of people condemned the sport for its 'barbarity'; and in the eighteenth century it gradually disappeared from the grammar schools.[53] In the mid seventeenth century the poet Robert Wild had treated cocking in a facetious, mock-heroic fashion in his *A notable, true, tragicall relation of a duel betwixt a Norfolk cock and a Wisbich cock*.[54] But by 1807 George Crabbe's tone was very different:

> Here his poor bird th'inhuman Cocker brings,
> Arms his hard heel and clips his golden wings;
> With spicy food th'impatient spirit feeds
> And shouts and curses as the battle bleeds.
> Struck through the brain, deprived of both his eyes,
> The vanquish'd bird must combat till he dies;
> Must faintly peck at his victorious foe,
> And reel and stagger at each feeble blow:
> When fallen, the savage grasps his dappled plumes,
> His blood-stain'd arms, for other deaths assumes,
> And damns the craven fowl that lost his stake,
> And only bled and perish'd for his sake.[55]

Cockpits became illegal in London in 1833 and in the whole country in 1835. Cock-fighting as such was finally prohibited in 1849, though like the other animal sports it survived in a clandestine way.[56]

Hunting, on the other hand, presented a much trickier issue. The medieval church had deemed it a carnal diversion, unsuitable for clergymen, and had (rather ineffectually) forbidden it to those in holy orders; the issue gained a new topicality in 1621, when the unfortunate Archbishop Abbot was unlucky enough to miss the stag and kill the gamekeeper.[57] The Puritans, like other moralists, thought that hunting wasted a great deal of time and money, as well as being destructive to poor farmers' crops.[58] In 1604 Archbishop Hutton told Robert Cecil that he would like to see 'less wasting of the treasure of the realm and more moderation in the lawful exercise of hunting, both that poor men's corn may be less spoiled and other of his majesty's subjects more spared' (a wistful hope with that obsessive huntsman, James I, just settled on the English throne).[59] But the archbishop was careful to describe hunting as a lawful exercise. So long as it could be represented as a necessary means of securing food or keeping down pests it was hard to attack it directly. If hunting was controversial, it was more because of the attempt to confine it to those above a certain social level than because of any doubts about man's right as such to pursue and kill the lower animals. Most Puritan casuists allowed it in moderation. In 1641 William Hinde published a lengthy discussion of the ethics of hunting. But he avoided coming down firmly against it. The most that he could say was that it was wrong to take pleasure in the pains of the hunted animal or to protract them unnecessarily and that consideration should be shown to the horses. Later on in the century a Yorkshire clergyman noted in his diary that, though some of his contemporaries questioned the lawfulness of hunting and coursing, he himself was satisfied that, even if Adam had not sinned, 'there would yet have been some creatures for food which man must have been forced to pursue with others as he doth'.[60]

From John Foxe to Edmund Ludlow, there was no shortage of godly figures who were passionately addicted to hunting and appear to have felt no pangs of conscience. In the seventeenth century the Puritan gentleman Nicholas Assheton chased happily after foxes, stags, otters, hares and every other kind of huntable wild life. Bulstrode Whitelocke had some doubts, but he soon allayed them with the reflection that God would never have given hounds their sharp noses and their speed if he had not intended them to be used for hunting; as for the hunted animals, they were 'creatures which by nature are continually in fear and dread, and that when they are not hunted, as well as when they are'. Hunting, he thought, did nothing to make hares or deer unhappier than they would have been anyway.[61]

Yet there had long been uneasiness on the matter. In the twelfth century John of Salisbury had thought that hunting had a brutalizing effect upon the character; and in the early Tudor period Sir Thomas More's Utopians took no pleasure in hunting, but felt pity for the innocent hare: hunting was 'the lowest, the vilest, and most abject part of butchery'. For More, as for John Foxe a few decades later, the Christian repudiation of the Jewish blood sacrifice was proof that God abhorred unnecessary bloodshed.[62] Archbishops Warham and Parker never went hunting. Neither did Bishop Jewel. What pleasure, he asked the hare-hunters, could be derived from pursuing with fierce dogs a timid animal that attacked no one and was put to flight by the slightest noise? Hunting for sport alone, thought Philip Stubbes, was wholly unlawful. There would be no hawking or hunting in heaven, a preacher told his congregation at St Paul's in 1603.[63] It was in response to such teaching that the godly Jacobean layman John Bruen was prevailed upon to lay aside his hounds, hawks and dogs and to dispark his game reserves.[64] There was a risk, warned a sporting writer,

> lest we be transported with this pastime, and so ourselves grow wild, haunting the woods till we resemble the beasts which are citizens of them, and by continual conversation with dogs, become altogether addicted to slaughter and carnage, which is wholly dishonourable.

Addiction to hunting, agreed Thomas Tryon, made men 'fierce, cruel and great devourers'.[65]

Yet only occasionally did a bold spirit directly challenge what was, after all, the chief recreation of the clergy's noble patrons. One such spirit was Thomas Bywater, who in 1605 presented his employer, Lord Sheffield, with a book reproving him for his sins, particularly 'hunting and hawking too much'. 'He would maintain to my face,' reported the indignant peer, 'that both hawks and hounds, which I did then and do now moderately delight in, were not ordained by God for man's recreation, but for adorning the world.' Bywater was a tutor in Lord Sheffield's household; it is not surprising that he failed to gain tenure.[66]

In the later seventeenth century many Dissenters grew worried about hare-coursing. To kill edible creatures was 'no doubt lawful,' thought Edward Bury in 1677, 'but to sport ourselves in their death seems cruel and bloody'. And, in a spirit anticipatory of much subsequent writing on the subject, he continued, 'suppose thou heardst such a poor creature giving up the ghost to speak after this manner (for it is no absurdity to feign such a speech), "Oh man, what have I done to thee? ... I am thy fellow creature".'[67]

The truth was that the doctrine of man's stewardship, strictly interpreted, was now making it impossible to condone killing animals for mere sport. One could take life for the sake of food or self-defence, but not for pleasure. 'It is lawful for man in his own defence and for his own safety to destroy serpents, hurtful beasts and noisome creatures,' ruled George Walker in 1641, 'yet to do it with cruelty and with pleasure, delight and rejoicing in their destruction, and without a sense of our own sins and remorse for them, is a kind of scorn and contempt of the workmanship of God.' Hunting as such was not unlawful, thought Sir Matthew Hale, but he could never approve of pursuing 'the harmless hare for no other end than sport'.[68] The Quakers would forbid hunting for sport altogether, while the *Tatler* in 1710 pronounced against hunting 'innocent animals which we are not obliged to slaughter for our safety, convenience or nourishment'.[69]

But it was hard to see that hares and deer were hunted for any other purpose than the pleasure of pursuit, despite the occasional half-hearted attempt to justify stag-hunting as a necessary form of self-protection. In William Browne's *Britannia's Pastorals* (1613) a swain asks,

> Is it not lawful we should chase the deer
> That breaking our enclosures every morn
> Are found at feed upon our crop of corn?[70]

But since deer survived only in protected deer parks the argument was patently disingenuous; and it had disappeared altogether by the 1780s, when the Royal Buckhounds at the stipulation of George III's second son, Frederick, Bishop of Osnaburg, abandoned the practice of killing the quarry and released the deer so that it could be hunted again.[71] In the eighteenth century it had become increasingly difficult to argue that either hare-coursing or stag-hunting served any necessary purpose; and moralists passionately denounced them accordingly. To some men, noted an observer in 1788, 'the delight found in pursuing a poor harmless hare, with a parcel of ugly roaring hounds ... may appear on consideration [as] inhuman and barbarous as bull-baiting'.[72] The shooting of harmless birds also began to be condemned.*

The fox, however, was a different matter. For he, as the poets put it, was a 'subtle, pilfering foe', a 'conscious villain' and the highly-organized sport of fox-hunting could be seen as 'just vengeance on the midnight thief'. The otter was also 'this midnight pillager'. The pursuit of such pests was accordingly represented as half battle, half morality-play.

> For these nocturnal thieves, huntsman, prepare
> Thy sharpest vengeance. Oh! how glorious 'tis
> To right th'oppress'd and bring the felon vile
> To just disgrace![73]

In 1776 Francis Mundy denounced hare-hunters:

> the murderous crew
> In harmless blood their hands imbrue...

But hunting the fox, he thought, was a different matter:

> Talk not of pity to such foes!
> Stern justice claims the life he owes.[74]

* See below, p. 280.

As the Jacobean preacher John Rawlinson had long ago explained, the beasts to which the Old Testament intended the righteous man to be merciful were the 'cattle or helpful beasts'; foxes, by contrast, were 'not helpful, but hurtful ... and therefore no pity [is] to be had of them'. The poet James Thomson could plead for the hare and the stag, but he too felt no sympathy for the fox, 'the nightly robber of the fold'.[75]

Indeed it was only in 1869, when the historian E. A. Freeman wrote a famous article attacking fox-hunting, that the modern agitation against the sport as cruel to the fox really got under way.[76] It is odd that it did not start much earlier, since most people had long known that hunters, far from trying to keep down foxes, which was the overt objective, were in fact carefully preserving them. Fox-hunting, although originally regarded as a socially inferior activity to deer-hunting, had gained steadily in popularity with the gentry during the sixteenth and seventeenth centuries, particularly when deer grew scarcer and hare-coursing was impeded by enclosures. William Harrison observed in 1577 that foxes would have been 'utterly destroyed ... many years agone' if gentlemen had not protected them to 'hunt and have pastime withal'; and Robert Reyce commented in 1618 that in Suffolk the fox would have been extinct if it had not been protected by the gentry for the sake of necessary warlike exercise 'against the time of a foreign invasion'. Indeed, as early as 1539 Robert Pye had informed Thomas Cromwell that foxes could easily be wiped out, if only the gentry would allow it; foxhounds, he added, did more harm to farmers' sheep and chickens than did foxes.[77] It was still possible in 1669 for John Worlidge to urge that, if foxes were hunted at breeding-time, they could be eliminated altogether; and regular payments were made by parish authorities to those who produced the carcasses of such vermin.[78] But by the beginning of the eighteenth century it had become common to preserve fox cubs, to import foxes from adjacent counties, to plant coverts for their shelter and even to chase 'bagged' foxes (that is, those brought along in a sack to be hunted).[79] Landlords preferred to pay compensation to farmers for the losses to their lambs and chickens, rather than give up the sport. In due course vulpicide (the secret killing of the fox) became one of the greatest moral offences a country gentleman could commit.[80] The owners of pheasant preserves waged a private war on foxes, but it was only an accredited huntsman who would dare openly claim payment from the parish for foxes' heads.[81]

This artificial preservation of foxes by fox-hunters did not stop the judges of King's Bench under Lord Mansfield in 1786 from reiterating the traditional doctrine that no action for trespass lay against fox-hunters who followed their quarry onto someone else's land, because the fox, unlike the hare, was a noxious beast which all men were at liberty to pursue and kill wherever they could; hunters could therefore chase across the country at large, regardless of who owned it.[82] But non-lawyers found it increasingly difficult to think of fox-hunting as the conscientious discharge of the painful duty of pest control; and by the later eighteenth century a small group of critics had begun to attack it on the grounds of its cruelty.[83]

iii. THE DETHRONEMENT OF MAN

Even within a fundamentally man-centred mode of thought, therefore, it was possible to condemn many of the ways in which animals had been customarily treated. The beasts had been created for Man's sake, but that was no reason for ill-treating them unnecessarily.

By the later seventeenth century the anthropocentric tradition itself was being eroded. The explicit acceptance of the view that the world does not exist for man alone can be fairly regarded as one of the great revolutions in modern Western thought, though it is one to which historians have scarcely done justice. Of course, there had been many ancient thinkers, Cynics, sceptics and Epicureans, who denied that men were the centre of the universe or that mankind was an object of special concern to the gods. In the Christian era a periodic challenge to anthropocentric complacency had been presented by sceptical thinkers like Celsus, who in the second century A.D. had attacked both Stoics and Christians by urging that nature existed as much for animals and plants as for man.[1] It was absurd to think that pigs were specially made to be eaten by men, said Porphyry a century later; one might as well believe that men were specially made to be eaten by crocodiles.[2] Moreover, the Old Testament contained many texts consistent with the view that God had created the inferior creatures for his sake and theirs, rather than for that of man alone. Some theologians accordingly taught that living creatures existed to reflect divine glory as well as to cater to human needs.

What is new about the early modern period is that, when Montaigne in the sixteenth century and the French *libertins* in the seventeenth revived the old attack of the classical sceptics upon man's 'imaginary sovereignty' over other creatures,[3] they found that now for the first time there were writers in the Christian tradition prepared to agree with them. In the mid sixteenth century the Marian martyr John Bradford explicitly challenged the scholastic doctrine that animals were made solely for human sustenance.[4] In the seventeenth century it became increasingly common to maintain that nature existed for God's glory and that he cared as much for the welfare of plants and animals as for man. And during the Civil War there were sectaries who took this view to its logical conclusion. 'God loves the creatures that creep on the ground as well as the best saints,' said one, 'and there is no difference between the flesh of a man and the flesh of a toad.'[5]

Although most contemporaries would have regarded this as an overstatement, it was quite usual in the later seventeenth century for relatively orthodox clergymen to urge that God was as concerned for beasts as for man. The Dissenter Samuel Slater even described it as a 'heathen' doctrine to say that God did not 'attend to the meaner and inferior creatures ... but only superintended the affairs and concernments of mankind'. Early Christian fathers like Jerome, he said, were quite wrong to think God indifferent to the welfare of, say, flies and gnats. Creatures were made 'to enjoy themselves' as well as to serve man, agreed Henry More, and it was 'pride', 'ignorance' or 'haughty presumption' to think otherwise.[6] The general shift in perspective during these years was well expressed by John Ray in 1691. 'It is a generally received opinion,' he wrote, 'that all this visible world was created for Man; [and] that Man is the end of the Creation, as if there were no other end of any creature but some way or other to be serviceable to man ... But though this be vulgarly received, yet wise men nowadays think otherwise.'[7]

It was above all the vast expansion in the size of the known world which was causing wise men to think differently. As the astronomers revealed not only that the earth was not the centre of the universe, but that there was an infinity of worlds, each perhaps inhabited by some unknown species, it became increasingly hard to maintain that creation existed for the exclusive benefit of the human denizens of one small planet. The old sublunary world was only a tiny fraction of the vast celestial universe now known to exist. 'The vulgar opinion of the unity of the world' was now 'exploded', wrote Henry Oldenburg in 1659; it was 'absurd' to think 'the heavenly hosts, which are so many times bigger than our earth, to be made only to enlighten and to quicken us'. There was no reason to think that either the earth or the human race was a particularly central part of the universe.[8] John Ray believed that 'in all likelihood' there were creatures living on the moon; and the possibility that not only it but other planets also might be inhabited was widely canvassed.[9] The great philosophers Galileo, Descartes, Gassendi and Leibniz all rejected the idea that the natural world was created for man alone.[10] Pondering his calculation that some stars were over twenty thousand times as large as the sun, William Gilbert reflected in 1636 on human insignificance. Poor man was a mere ant upon the face of the earth, who, by comparing himself only with those beneath him, had foolishly swollen 'into a conceit of being somebody'. The full extent of human ignorance was revealed, thought John Locke, 'when we consider the vast distance of the known and visible parts of the world, and the reasons we have to think that what lies within our ken is but a small part of the universe'.[11]

Even on the earth itself, the microscope had by the end of the seventeenth century begun to reveal millions of animated beings, protozoa and bacteria, pursuing their existence in utter indifference to human concerns, occupying a world of beauty and intricacy on which no men had ever previously set eyes. The Dutchman Anton van Leeuwenhoek found 8,280,000 living creatures in a drop of water and declared in 1683 that there were more animals in his own mouth than there were people in the United Provinces.[12] At the same time, explorers were daily stumbling upon inhabited tracts of the earth's surface, forests and deserts created for no apparent human purpose, swarming with hitherto unknown forms of life for which there was no obvious human use. By the time of Linnaeus the number of known plants was ten times that which had been recorded in classical antiquity and the range of known animal life had been similarly extended.[13] Many things, observed Descartes, existed or formerly existed and had ceased to be; yet they had never been seen or known by man and were never of use to humans. We should therefore 'beware of presuming too highly ourselves' so as to think 'that all things were created by God for us only'. Or, as the poet Gray put it more succinctly a century later,

> Full many a flower is born to blush unseen
> And waste its sweetness on the desert air.[14]

The destruction of the old anthropocentric illusion was thus begun by astronomers, botanists and zoologists. It was completed by the students of geology. As early as 1738 the observations made by the Cumbrian Quaker Thomas Story of the strata of the cliffs near Scarborough had convinced him that the earth was 'of much older date than the time assigned in the Holy Scriptures'.[15] In the later eighteenth century the accumulated work of similar observers had encouraged the French naturalist Buffon to abandon biblical chronology to the extent of allowing that the earth had existed for 'some seventy thousand years' before the appearance of man. By the 1820s the geologists were certain that the earth's prehistory was a matter not of thousands of years but of millions. Whole species of animals and plants had come into existence, lived, and been obliterated, long before humanity appeared. As Charles Lyell explained in 1830, man's arrival upon the planet was relatively recent: 'at periods extremely modern in the history of the globe, the ascendancy of man, if he existed at all, had scarcely been felt by the brutes'; earlier writers had erred by 'undervaluing greatly the quantity of past time'.[16] In the later seventeenth century most of John Ray's contemporaries had been unwilling to admit that any former species had been extinguished before the appearance of man. But even then Robert Hooke knew that there had; and in the eighteenth and early nineteenth centuries the accumulating evidence of the fossils would prove irresistible.[17] The Fall of Man could no longer be held responsible for nature's physical characteristics; the earth and the species on it had not been created for the sake of humanity, but had a life and history independent of man. In 1780, accordingly, the atheistic geologist

G. H. Toulmin declared that man was merely a small part of nature and rejected anthropocentric religious myths as mere figments of human pride.[18]

Of course this was too much for most people; as the nineteenth-century debates on evolution would show, anthropocentrism was still the prevailing outlook. Darwin upset many by his polemical rejection of the argument from design and by his demonstration that the features of, say, the orchid derived from their advantages in the plant's struggle for existence, not from God's desire to provide man with an object of beauty and interest. One of Samuel Wilberforce's objections to Darwin was that 'the degrading notion' of human descent from the brutes was 'utterly irreconcilable' with 'man's derived supremacy over the earth'.[19] In the eighteenth century most of the writers who paid lip-service to the doctrine that the world was not made for man alone usually moved on quickly to demonstrating that, even so, it had been remarkably well designed to receive humanity. Nevertheless, it had become by early Hanoverian times a commonplace to concede that it was not only human purposes which the world was designed to serve. It was repugnant to reason 'to affirm that the world was made for the sake of man alone,' remarked the sceptical Bolingbroke, adding with sardonic pleasure, 'some modern divines have been candid enough to give up the point.'[20] Man was now but one link in Nature's mighty chain and no more indispensable than any other link. As Henry Baker put it in *The Universe* (1727), subtitled 'A Poem intended to Restrain the Pride of Man':

> Each hated toad, each crawling worm we see,
> Is needful to the whole as well as he.

It was comically vain on man's part to imagine that it was for him that the earth had been made.

> As well may the minutest emmet say
> That Caucasus was raised to pave his way.[21]

It was only the 'arrogance of humanity', observed Edward Bancroft in 1769, that had generated the delusion that the whole of animate nature had been created solely for its use. To regard the happiness of man as the only object of creation, agreed his fellow-naturalist George Gregory, was 'narrow-minded' and 'absurd'. Each part of the natural world was an end in itself.[22]

Some philosophers indeed had already begun to move from the mere denial that man was uppermost in nature's intentions to the more drastic assertion that nature had no intentions at all, or at least that it was impossible to know what they were. Both Bacon and Descartes had regarded the appeal to final causes as inappropriate in the study of natural history, on the grounds that it was preposterous for man to claim to know God's ultimate intentions. Some of their contemporaries were said to be 'epicureans' or 'libertine' sceptics who denied the role of providence altogether, seeing the world as the product of the chance collision of atoms.[23] Such views came unambiguously into the open with Spinoza, for whom nature had no end and all final causes were human fictions. In his *Dialogues concerning Natural Religion* (published posthumously in 1779) David Hume described 'a blind nature, impregnated by a great vivifying principle, and pouring forth from her lap, without discernment or parental care, her maimed and abortive children'. Immanuel Kant in his *Critique of Teleological Judgement* (1790) professed himself unable to find 'any being capable of laying claim to the distinction of being the final end of creation'. Man was as much a means as an end: 'nature has no more exempted him from its destructive than from its productive forces, nor has it made the smallest exception to its subjection of everything to a mechanism of forces devoid of an end.' The emerging concept of an ecological system would make obsolete the old language of means and ends.[24]

It took a long time for these new currents of thought to reveal their full implications. But even in the seventeenth century the growing challenge to the old anthropocentrism had begun to affect contemporary thinking about the treatment of animals; for if the beasts were no longer to be thought of as created solely for the sake of man, then human conduct towards them appeared in a new and much less favourable light. Long before Hume and Kant there were individuals prepared to concede a parity to all parts of creation. 'The inferior creatures,' Thomas Tryon told his contemporaries in 1684, 'groan under your cruelties. You hunt them for your pleasure, and overwork them for your covetousness, and kill them for your gluttony, and set them to fight one with another till they die, and count it a sport and a pleasure to behold them worry one another.' Man, thought Margaret Cavendish, behaved as if

> all creatures for his sake alone
> Were made for him to tyrannize upon.

What right, she asked in her *Dialogue betwixt Birds* (1653), did human beings have to shoot sparrows for taking cherries and then eat the fruit themselves? And the question was repeated by Tryon in 1683 in his defence of wild birds: 'What right, I pray, has man to all the corn in the world?'[25]

The only answer to that question now was the one given in the mid seventeenth century by Thomas Hobbes. Man, like any other living being, was entitled by the right of nature to take those steps which he thought necessary for his preservation and subsistence. He could therefore kill other creatures 'for his safety and benefit'. Useful animals could be reduced to servitude and noxious ones destroyed, just as an individual in the state of nature was entitled to kill another human being if he felt he was a threat to his preservation. Human rule, therefore, reflected merely the naked self-interest of the human species. 'If we have dominion over sheep and oxen,' wrote Hobbes, 'we exercise it not as dominion, but as hostility; for we keep them only to labour, and to be killed and devoured by us; so that lions and bears would be as good masters to them as we are.' Man's rule over other creatures rested merely on his superior power, stemming from his manual skills and his use of speech, not on divine law or a special grant from God. Hobbes mocked the old notion that the world was made for man, echoing the words of Porphyry fourteen hundred years earlier: 'I pray, when a lion eats a man and a man eats an ox, why is the ox more made for the man than the man for the lion?'[26] These sentiments were as revolutionary as anything Hobbes ever wrote about politics; and they evoked a corresponding warmth of protest. 'I am sorry,' wrote Bishop Bramhall, 'to hear a man of reason and parts ... compare the murdering of men with the slaughtering of brute beasts.' For Bramhall held to the older view in which men occupied a divinely-privileged position: their dominion over the animals rested on God's grant as set out in Genesis. A man who killed a dangerous lion had divine authority for doing so, whereas the lion had no such authority to eat the man. Hobbes's position, thought the bishop, was perverse: 'He acquitteth the beasts from the dominion of man and denieth that they owe him any subjection.' He had turned himself into an 'attorney-general for the brute beasts'.[27]

Even for more conventional thinkers than Hobbes mankind was no longer the sole object of creation. Animals were to be regarded as their fellow-creatures – 'under-graduated fellow-creatures' perhaps, as Thomas Tryon called them,[28] but still fellow-creatures; and they should be treated accordingly. As the Presbyterian minister John Flavell wrote of a tired horse in 1669,

> What hath this creature done that he should be
> Thus beaten, wounded and tired out by me?
> He is my fellow-creature.

We ought, urged Benjamin Parker in 1745, 'to have more regard and esteem for our fellow creatures than to imagine them made for no nobler ends than to become our vassals'. To Christopher Smart animals and birds were 'my fellow subjects of th'eternal King'.[29]

In late-eighteenth-century Romanticism this theme of universal brotherhood grew very insistent. Burns was

> truly sorry man's dominion
> Has broken nature's social union.

Blake asked the fly,

> Am not I
> A fly like thee
> Or art not thou
> A man like me?

And Coleridge, moved by revolutionary ideals of fraternity, addressed 'A Young Ass':

> I hail thee BROTHER.

Animals had thus moved from being mere 'brutes' or 'beasts' to being 'fellow beasts', 'fellow mortals' or 'fellow creatures' and finally to being 'companions', 'friends' and 'brothers'.[30]

iv. NEW SENSIBILITIES

From the later seventeenth century onwards it had thus become an acceptable Christian doctrine that all members of God's creation were entitled to civil usage. Moreover, the area of moral concern had been widened to include many living beings which had been traditionally regarded as hateful or noxious. 'Even to the reptile,' wrote John Dyer in 1757, 'every cruel deed/Is high impiety.' Christopher Smart sang of the beetle, 'whose life is precious in the sight of God, tho' his appearance is against him', and the crocodile, 'which is pleasant and pure when he is interpreted, tho' his look is of terror and offence'. Parents should not let their children cause needless harm to any living thing, declared John Wesley, for the golden rule applied to all creatures – snakes, worms, toads and flies included. It was criminal, the Rev. James Granger told his rural congregation at Shiplake, Oxfordshire, in 1772, to destroy the 'meanest insect' without good reason. Worms, beetles, snails, earwigs and spiders all found their advocates; and naturalists began to seek for more humane methods of killing them.[1]

The eighteenth century abounds in these new susceptibilities. For if in literature we have Tristram Shandy's Uncle Toby, reluctant to kill the fly, then in actual life there was the Norwich doctor Sylas Neville, who in 1767 caught two mice in a trap, but then released them, being 'unwilling to kill the troublesome little vermin'. There were also the author William Melmoth, worried about the cruelty of destroying the snails on his garden peaches, and the Calvinist divine Augustus Toplady, deploring the cruelty of digging up ant-hills, and the writer William Chafin, lamenting the callow theft by schoolboys of nuts hoarded by industrious mice.[2]

Of course, spontaneous tender-heartedness, as such, was not new. Some medieval examples have already been quoted.* In sixteenth-century France Montaigne had denounced cruelty to animals, partly because they shared some qualities with men and partly because they were God's creatures worthy of respect, but also because such cruelty offended his innate sensibilities: 'If I see but a chicken's neck pulled off or a pig sticked, I cannot choose but grieve; and I cannot well endure a silly dew-bedabbled hare to groan when she is seized upon by the hounds.' It is not difficult to detect his influence upon Margaret Cavendish, who in 1667 described herself as 'tender-natured, for it troubles my conscience to kill a fly and the groans of a dying beast

* Above, p. 152.

strike my soul'.[3] Shakespeare had also displayed explicit sympathy with hunted animals, trapped birds, tired horses, even flies, snails and 'the poor beetle that we tread upon'; and the Elizabethan John Stubbs (of *The Gaping Gulf*) alluded to the 'common compassion' which some men felt when they saw overburdened beasts.[4] The Seeker Thomas Taylor in 1661 contrasted the devotees of cruel animal sports with 'the tender nature of Christ and all Christians, truly so-called, who could never rejoice in any such things by reason of their tender, pitiful and merciful nature'. In his textbook on Greek history Francis Rous praised the Athenians because they showed themselves 'tender-hearted', not just to men, 'but even to brute beasts'. 'Tender compassion to the brutes,' agreed Richard Baxter, had been put by God into 'all good men.'[5]

These were not merely pious assertions, for there is no shortage of well-attested instances of such tender-heartedness at work. When in 1614 'a cur dog' was thrown into a London privy by 'an untoward lad . . ., taking delight in Knavish pastimes', and left to remain there 'starving and crying for food' for three days, it was eventually rescued by 'the good man of the house, who grieved to see a dumb beast so starved and for want of food thus to perish'. Among the Diggers in 1649, says Gerrard Winstanley, 'tender hearts' grieved to see their cows bruised and swollen after being beaten by the lord of the manor's bailiffs. Two decades later Samuel Pepys was maddened to see the son of Sir Heneage Finch beating a little dog to death and letting it lie in pain. At Florence in 1672 an experiment upon a live dog in the presence of some Englishmen was ruined when its agonies 'moved the compassion of one of the servants . . . out of an untimely charity to rid her from these torments by striking her on the head with a stick'; even the vivisectors Hooke and Boyle decline to experiment upon the same animal twice 'because of the torture of the creature'.[6] When Colonel Abraham Holmes, a supporter of Monmouth, was executed with some of his companions at Lyme Regis in 1685, the horses could not pull the sled carrying the condemned men to the scaffold. The attendants began to whip them furiously, whereupon Colonel Holmes, with one of those superb gestures of which the men of the seventeenth century were so frequently capable, got out to walk, saying, 'Come, gentlemen, don't let the poor creatures suffer on our account. I have often led you in the field. Let me lead you on in our way to Heaven.'[7]

All this was before 1700. In the eighteenth century the sensibilities were not different in kind, but they seem to have been much more widely dispersed, and they were much more explicitly backed up by the religious and philosophical teaching of the time. As Christopher Smart put it, kind acts to animals were not trivial matters, reflecting personal inclination:

> Tho' these some spirits think but light,
> And deem indifferent things;
> Yet they are serious in the sight
> Of CHRIST, the King of Kings.[8]

The intellectual genealogy of the so-called 'man of feeling' has been traced back to the Latitudinarian divines of the Restoration period, who, in reaction to Hobbes, taught that man's innate instincts were kindly and that it was unnatural to take pleasure in cruelty.[9] As early as 1654 John Hall maintained that children were naturally averse to killing anything: 'Had we not by incogitancy and custom been led there-unto, a man should no more kill a fly than a soldier without other engagement would kill a man.' For Samuel Parker, Isaac Barrow, John Tillotson, Robert South, Gilbert Burnet and Samuel Clarke, virtue was benevolence, and kindliness the most refined source of pleasure. This teaching was subsequently incorporated in the moral philosophy of Richard Cumberland, of the third Earl of Shaftesbury and of Francis Hutcheson; by the 1720s 'benevolence' and 'charity' had become the most favoured words in literary vocabulary. There was something in human nature, said William Wollaston, which made the pains of others obnoxious to us. 'It is grievous to see or hear (and almost to hear of) any man, or even any animal whatever, in torment.' The mid eighteenth century saw a cult of tender-heartedness, a vogue for weeping and a widespread acceptance by the middle classes of the principle that 'to communicate happiness is the characteristic of virtue'.[10] Kindliness and benevolence had become official ideals.

It was this mode of thought which gave rise to later utilitarianism, for the benevolent, as Cowper put it, wished 'all that are capable of pleasure pleased';[11] and, although its main implications were for the human species, whether slaves, children, the criminal or the insane, its relevance to animals was inescapable. As the editor of Bishop Cumberland's *Laws of Nature* commented in 1727: 'The author's scheme would have been more complete had he included benevolence towards brutes ... because we can't imagine but that Deity takes pleasure in the happiness of all his creatures that are capable thereof ... A truly benevolent man receives pleasure even from the happiness of the brute creation.' Gentle usage of the creatures, thought David Hume, was required 'by laws of humanity'. The law of universal benevolence, said the Nonconformist Philip Doddridge, applied even to brutes, since they were capable of sensation and therefore of pleasure and pain. 'My cat Jeoffry,' wrote Christopher Smart, 'is an instrument for the children to learn benevolence upon ... God be merciful to all dumb creatures in respect of pain.'[12]

What this new mode of thinking implied was that it was the *feelings* of the suffering object which mattered, not its intelligence or moral capacity. 'Pain is pain,' wrote Humphry Primatt in 1776, 'whether it be inflicted on man or beast.' As Rousseau had said twenty years earlier, neither animals nor men should be unnecessarily ill-treated; they were equally sentient beings.[13] Or, as Jeremy Bentham observed in 1789 in a famous passage, the question to be asked about animals was neither 'Can they *reason*?' nor 'Can they *talk*?', but 'Can they *suffer*?'[14] This was a new and altogether more secular mode of approach. It was now possible to attack cruelty to animals without invoking God's intentions at all. The ill-treatment of beasts was reprehensible on the purely utilitarian grounds that it diminished their happiness. Animals had feelings and those feelings ought to be respected. Whether animals had reason was irrelevant. After all, as one of John Wesley's correspondents observed, if pity was to be extended only to those with reason, then 'those would lose their claim to our compassion who stand in the greatest need of it, namely children, idiots and lunatics'.[15] Neither was it necessary to prove that animals had souls, for, if they did not, then their lack of future recompense was all the greater argument for treating them considerately in this world. The emphasis on sensation thus became basic to those who crusaded on behalf of animals, like James Granger, who in 1772 invoked 'the great law of humanity, which comprehends every kind of being that hath the same acute sense of pain which he finds in his own frame', and John Oswald and John Lawrence, both of whom wrote in the 1790s 'to lessen the sum of animal misery in the world'. In 1826 Sir Richard Phillips applied the golden rule to 'any sentient or suffering being', including the 'meanest animals'. In 1839 William Youatt urged that the only criterion for judging an action affecting other creatures, whether human or animal, should be 'its increasing or diminishing the general sum of enjoyment'; and in 1846 a Gloucester surgeon advocated respect for 'the interest and feelings of every sentient being that holds life'. It was in this spirit that Wordsworth had urged his contemporaries

> Never to blend our pleasure or our pride
> With sorrow of the meanest thing that feels.[16]

Inevitably the creatures which excited most sympathy were those who communicated their sense of pain in most recognizably human terms. As a commentator wrote in 1762, 'we are moved most by the distressful cries of those animals that have any similitude to the human voice, such as the fawn, and the hare when seized by dogs.'[17] Thomas Bewick's feelings of humanity to animals were first aroused when, as a boy, he caught a hare in his arms, while it was surrounded by the hunters and their dogs; 'the poor terrified creature screamed out so piteously, like a child, that I would have given any thing to save its life.' Yet despite his repugnance for blood sports, Bewick retained an enthusiasm for angling: 'I argued myself into a belief that fish had little sense, and scarcely any feeling.'[18] It was this uncertainty as to whether fish had sensation, since, as well as being virtually bloodless, they did not cry out or change expression, which enabled angling to retain its reputation as a philosophical, contemplative and innocent pastime, given impeccable ancestry by the New Testament and particularly suitable for clergymen. Fishing, unlike hunting, had never been forbidden to clerics by the medieval church; and in the sixteenth and seventeenth centuries it was the favourite recreation of many godly divines, including William Perkins, who had thundered against other animal sports.[19] Commentators objected to some details, like the use of live bait or the practice of spearing fish,[20] but they seldom attacked the sport itself, partly, of course, because fish were usually caught to be eaten. Yet before the beginning of the nineteenth century, fishing too was sometimes attacked because it inflicted pain. In 1799 Charles Lamb described anglers as 'patient tyrants, meek inflictors of pangs intolerable, cool devils', while for Byron angling was 'that solitary vice':

> Whatever Izaak Walton sings or says:
> The quaint, old, cruel coxcomb, in his gullet
> Should have a hook, and a small trout to pull it.[21]

Yet even today there are no laws for the humane treatment of fish.*

It was in keeping with the new emphasis upon sensation that the eighteenth century witnessed growing criticism of some forms of cruelty

* In 1980, however, a High Court decision established that a goldfish is entitled to the law's protection under the Protection of Animals Act (1911); *The Times*, 26 June 1980.

which had passed without much comment in earlier periods. Vivisection, which Isaac Barrow in 1654 had described as 'innocent' and 'easily excusable', was castigated a century later by Dr Johnson as the work of 'a race of men that have practised tortures without pity and related them without shame and are yet suffered to erect their heads among human beings'.[22] Many of his contemporaries shared Johnson's disgust. In 1816 a Dr Wilson Philip was rejected by the ballot of the Royal Society because of the great offence given to Fellows by the cruelty of his experiments upon animals.[23] Conventional methods of meat production also came under attack. 'What rapes are committed upon nature,' exclaimed Defoe, 'making the ewes bring lambs all the winter, fattening calves to a monstrous size, using cruelties and contrary diets to the poor brute, to whiten its flesh for the palates of the ladies!' The eighteenth century saw much protest against practices like that of crimping fish (i.e. cutting their live flesh to make it firmer) or plucking poultry before they were dead. Even William Cobbett, whose general attitude was nothing if not realistic, held that to cause pain to animals in order to heighten the pleasure of the palate was an abuse of man's God-given authority.[24] In the later part of the century methods of slaughter also came under critical scrutiny. The treatment of cattle at Smithfield market was put under statutory surveillance in 1781. In 1786 slaughterhouses had to be licensed and there was much discussion of humane-killers.[25] Meanwhile demands increased for legislative action against all kinds of cruelty to animals.[26] In the later eighteenth century some grammar schools introduced rules against the ill-treatment of animals;[27] and even before Parliament began to act there were prosecutions for cruelty brought under the head of trespass, nuisance or malicious damage.[28]

In such ways the notion that the feelings of all sentient beings should be regarded began to affect educated opinion. Of course, there were dissenters from the general trend, like William Whewell, the mid-nineteenth-century Master of Trinity College, Cambridge, who dismissed the idea that man's happiness should sometimes be sacrificed to increase the pleasure of animals as a *reductio ad absurdum* of Benthamite teaching; a view for which he was sharply reprimanded by John Stuart Mill, who, unlike Whewell, was quite certain that it was not only humanity which was entitled to humane treatment.[29]

At one point in the late eighteenth century there was a move to suggest that plants, no less than animals, were entitled to consideration, on the same utilitarian grounds. This was not an altogether novel idea. Montaigne had urged that trees and plants should be treated with humanity;[30] and various seventeenth-century English writers had embroidered upon the implications of the supposed antipathies and sympathies of vegetation. It was not hard to believe, thought Nathanael Homes in 1661,

> that plants and trees and herbs have their passions or affections; their love appearing in their sympathy as ... in the ivy and oak, etc.; their hatred in their antipathy, as in the vine and colewort, that will not prosper if near each other; their sorrow in pining and withering; their joy in blossom and flowering.

The agriculturalist John Worlidge in 1677 detected 'a kind of perception in them, tending themselves to that which nourisheth and preserves them, and eschewing and voiding that which injureth them'. Plants and herbs were 'our fellow-creatures', observed Lord Herbert of Cherbury.[31] The idea of plant sensitivity was propounded in the mid seventeenth century by Henry Power and Sir Thomas Browne. It was rejected by those who took a mechanistic view of the nature of vegetation, but by the later eighteenth century botanical science had come round to the idea that plants were sentient creatures. After all, they too breathed, slept, felt irritability and, as Linnaeus had emphasized, had a sexual life. Plants, claimed the author of *A Philosophical Survey of Nature* (1763), 'feel pain in degree proportionate to the delicacy of their construction.' 'I doubt whether we are right in confining the capacity of pleasure and pain to the animal kingdom,' agreed the botanist George Bell in 1777; and in 1784 Thomas Percival told a Manchester audience that 'plants, like animals, are endowed with the powers, both of perception and enjoyment'.[32] 'And 'tis my faith,' wrote Wordsworth, 'that every flower enjoys the air it breathes.' 'I would not strike a flower,' he added, 'as many a man would strike his horse.'[33]

But most contemporaries dismissed such sentiments as poetical fancies and in the nineteenth century botany and poetry went their separate ways. It was animals, not plants, who gained most from the new disposition to found the rights of man upon his capacity for happiness. Indeed, if men had rights, so too did they. In a posthumous work published in 1755 the philosopher Francis Hutcheson declared that brutes 'have a *right* that no useless pain or misery should be inflicted on them'.* Flies were as capable of pain as men, said Thomas Percival in 1775, and had no less a 'right' to life, liberty and enjoyment. 'The day may come,' wrote Jeremy Bentham in 1789, 'when the rest of the

* Though the key word was 'useless', for Hutcheson emphasized that 'the brutes ... can have no right or property valid against mankind in any thing necessary for human support'; *A System of Moral Philosophy* (1755), i. 313.

animal creation may acquire those rights which only human tyranny has withheld from them.' In 1798 John Lawrence proposed that the rights of beasts should be acknowledged by the state: 'the *ius animalium* ... surely ought to form a part of the jurisprudence of every system founded on the principles of justice and humanity.' Cruelty now was not merely inhumane; it was unjust.[34]

There were also hints that the rights of animals extended to something more than mere protection from physical pain. Man's stewardship of creation, as Thomas Tryon had stressed, involved 'assisting those beasts to the obtaining of all the advantages their natures are by the great, beautiful and always beneficent creator made capable of'. They were capable of pleasures of the mind, Lord Monboddo would urge, of fellowship of the herd and affection for offspring.[35] Bernard Mandeville raised doubts about the ethics of castrating domestic animals as early as 1714. A hundred years later Shelley denounced the practice as 'unnatural and inhuman', while the physician William Lambe called it 'a shocking outrage on the common rights of nature'.[36] In the late Victorian age some humanitarians would urge the right of animals to 'self-realization'.[37]

So much then for the intellectual origins of the campaign against unnecessary cruelty to animals. It grew out of the (minority) Christian tradition that man should take care of God's creation. It was enhanced by the collapse of the old view that the world existed exclusively for humanity; and it was consolidated by a new emphasis on sensation and feeling as the true basis for a claim to moral consideration. In this way the anthropocentric tradition was, by a subtle dialectic, relentlessly adjusted to bring animals within the sphere of moral concern. The debate on animals thus furnishes yet another illustration of that shift to more secular modes of thinking which was characteristic of so much thought in the early modern period. Yet the initial impulse had been strongly religious. The teachings of classical writers like Plutarch or Porphyry had not been forgotten, but the Old Testament was the authority which was most frequently cited by the propagandists. Clerics were often ahead of lay opinion and an essential role was played by Puritans, Dissenters, Quakers and Evangelicals;* though the new feelings about animals were in no sense confined to those within the Protestant tradition. Obadiah Walker, who in 1673 urged that children should be restrained from tormenting small animals,[38] was a Roman Catholic.

* As Horace Walpole wrote to a friend on 20 June 1760: 'I met a rough officer at his house t'other day, who said he knew such a person was turning Methodist; for, in the middle of conversation, he rose, and opened the window to let out a moth'; *Letters*, ed. Helen and Paget Toynbee (Oxford, 1903–18), iv. 399.

Lewis Gompertz (1779–1865), the first secretary of the S.P.C.A., was a Jew. It is wrong to suggest that sceptics initiated the movement and that Christians only tagged on at the end of the eighteenth century.[39] It must also be emphasized that the Tudor and Stuart period, far from being a relative blank in the story, was the one during which all the essential ingredients of later feelings were assembled. By 1700 all the key arguments were there.

V. NEW CONDITIONS

Yet to suggest that concern for animals' rights developed logically out of elements latent in the Judaeo-Christian tradition is merely to beg the question. For, if the intellectual possibility had always been present, why was it only in the early modern period that it was realized?

The answer seems to be that these purely intellectual developments had to be stimulated by external social change. The triumph of the new attitude was closely linked to the growth of towns and the emergence of an industrial order in which animals became increasingly marginal to the processes of production.[1] This industrial order first emerged in England; as a result, it was there that concern for animals was most widely expressed, though the movement was very far from being peculiar to this country.* Of course, working-animals of every kind were extensively used during the first century and a half of industrialization. Horses, donkeys, even dogs, were employed in woollen mills, breweries, coal mines and railway shunting-yards. Horses did not disappear from the streets until the 1920s or from the farms until the 1940s.[2] But, long before that, most people were working in industries powered by non-animal means. The shift to other sources of industrial power was accelerated by the introduction of steam and the greater employment of water power at the end of the eighteenth century; and the urban isolation from animals in which the new feelings were generated dates from even earlier.

For the agitation did not begin among butchers or colliers or farmers, directly involved in working with animals. As the poet Seamus Heaney recently put it:

> 'Prevention of cruelty' talk cuts ice in town
> Where they consider death unnatural,
> But on well-run farms pests have to be kept down.[3]

Neither did the pressure emanate from those most accustomed to handling animals for working purposes. Grooms, cab-drivers and other servants did not own the animals themselves and were usually concerned only to get their particular job done as quickly as possible. The new sentiment was first expressed either by well-to-do townsmen, remote from the agricultural process and inclined to think of animals as pets rather than as working livestock, or by educated country clergymen, whose sensibilities were different from those of the rustics among whom

* The development of new feelings towards animals in European countries is outside the scope of this book. But it is evident that the conditions which shocked Victorian travellers to Spain or Italy were not to be found everywhere on the Continent. In the seventeenth and eighteenth centuries nearly all European countries had their advocates of greater humanity to beasts; and in the nineteenth century most of them enacted legislation on the subject. In the Jacobean period one of the most striking statements on the topic was the chapter 'Of gentleness towards brute beasts' in John Molle's translation (1621) of *The Walking Librarie* by Philipp Camerarius, counsellor to the free state of Nuremberg. At the end of the eighteenth century it was Arnaud Berquin, author of *L'Ami des Enfants* (1782–3), translated by Mark Anthony Meilan in 24 volumes in 1786 as *The Children's Friend*, who, along with Mrs Barbauld, was thought by Coleridge to have done most to make compassion towards animals 'universally fashionable'; *The Watchman*, ed. Lewis Patton (1970), 313. Earlier in the century the French clergyman Jean Meslier (1664–1729) had deplored even the crushing of spiders. His sensibilities towards the sufferings of animals were so great as to lead him to atheism, on the grounds that the natural order was demonstrably imperfect if it could permit such cruelty; Maurice Dommanget, *Le Curé Meslier* (Paris, 1965), 62–3, 249.

they found themselves. In the seventeenth century it was noticed that hunting was too demanding of time and money to be a suitable recreation for men of business; and the reformist movements for the abolition of the cruel sports were firmly based on the towns.[4] For evidence of the new urban sensibilities we need look no further than Samuel Pepys, who records how in 1665 he met a great crowd of hunters at the King's Head, Deptford, 'and a great many silly stories they tell of their sport, which pleases them mightily and me not at all, such is the different sense of pleasure in mankind'.[5]

Such feelings were not just urban. They were those of the professional middle classes, unsympathetic to the warlike traditions of the aristocracy. For hunting was notoriously a military exercise and a training ground for cavalry. It taught men and horses to endure hardships, to cross difficult terrain and to become expert in battle tactics. It was 'the very imitation of battle', wrote Sir Thomas Elyot. For the hunting poet William Somervile it was the 'image of war, without its guilt' (and, as Jorrocks added, 'only five and twenty per cent of its danger').[6] The Elizabethan author Sir Thomas Cockaine listed several English commanders who had learned their military skill on the hunting field; and only the occasional sceptic like the Puritan William Hinde dared to suggest that 'many of our gentlemen which are most addicted to this exercise of hunting do not always prove the best soldiers'.[7] The game laws reserving the beasts of the chase to the upper classes had not been designed merely to secure material privileges. They were there because the symbolism of hunting was military and aristocratic; like riding the great horse, the sport was in itself an assertion of social superiority.

Just as hunting had been valued because it simulated warfare, so cock-fighting and bear-baiting had been esteemed as representations of private combat. The cock was a symbol of masculine fortitude and sexual prowess (the *double entendre* being important). He fought to the death, even when blinded and hideously wounded; he was praised for his 'invincible courage' and 'resolution' and the sport itself eulogized as a 'noble and heroic recreation'.[8] The vogue of cock-fighting among the gentry paralleled the late-seventeenth-century taste for the heroic in literature; and in the schools the sport was justified as a means of inculcating bravery into schoolboys.[9] When Prince Lewis of Baden saw a cock-fight at the English court in 1694, he remarked that he would never have guessed that a bird could display so much valour and magnanimity.[10] It was no coincidence that the opponents of animal sports were often the enemies of the aristocratic duel (and later of pugilism). Military values were anathema to all who wished to tame the aristocracy and jettison the legacy of the feudal past: whether polite writers like Addison and Chesterfield, who regarded rural sports as fit only for bumpkins and boobies, or opponents of gothic barbarism like Alexander Pope, who were appalled by the hunters' custom of presenting aristocratic ladies with a knife to cut the stag's throat, as it lay 'helpless, trembling and weeping'.[11] 'To read the history of kings,' remarked the democrat Tom Paine, 'a man would be almost inclined to suppose that government consisted of stag hunting'; and in the nineteenth century radicals like Richard Cobden denounced hunting as a 'feudal sport'.[12]

Conversely, the age-old association of hunting with class privilege made some aristocrats resistant to the new sentiments when they might otherwise have been sympathetic. Cruel or not, thought the Hon. John Byng in 1794, field sports were an important right of the gentry which had been handed down since the Conquest; they should be preserved against *sansculottes* preaching the rights of man or animals. Social pretension led many Hanoverian towns to keep packs of hounds; and in Victorian times there were many tradesmen who took up hunting as a means of symbolizing their social ascent.[13]

In practice, it was almost impossible to reflect on animals without being distracted by the conflicting perceptions imposed by social class. Many of the poor saw the gentry's dogs, horses and deer as symbols of aristocratic privilege, threats to their customary rights, to be callously mutilated in some defiant gesture of social protest. Hence the long series of statutes making the 'malicious wounding' of animals a felony.[14] Conversely, many of those who defended the animals were very unsympathetic to the lower classes. Usually, however, the concern for animal welfare was part of a much wider movement which involved the spread of humane feelings towards previously despised human beings, like the criminal, the insane or the enslaved. It thus became associated with a more general demand for reform, whether the abolition of slavery, flogging and public executions or the reform of schools, prisons and the poor law. The pamphleteer who in 1656 called for a law against cruel sports also denounced torture and pressing to death as barbarous, and condemned hanging, drawing and quartering as 'an act of cruelty and too much insulting over a poor fellow-creature in misery'.[15] In the late eighteenth century many of the champions of animals were simultaneously active in other spheres. William Cowper, though indifferent to politics, was a passionate enemy of the slave trade. Humphry Primatt and Mrs Barbauld objected to negro slavery and religious intolerance. James Granger ended his *An Apology for the Brute Creation* (1772) with a plea for the kinder treatment of agricultural labourers in their old age.[16] Thomas Bewick opposed Pitt's war against the French Revolution as 'superlatively wicked', and George Nicholson was against wars of every kind.[17] In 1791 the vegetarian Scot John Oswald hailed the French Revolution as offering hope for animals as well. Now that 'the barbarous governments of Europe' were 'giving way to a better system of things', he felt, 'the day is beginning to approach when the growing sentiment of peace and goodwill towards men will also embrace, in a wide circle of benevolence, the lower orders of life'. Other contemporary radicals enthusiastically embraced the cause of humanity to animals.[18] No wonder that in 1802 the politician William Windham, speaking in Parliament, defended bull-baiting as a preferable alternative to Jacobinism.[19] At the very end of the nineteenth century the Humanitarian League put forward a programme for reform of prisons, punishments, wages, the poor law and the position of women as part of 'a comprehensive doctrine of humaneness, to be applied to all sentient beings'. In 1894, H. S. Salt wrote:

> It is only by the spread of the same democratic spirit that animals can enjoy the 'rights' for which even men have for so long struggled in vain. The emancipation of men from cruelty and injustice will bring with it in due course the emancipation of animals also. The two reforms are inseparably connected, and neither can be fully realized alone.[20]

Yet not all animal-lovers were either social reformers or lovers of humanity. 'I knew an old maiden lady,' recalled a correspondent in 1787, 'whose tears would tenderly flow at the relation of the sufferings of a cat, but who did not exhibit any active benevolence at the call of the wants of her poor or suffering neighbours.'[21] Like present-day legacies to cats' homes, a concern for animal welfare could be an alternative to charity rather than a form of it. In the nineteenth century we have the remarkable confession of the anti-vivisectionist Anna Kingsford: 'I do not love men and women ... It is not for them that I am taking up medicine and science ... but for the animals and for knowledge generally ... I cannot love both the animals and those who systematically mistreat them.'[22]

Similarly, in the seventeenth and eighteenth centuries much of the pressure to eliminate the cruel sports stemmed from a desire to discipline the new working class into higher standards of public order and more industrious habits. It is often remarked (and it was noticed at the time) that it was the sports with a strong proletarian following which were outlawed – cock-throwing, bull-baiting and cock-fighting – whereas the gentlemen's fox-hunting, fishing and shooting survived unscathed.[23] This criticism is not wholly fair, for that annual recreation of aristocratic boys, the Eton Ram Hunt, in which the animal was ritually clubbed to death in Weston's Yard, was stopped in 1747,[24] nearly a hundred years before the reformers managed to suppress the proletarian Stamford bull-running (1840); and anyway, there were, as we have seen, plausible intellectual reasons for allowing the pursuit of the fox to continue.* But it is true that love for the brute creation was frequently combined with distaste for the habits of the lower orders; and middle-class opinion was as outraged by the disorder which the animal sports created as by the cruelty they involved. The ill-treatment of animals, thought William Baker in 1770, was 'a vice to which the illiterate vulgar are greatly addicted'. It was, agreed William Jones of Nayland in 1771, 'one of the distinguishing vices of the lowest and basest of the people.'[25] Throughout the nineteenth century it remained axiomatic that it was 'among the lower classes of the community that cruelty mostly abounded'. The villains were 'the hardened and profligate' cab-drivers, 'the rascally and insensible blackguards of Smithfield market', and the unspeakable bargees. The Act of 1835 against cruelty to animals declared its intention to reduce both the sufferings of dumb creatures and 'the demoralization of the people'.[26]

The S.P.C.A. can thus be seen as yet another middle-class campaign to civilize the lower orders. In the early years those whom it prosecuted for cruelty to animals sprang almost exclusively from the working classes. Lewis Gompertz, its first Secretary, even offered the well-to-do some advice on how to deal with cases of cruelty they met in the street, recommending 'expertness in pugilism' and sensibly warning them against reproving an offender 'before numerous others of his class'.[27] In 1868 John Stuart Mill declined the Vice-Presidency because the Society's operations were limited 'to the offences committed by the uninfluential classes of society'. Yet he himself could speak of animals as 'those unfortunate slaves and victims of the most brutal part of mankind'.[28]

Behind the apparent class bias lay a genuine gulf in sensibilities. Kindness to animals was a luxury which not everyone had learnt to afford. Just as the early-nineteenth-century working classes, dependent on the labour of their children, were reluctant to adopt the now-fashionable middle-class view that the growing child should be protectively insulated

* Above, pp. 163–5.

from the world, so most workers continued to regard animals in a functional light, untinged by sentiment. When the S.P.C.A. established a fund to reward humane cab-drivers, 'so few cases ... of exemplary conduct' were discovered that the balance had to be transferred to another purpose. When it endeavoured to prosecute those who staged dog-fights, it found that 'no witness would give evidence, there being a repugnance in the lower orders to do away with dog-fights'.[29]

Those recent historians who see the anti-slavery movement of the late eighteenth and early nineteenth centuries as a means of diverting radical energies away from the miseries of the English working class could say the same about the campaign against animal cruelty.[30] It too buttressed the new industrial system by representing inhumanity to beasts as something which belonged to more uncivilized regimes in the past, along with judicial torture, mutilations and similar barbarities, by contrast with the unprecedented sensibilities of the new order. Just as John Lawrence could comment in 1798 on 'the superior humanity of the present over any former period and ... its probable or rather certain increase, with increasing light', so Lord Erskine in 1809 was confident that a law against animal cruelty would make benevolence 'habitual' and thereby open up 'an aera in the history of the world'. The Stamford bull-running, thought *The Stamford Mercury* in 1814, was more appropriate to 'the barbarism and darkness of past ages'.[31] In this way the movement for animal welfare could, like the anti-slavery campaign, be cynically said to have helped to 'give legitimacy to an emerging British ruling class by incorporating "benevolence" into its ideology, while at the same time carefully limiting the scope of that benevolence so that it could not threaten class hegemony'.[32]

It certainly enhanced British complacency. For it was common to champion animals by comparing them unfavourably with those inferior races of mankind who, as Lewis Gompertz put it, bore 'great resemblance, not only in looks, but in manners and intellect, to the monkey tribe'. Greenlanders were less civilized than English sheep and oxen, thought John Wesley, while to compare the savages of Siberia or Tartary with 'horses or any of our domestic animals would be doing them too much honour'. In his *Rural Rides* (1830) William Cobbett described a dog which had more reason 'than one half of the Cossacks' and 'a great deal more than many a negro'. That great Victorian popularizer of love for animals, the Rev. J. G. Wood, regularly portrayed dogs and horses as morally superior to natives and savages; and many of his contemporaries compared their domestic pets favourably with Hottentots and Bushmen.[33] Of course, such attitudes to other races were far from characteristic of all nineteenth-century Englishmen. But the tendency was there; and Darwin himself encouraged it:

> He who has seen a savage in his native land will not feel much shame, if forced to acknowledge that the blood of some more humble creature flows in his veins. For my own part I would as soon be descended from that heroic little monkey who braved his dreaded enemy in order to save the life of his keeper ... as from a savage who delights to torture his enemies, offers up bloody sacrifices, practises infanticide without remorse, treats his wives like his slaves, knows no decency, and is haunted by the grossest superstitions.

It was grotesque, agreed the anti-vivisectionist Frances Power Cobbe, to label Landseer's dog 'a thing' and to give the title of 'person' to 'a Fuegian who eats his grandmother and can barely count his fingers'.[34]

No wonder that cruelty to animals was so often described as 'barbarous'. Pity, compassion and a reluctance to inflict pain, whether on men or beasts, were identified as distinctively civilized emotions. Vivisection for curiosity's sake, thought Thomas Percival in 1775, was a fit amusement 'only for the cannibals of New Zealand'. Humanity to animals, explained Charles Darwin, was one of the noblest moral qualities, and one of the last to be acquired, for savages did not possess it.[35]

Observers had often remarked that the love of animals did not necessarily conduce to the love of humanity. 'They which love beasts in a high measure,' remarked Edward Topsell in 1607, 'have so much less charity to men.' Excessive love of dumb creatures was no sign of humanity, agreed Thomas Fuller: 'Some nice consciences that scruple the baiting of bulls will worry men with their vexatious cruelties.'[36] Or, as George Bernard Shaw would remark, 'I know many sportsmen and none of them are ferocious. I know several humanitarians and they are all ferocious.'[37]

Neither was the love of animals often taken to the point at which it threatened human interests. It was no accident that so much of the campaign related to the treatment of those domestic beasts on whom society was economically dependent and in whose case the coincidence of charity and self-interest was most obvious. Sheep, noted Edward Topsell in 1607, were loved 'not because they are God's creatures, but for that they are profitable and serviceable for the necessities of men'. Domestic animals, agreed John Locke, were kept for man's 'pleasure and advantage and so are taken care of, not out of any love the master has for them but love of himself and the profit they bring him'.[38]

The development in the early modern period of more humane methods of horse-breaking can be plausibly linked to a rising awareness of the animal's economic value. Certainly it was concern to husband valuable resources which underlay the protests of subsequent agricultural writers like Gervase Markham against the 'tyrannical martyring of poor horses' by ferocious breaking-in, cruel bits and brutal punishments.[39] It also explained his recommendation that the ox-keeper should be 'gentle and loving with his oxen'. 'Gentle usage with these slow and dull creatures is best,' repeated John Laurence in 1726, 'for that is observed to bring them to the yoke better than even repeated severities.'[40] 'Enlightened self-interest,' agreed a Victorian horse-trainer, 'is the most powerful, and by far the most generally applicable antidote to cruelty.' Valuable horses, he explained, were broken more humanely than undistinguished ones, because it was worth spending the time on them. As Lord Erskine told Parliament in 1809, the duties and the interests of the owners of domestic animals were inseparable.[41]

It was also self-interest which provoked the first protests against battery farming. In Elizabethan times Thomas Muffett objected that 'to cram capons ... and to deprive them of all light is ill for them, and us too; for though their body be puffed up, yet their flesh is not natural and wholesome; witness their small, discoloured and rotten livers'. 'The only way of having sound and healthy animal food,' repeated George Cheyne in the early eighteenth century, 'is to leave them to their own natural liberty, in the free air ... with plenty of food and due cleanness and a shelter from the injuries of the weather.'[42] For the same reason it was urged that slaughter should be quick and humane; 'beasts killed at one blow are tenderest and most wholesome'.[43]

The same element of self-interest runs through all the legislation against animal cruelty. It appears in the first modern enactment on the subject, the law of 1641 in Puritan Massachusetts which prohibited 'tyranny or cruelty towards any brute creatures *which are usually kept for the use of man*' (i.e. not towards others).[44] It can be seen earlier in Jacobean regulations limiting the burden and speed of the carrier and postal services, out of consideration not just for 'the beasts themselves', but also for 'all owners of horses'.[45] And it underlay attempts in the same period to prohibit the Irish practice of yoking horses to the plough by their tails, a 'barbarous custom' which, 'besides the cruelty used to the beasts', ruined the horses.[46] So close was the overlap of kindness and self-interest that in the seventeenth century it was believed that the 'law of England' provided 'that any man's person and estate should be seized into the king's hands in case of some wild cruelty to his beasts; for he would appear in the eye of the law an idiot or a lunatic that should put his horses or asses to the sword'.[47]

Animals, in short, were like servants: they responded best to reasonable treatment. As Lord Erskine told Parliament in 1809, an owner who was in doubt as to whether it was cruel to strike a 'lazy or refractory' beast could easily resolve his dilemma: the criterion was exactly the same as that which determined whether or not it was cruel to strike an apprentice.[48]

Of course, many early protests against cruelty to domestic animals went well beyond considerations of mere utility. There was no economic motive for Sir Matthew Hale's decision to put his aged horses out to graze, rather than sell them to the knackers, or for the actions of the second Duke of Montagu (1690–1749) ('one of the most feeling men' Horace Walpole ever knew), who turned Boughton into a geriatric home for old cows and horses.[49] Nor does self-interest explain the increased sensitivity of eighteenth-century passers-by to the cruel treatment of horses in the street or the mounting volume of protest against the traditional practices of docking the animal's tail or cropping its ears or tying up its head so as to make it look more imposing.[50]* Even the unceasing (and unsuccessful) attempts of bee-keepers throughout the seventeenth and eighteenth centuries to devise an effective method of taking honey without first destroying the bees (which was the normal practice) were motivated as much by distaste for cruelty as by dislike of waste.[51]†

Nevertheless, legislation at first proved possible only where economic interest was involved. The feelings so frequently expressed about wild birds, hares or insects went unheeded, and the statutes passed in the early nineteenth century related exclusively to horses, cattle, dogs, poultry and other domestic animals. Nor did concern for animal welfare stop many people from continuing to eat meat.‡ If the animal was edible, then it was only 'unnecessary' cruelty which was forbidden.

The late eighteenth century thus abounded in apparent contradictions.

* Increasing prudery, however, may account for the abandonment of the practice of docking tails, which was thought 'offensive both to humanity and decency'; Thomas Bewick, *A General History of Quadrupeds* (1807; reprint, 1970), 11.

† The problem was not solved until the invention in 1851 of a frame which could be removed from the hive at will; H. Malcolm Fraser, *History of Beekeeping in Britain* (1958), 77–8.

‡ Though see below, pp. 290–97.

Some animals were pets, others were 'vermin'. Opponents of hunting were well disposed to fishing. The huntsmen themselves combined zest for destroying wild animals with great tenderness for dogs and horses. Pamphlets against cock-fighting were bound up in calf-skin. Dr Johnson's kindness to animals is normally thought to have been proved by his readiness to go out shopping for live oysters to feed to his cat Hodge. That great pet-lover, the poet William Cowper, felt no compunction about slaying a viper which threatened his kittens. His contemporary, the Rev. Joseph Greene of Stratford-upon-Avon, was a vigorous opponent of cock-fighting and cock-throwing. But in 1778 he bought a large quantity of Poultey's Paste, a poison to kill rats, 'those mischievous and destructive vermin'.[52] Certain creatures were now admitted into the sphere of moral consideration, but the line had to be drawn somewhere. Most people still excluded fish, predators, pests and insects. The needs of human survival seemed to require such an exclusion, just as in practice they also involved excluding some sections of humanity. The correspondence of E. W. B. Nicholson, Bodley's Librarian at Oxford and a great champion of animal rights, contains a touching letter of 1879, written by an animal-lover whose house had been overrun by black beetles: 'I hate making war even upon black beetles,' it runs, 'they have as much right to live as black Zulus. But what can one do in either case?'[53]

It was not a question which it was easy to answer. It was now agreed that it was wrong to cause unnecessary suffering to certain animals, but it was not clear to which animals, or where the point was at which the suffering became 'unnecessary'. What was clear was that the gulf was now very much wider between human needs on the one hand and human sensibilities on the other.

V

TREES AND FLOWERS

One of the most treasured memories of an old lady friend of mine, recently deceased, was of her visits, some sixty years or more ago, to a great country-house where she met many of the distinguished people of that time, and of her host, who was then old, the head of an ancient and distinguished family, and of his reverential feeling for his old trees. His greatest pleasure was to sit out of doors of an evening in sight of the grand old trees in his park, and before going in he would walk round to visit them, one by one, and resting his hand on the bark he would whisper a good-night. He was convinced, he confided to his young guest, who often accompanied him in these evening walks, that they had intelligent souls and knew and encouraged his devotion.

W. H. Hudson, *Far Away and Long Ago* (1918; Everyman's Library, 1939), 202.

i. THE WILD WOOD

The brute creation was not the only part of the natural world to be regarded with new feelings in the early modern period. By the eighteenth century tree-planting and landscape-gardening had become characteristic pastimes of the well-to-do, while a passion for flower-cultivation was spreading through the population at large. Just as animals were viewed by many with increasing sympathy, so trees and flowers steadily acquired a new emotional importance. Moreover, the categories into which men unconsciously classified them closely paralleled those which they had used for animals. The beasts had been divided into the wild, to be tamed or eliminated, the domestic, to be exploited for useful purposes, and the pet, to be cherished for emotional satisfaction. The early modern period had duly seen the elimination of many wild animals, the increased exploitation of domestic ones, and a rise in interest in the third category, the pet, maintained for non-utilitarian reasons. Almost exactly the same development occurred in the case of trees.

For although orchards and domestic groves had always been looked on favourably, the forest had initially been seen as wild and hostile. Since Mesolithic times, human progress had depended upon grubbing up and demolishing the trees with which much of the land had originally been covered. The process accelerated in the Neolithic age, when stone axes made it possible to fell timber as well as to graze or burn it down; and it continued in successive waves of advance and retreat under the Romans, Saxons and Danes. Modern historical geographers have shown that the wild wood had gone from much lighter land even before the arrival of the Romans and that by the end of the Anglo-Saxon period the bulk of forest clearance in England had been accomplished. Probably no more than twenty per cent of the country was wooded at the time of Domesday Book. By the late thirteenth century most of today's human settlements had been established, often in a landscape of winding lanes and irregularly-shaped fields, painfully carved out of the woodland, though in some parts of the country, like the Brecklands or the area around Cambridge, the trees had gone altogether. Little virgin forest remained and most of the surviving woods were to some extent coppiced, pastured or otherwise exploited. In the late fourteenth and fifteenth centuries some of the trees spread back, but, as the population rose again, the onslaught was renewed. In Tudor and Stuart times woodlands continued to give way, primarily to grazing and cultivation, but also to meet the expanding demand for building materials and industrial fuel, whether for iron manufacture or salt-boiling or the production of glass and pottery. Disparking, enclosure of chases, encroachment on the commons, the lax administration of the royal forests and the steady reduction in their extent: all meant the clearing of woodland and the felling of trees. It was not on Tower Hill that the axe made its most important contribution to English history.[1]

Of course, contemporaries exaggerated the depredations on the woods. We know now that the iron industry, although initially leading to much destruction of woodland, had in the end the opposite effect, by providing a stimulus to the regular cultivation of coppices for charcoal, a paradox which several contemporaries perceived. The timber shortage was never more than local.[2] Nevertheless, the general contraction of woodland for the sake of pasture or tillage in response to market forces was a visible process of which everyone was intensely aware. The Elizabethan poet Michael Drayton wrote nostalgically of the time 'when this whole country's face was forestry'; and in Pembrokeshire his contemporary, George Owen, found 'by matters of record that divers great cornfields were, in times past, great forests and woods'.[3] In many areas aged inhabitants told stories of how the trees had once been so numerous that small boys and squirrels could travel many miles without touching the ground.[4] The process of destruction cannot be quantified, but there is no doubt that between 1500 and 1700 the number of trees was substantially reduced, particularly in the Midlands and the

North.[5] In the 1690s Gregory King estimated that there were only three million acres of cultivated woodland left (about 8 per cent of England and Wales) and another three million of forests, parks and commons. And the contraction was still going on. There was hardly a county in the kingdom, thought a contemporary in 1764, where one would not find places called 'forest', 'grove' or 'park' which were now arable or pasture or bare heath. In the 1790s, when John Byng sat down to list the changes which had occurred in England since the seventeenth century, he put high on his list the further erosion of the old woods. By 1800 there were no more than two million acres of woodland in England and Wales, and by the beginning of the twentieth century the percentage of the United Kingdom occupied by woodland (4 per cent) would be the lowest in Europe.[6]

To many, this development symbolized the triumph of civilization. Forests had originally been synonymous with wildness and danger, as the word 'savage' (from *silva*, a wood) reminds us. Early man, it has been plausibly suggested, preferred open country to woodland because it was safer: he could see what was coming and guard against it in advance.[7] When Elizabethans spoke of a 'wilderness' they meant not a barren waste, but a dense, uncultivated wood, like Shakespeare's Forest of Arden, 'a desert inaccessible under the shade of melancholy boughs'. A mid-seventeenth-century poetical dictionary suggests as appropriate epithets for a forest: 'dreadful', 'gloomy', 'wild', 'desert', 'uncouth', 'melancholy', 'unpeopled' and 'beast-haunted'.[8] In New England, Plymouth Colony was founded in a 'hideous and desolate wilderness ... full of wild beasts and wild men ... and the whole country full of woods and thickets'. The colonists were aghast at the sight of a countryside covered by 'wild and uncouth woods'; and they set about destroying trees so as to make 'habitable' what Cotton Mather regarded as 'dismal thickets'.[9] Only 'wild creatures' they thought, would 'ordinarily love the liberty of the woods'.[10] Old England, explained the Elizabethan lawyer John Manwood, had also originally been 'a wilderness', but the early inhabitants had destroyed 'the woods and great thickets' near places of human habitation so that they would provide no shelter for dangerous wild animals: 'by that means the wild beasts were all driven to resort to those places where the woods were left remaining to make their abode in them ... so that ... the first beginning of forest in England was *propter defectum inhabitantis populi*, for want of people to inhabit those vacant places wherein wild beasts were.'[11]

The woods, therefore, were homes for animals, not men. Hence the poet William Browne could describe wild beasts as 'forest citizens'.[12]

Hence also the assumption that any men who lived in the woods must be rough and barbarous. The first human beings, it was widely believed, were 'woodland men', *homines sylvestres*. The progress of mankind was from the forest to the field. The ancient Britons, thought the eighteenth-century antiquary John Woodward, were barbarous and savage and their towns were 'groves and thickets', surrounded by a hedge or ditch.[13] The Irish, said an Elizabethan, remained 'wood-born savages', while John Locke contrasted the 'civil and rational' inhabitants of cities with the 'irrational, untaught' denizens of 'woods and forests'. The ancient Hindus, thought Edmund Burke in 1783, had developed a civilization possessing 'all the arts of polished life, whilst we were yet in the woods'. Only by being drawn out of forests would men be led to civility.[14]

Literary convention as well as actual experience thus underlay the seventeenth-century commonplace that forest-dwellers tended to be lawless squatters, poverty-stricken, stubborn and uncivil. But it was undeniable that the woodland areas really did contain cabins erected by beggarly people, who had gone in search of space, or employment in the charcoal industry, and had squatted illegally, often free from the normal social restraints of church and manor courts, and subsisting by pilfering timber and game.[15] The seventeenth and eighteenth centuries would see many bitter disputes between forest-dwellers and the officials of the Crown and the larger landlords, who tried to impose more efficient control upon the resources and inhabitants of the forest areas. Competing claims to land use led to mutual hostility and misunderstanding; and it is not surprising that those concerned to maintain the social hierarchy disliked woodlands as potential black spots: the forest, said the agriculturalist Charles Vancouver in 1813, was a 'nest and conservatory of sloth, idleness and misery'.[16]

As well as being the scene of protracted social conflict, the forests were also disliked because they provided a refuge for outlaws and a base for dangerous criminals. Selwood Forest, for example, was a notorious haunt of bandits and coiners until Thomas, Viscount Weymouth, built a church there in 1712 and began to cut the woods down. Cranborne Chase harboured smugglers and deer-stealers; and many other forest areas had a similar reputation.[17] Even clumps of trees on the roadside were disliked because they provided a hiding-place for robbers. Fulbrook in Warwickshire had once been a safe route for travellers, lamented the fifteenth-century historian John Rous, but when it was imparked by its noble owner the hedges and pales provided shelter for dangerous thieves.[18]

Untamed woodlands were thus seen as obstacles to human progress;

and some of the moralists who condemned enclosures made an exception for taking in and grubbing up trees.[19] The attitude of agricultural improvers to woods and trees was usually extremely hard-nosed. As one Elizabethan put it, a man who owned 'a fair great tree' which yielded no fruit and whose boughs kept the sun from the ground beneath would do better to hew it down than allow it for its beauty's sake to grow to his continual loss and hindrance.[20] In 1587 Thomas Churchyard felt that there was hope for the Welsh because

> They have begun of late to lime their land
> And plough the ground, where sturdy oaks did stand
> . . .
> They tear up trees and take the roots away.

In England also many former woods had become 'gallant corn countries', reflected Walter Blith in 1653 with satisfaction.[21] Hedgerow trees were unpopular with some agriculturalists because they weakened the fence and led to mildew in the corn. Some even objected to the hedges themselves, since they shut out the sun, required expensive maintenance and attracted birds which raided the corn.[22]*

Of course, everyone agreed that timber reserves were needed and that wood was wanted for fuel and other purposes. But it should be grown on inferior land as a crop, in coppices or high timber woods which were regularly cut and harvested. For woodland of any other kind there was no room. In the 1680s John Houghton wrote an essay to prove that it would be a very good thing if there were no trees at all within twelve miles of any navigable river. In 1712 John Morton observed with complacency that there were very few woods in Northamptonshire: 'In a country full of civilized inhabitants,' timber could not be 'suffered to grow. It must give way to fields and pastures, which are of more immediate use and concern to life.' John Dyer in 1757 wrote of a 'careful swain' that

> haughty trees, that sour
> The shaded grass, that weaken thorn-set mounds
> And harbour villain crows, he rare allow'd.[23]

* In the 1790s Coleridge's friend the Somerset tanner Thomas Poole was busy on his farm 'taking down useless hedges', an anticipation of the practice which in the later twentieth century has brought back open-plan farming; Mrs Henry Sandford, *Thomas Poole and his Friends* (1888), i. 170–71.

As Dr Thomas Preston told a Commons committee in 1791, the decline of oak trees in England was not 'to be regretted, for it is a certain proof of national improvement; and for Royal Navies countries yet barbarous are the right and only proper nurseries'.[24] In England, as in the Book of Psalms, a man was famous according as he had lifted up axes upon the thick trees. Indeed in 1629 an intrepid figure living twenty miles from Durham was said to have personally felled over 30,000 oaks in his lifetime and to be still at it. In the nineteenth century Mr Gladstone's much publicized tree-felling exhibitions were the last relics of the long tradition that to cut down trees was to strike a blow for progress.[25]*

Nowadays, when the woods have dwindled until they cover less than half the space given over to urban development, our attitude is very different: we think it a better thing to plant trees than to cut them down. It is in the early modern period that the origins of this new attitude lie. Of course, there was no simple *volte-face*, no dramatic shift from tree-destruction to tree-preservation. Nevertheless, the rise of a more sympathetic attitude is unmistakable.

* Which is why, when Gladstone visited Germany in 1895, Bismarck pointedly presented him with a young oak to plant when he got home.

In part the origins of this change were practical, stemming from the continuing need of wood for building, for domestic use and for fuel. As several historians have recently emphasized, English woodlands, particularly in the Lowland Zone, had been intensely managed as a valuable self-renewing resource since at least early Norman times. Selected woods were deliberately fenced off from grazing animals; coppice-farming was widely practised; trees were cropped, pollarded and selectively felled; and by at least the thirteenth century there was a well-established trade in timber and wood.[1] Unlawful tree-felling had been penalized as early as the seventh century; and later manorial customs and village by-laws reveal that access to wood, bushes and undergrowth was often carefully rationed. When felling was permitted in the royal forests, there would be temporary enclosure to protect the new replacement growth.[2] In many areas, in other words, woods had already ceased to be wild and hostile and had become domestic, an essential part of the rural economy.

The timber legislation of the Yorkist and Tudor periods merely put these medieval practices on to the statute book. Between 1483 and 1585 the fear that wood was being wasted inspired various Acts of Parliament which permitted the protective enclosure of young trees, prohibited the conversion of coppice and underwood into pasture or tillage, ordered the preservation of a stated number of timber trees ('standards') per acre and forbade their use in the Wealden iron industry.[3] From Fitzherbert in the early Tudor period onwards, scores of agricultural writers urged a rational policy of woodland management; and as the needs of the shipping industry became more pressing, the chorus grew insistent.[4] As James I remarked in 1610, 'If woods be suffered to be felled, as daily they are, there will be none left.' There was a time, wrote Captain John Smith in 1670, when England had been overgrown with woods and it had been beneficial to grub them up. But that time was past.[5]

The shift from mere preservation to actual planting is often said to have followed the publication of John Evelyn's famous work *Sylva* (1664), which was written after the Commissioners of the Navy had consulted the Royal Society about the timber shortage. The book attacked 'the disproportionate spreading of tillage' and enjoyed great literary success, going into four editions in Evelyn's lifetime, and, according to its author's not uncorroborated boast in 1678, leading directly to an act for planting in the Royal Forest of Dean (1668) and to the sowing of 'millions of timber trees' by private landowners.[6] Yet Evelyn's originality should not be exaggerated. His rather mannered compilation was heavily indebted to the arguments of earlier propagandists; and he himself was able to cite many landowners who had planted timber before he wrote.[7]

In fact there was nothing new about planting as such. It is true that the administrators of the royal forests in the Middle Ages had merely pursued a conservationist policy, forbidding unauthorized felling and protecting young trees. Only in the sixteenth century, as the government grew more interested in woodlands as a resource for shipbuilding, is there direct evidence of some fitful and erratic attempts at planting; the first documented instance is as late as 1580, when Burghley had thirteen acres of oaks sown in Windsor Park.[8] Not until the reigns of Charles II and his successors was a more ambitious planting policy attempted, first in the Forest of Dean (1668) and then in the New Forest (1698).[9] It is therefore not surprising that some seventeenth-century writers thought that tree-planting for the sake of timber was a very recent development. Yet on private estates it may have been long practised. Not all the hedgerow trees in which England abounded had got there by natural means; and, although documentation is scanty, there is enough circumstantial evidence to suggest that trees had been planted in England from Norman times and probably earlier.[10] In the early fourteenth century the chronicler of Pipewell claimed that, when the abbey was first founded, the monks had planted trees daily, a romantic legend, no doubt, but a revealing one.[11]* In early Tudor times the agricultural writer Fitzherbert assumed that farmers would gather acorns and ash keys to plant; he also gave advice on transplanting.[12] The duty of tree-planting was written into many sixteenth-century leases and manorial regulations.[13]

The motives for this activity were predominantly economic. Timber was wanted for use, and for profit. In the seventeenth century the agricultural improvers set out to show that by planting trees landowners could raise the value of their estates. Their works are full of calculations about costs, returns and rates of interest, all designed to show that planting for the Navy would be a profitable venture as well as a patriotic one. Evelyn himself quoted Sir Richard Weston's estimate that £30 laid out on a plantation of white poplars would bring a return of £10,000 in eighteen years' time, while the agricultural

* When nostalgic Elizabethans lamented the dissolution of the abbeys, they recalled not only the monks' charity but also 'their planting of woods, their setting of trees'. (Francis Trigge), *An Apologie* (1589), 7.

propagandist John Beale heard a farmer cite a proverb to the effect that in most of England an acre of coppice was as valuable as an acre of wheat, if not more so.[14] It seems clear that most of these calculations were on the optimistic side. Forestry seldom paid unless the land was too poor for other uses. Yet, in the eighteenth century, estate accounts suggest that a respectable revenue could be raised from well-managed park woodlands, while a definite search for profit underlay the switch during the same period to quicker-growing conifers. In several parts of Great Britain, thought Adam Smith, the profit from planting timber was equal to that from either corn or pasture. It was also possible to raise large sums of money by bulk felling, many landowners, in the words of the second Earl of Carnarvon, regarding trees as 'an excrescence of the earth, provided by God for the payment of debts'.[15]

Economic considerations also lay behind the opposition and sometimes violent resistance of commoners to disafforestation. Labourers were notoriously better off in woodland areas and they resented the loss of opportunities for grazing and woodcutting, buttressed as they often were by customary rights.[16] Their needs conflicted with the rational exploitation of woods for the production of timber, which involved fencing off enclosures to prevent young growth. Evelyn wanted a law banning access to the woods by cattle for a longer period than that conventionally observed; there was too much indulgence to 'a few clamorous and rude commoners', he thought.[17] Woods and hedges were constantly raided by the poor in search of firewood. If a man planted trees one day, complained Arthur Standish in 1613, 'the poor plucketh or cutteth them up the next day, if not the same night.' It was almost impossible to raise a quickset hedge near a city, declared Timothy Nourse in 1700, for the poor folk living on the outskirts would plunder it for kindling to keep them warm in cold winters.[18] Yet the very frequency of wood thefts was testimony to the importance of hedges and woodlands in the domestic economy of the poor. Economic reasons impelled cottagers living in forest areas to resist the extension of enclosure and the plough.

But along with these practical arguments for the preservation of old woods and the planting of new woods went other, less narrowly utilitarian considerations. When large-scale forest preservation began in the early Middle Ages its avowed purpose had been to provide a favourable environment for beasts of the chase. In the thirteenth century perhaps a quarter of England was deemed to be royal forest and under special forest law. The forest, said the lawyers, was 'a certain territory of woody grounds and fruitful pastures, privileged for wild beasts and fowls of forest, chase and warren to rest and abide in, in the safe protection of the king, for his princely delight and pleasure'.[19] As this definition reveals, the so-called 'forest' was not all covered with trees. It harboured cattle as well as deer. 'Forest,' it has been justly observed, 'was not necessarily woodland and woodland was not necessarily forest.'[20] But the needs of game-preservation made it necessary to keep up some woods and coverts within the forest, even at the price of forgoing good timber. The forest law prohibited *purpestre* (encroachment), waste (damaged trees) and assarting (rooting them up). A freeholder living under its jurisdiction could not fell his own trees without permission.

It is doubtful whether this system was a serious check to the conversion of woodland to arable or pasture at times of population pressure, particularly as the Crown was usually prepared to regard forest law as a profitable jurisdiction rather than a serious means of woodland conservation. From the thirteenth century onwards monarchs showed themselves ready to disafforest certain areas altogether so as to relieve their financial difficulties; and this process of disafforestation continued actively during the sixteenth and early seventeenth centuries. Charles I's attempt to push back the boundaries of the forest proved unsuccessful, and an Act of 1641 firmly contracted the area under forest jurisdiction to that obtaining in 1625 (and not even that if no forest court had been held in the locality for the previous fifty years).[21] In any case the ineffectual forest administration had never been able to stop much of the unlicensed encroachment by commoners grazing their animals and stealing the wood. Even so, the royal forests represented a notable attempt to preserve a large part of England for recreational use, albeit for the recreation of a privileged few. It was only in the later seventeenth century that the royal forests were regarded primarily as mere timber reserves. Previously the woods had been preserved more for the sake of the game (*venison*) than for that of the timber (*vert*). As a keeper in the New Forest said in 1609, it was wrong to fell timber if the trees were needed as a covert for the game; and in the Restoration period the demands of serious timber-planting were seen as in conflict with the maintenance of the forest as a place for the king's deer.[22]

The same motive had accounted for the proliferation of private deer parks from early Norman times onwards. In Saxton's Elizabethan maps there are over eight hundred of them; and they were a distinctive feature of the English landscape, equipped, as they were, with oaken paling and often reinforced by an overhanging plantation of thick trees.[23] These parks were an important symbol of social rank. The agriculturalist John Houghton hated trees and thought parks a sign of bad husbandry. But he recognized that they were of some value to their owners, in that 'they make or preserve a grandeur, and cause them to be respected by their poorer neighbours'. He therefore reluctantly conceded the case for some parks 'till ... want of good ground doth urge the contrary'.[24] In fact, parks were not all woodland, any more than were the royal forests, and they offered grazing for cattle as well as for deer. But it was common to plant trees in them both for timber and for the sake of the game. At Althorp in Northamptonshire, for example, there are still date-stones commemorating the planting of oak trees by successive members of the Spencer family from 1567 onwards.[25] In the eighteenth century the association between hunting and trees was reaffirmed when the passion for fox-hunting led to the planting of coverts in many parts of rural England.

These deer parks and royal forests generated a further and more enduring reason for tree conservation, namely the belief that wood added beauty and dignity to the scene. As Manwood, the forest lawyer, put it, trees were grown to provide deer with food and cover; but there was another motive and that was

> *propter decorum*, that is for the comeliness and beauty of the same ... For the very sight and beholding of the goodly green and pleasant woods in a forest is no less pleasant and delightful in the eye of a prince than the view of the wild beasts of forest and chase, and therefore the grace of a forest is to be decked and trimmed up with store of pleasant green coverts, as it were green arbours of pleasure for the king to delight himself in.[26]

The pleasures of a deer park were equally great, thought the fourth Lord North, not just because of the sight of the deer, 'but in having so much pasture ground at hand lying open for riding, walking or any other pastime'. Parks were even used for picnics, as in 1528, when the monks of Butley entertained the Duke of Suffolk and his wife to a meal under the oaks after hunting in Staverton park.[27]

In the sixteenth and seventeenth centuries many of these reserves were disparked or given over to cattle, as an expanding market provided incentives for the more profitable use of the land. But those which remained became increasingly ornamental as owners demonstrated their wealth by deliberately rearranging the landscape and converting good arable land into a pleasure ground dotted with trees. This was the period when a new type of country house emerged. Many gentlemen ceased to live in the middle of the village, and distanced themselves by removing to the centre of a landscaped park, if necessary obliterating or removing the village so as to provide a sense of space and separation. At Hatfield, for example, Robert Cecil bought out the copyhold tenants and turned their farms into parkland.[28] There were probably as many villages swept away to make room for aristocratic houses and parks in this period as in the eighteenth century: four instances are recorded in Staffordshire alone.[29] As the gardening writer Stephen Switzer would later observe, these Tudor parks needed only water to become what the eighteenth century would recognize as a landscape garden.[30] In achieving this effect trees played an essential part.

Even the agricultural improvers recognized that tree-planting had satisfactions which were aesthetic as well as economic. Since Anglo-Saxon times there had been monks who planted trees for the sake of their beauty. In the Middle Ages orchards were thought to be places of delight; and in the reign of Henry VIII John Leland could write of what he called the 'pleasure of orchards'.[31] The ensuing century and a half witnessed a vast increase in the amount of fruit-tree planting and in the varieties of fruit grown. More fruit trees had been planted in the last thirty or forty years, thought Ralph Austen in 1676, than 'in several hundreds of years in former ages'.[32] In the countryside hedgerow trees had many defenders. They provided shade for cattle and a protective shelter under which men could stand in rainy weather. They demarcated property and furnished lops for fuel and fencing. But the hedgerow tree, as Walter Blith conceded in 1653, was not just useful: it was also 'a thing of delight'. He therefore recommended planting here and there an oak, ash or elm, not just for profit, but to be 'most delightful and honourable unto men of ingenious spirits'. Hedgerow trees, agreed Leonard Meager in 1697, could afford 'pleasant and delightful prospects to the eye'. In the eighteenth century some agricultural authors urged that hawthorn quicksets should be diversified for ornament's sake by being mixed with crab, Lombardy poplar, rowan, alder, eglantine and woodbine.[33] Foreign visitors thought that English hedges made the whole countryside seem a beautiful garden.[34]

It was in accordance with this feeling for the aesthetic value of trees that the poet Drayton, lamenting the erosion of the old forests, deplored not just the loss of the timber, but also the vanished beauty of the woods themselves. In 1625 a country gentleman, Sir George Heneage, could remark wistfully of a Lincolnshire property at South Willingham that 'in antique time ... the place itself and the neighbouring hills ... were much beautified with groves and thickets,' whereas now they were 'naked and ... wholly disfurnished thereof'.[35] James I's commissioners for the sale of woods ran into opposition based on similar aesthetic grounds, while in 1650 the parliamentary surveyors of the Commonwealth urged that two hundred timber trees at Nonsuch House should be spared because they were essential to 'the magnificence of the structure' and 'the pleasantness of the seat'.[36] Trees, in other words, were now planted and cherished for amenity's sake alone.

It seems impossible to tell when such planting had begun. From time immemorial, clumps had been preserved near houses to provide shelter against the wind and to give shade in summer. The forest may have been intimidating, but the grove was a recognized symbol of human habitation. Trees were deliberately planted round medieval churches to act as a wind-break, while Chaucer wrote of his reeve that

> His woning was ful faire upon an heeth;
> With grene trees yshadwed was his place.[37]

In Elizabethan times William Harrison remarked that in many parts of the country a man could ride for ten or twenty miles and find very few trees, save 'where the inhabitants have planted a few elms, oaks, hazels or ashes about their dwellings for defence from the rough winds'.

In mid-seventeenth-century Herefordshire there were rows of elms along the highways and at every cottager's door, while in 1748 a visitor noted that, though there was little wood in Cambridgeshire, the villages were usually adorned with groves of elms. Everywhere cottagers planted small trees or bushes on which to hang out the washing to dry.[38] Since medieval times lawyers had ruled that a tenant was guilty of waste if he felled not just timber trees, but also those trees which stood, as they put it, 'in the defence and safeguard of the house'.[39] Even the agriculturalists conceded that 'groves or plumps of trees' might be erected around houses 'for delight and pleasure'; and it was near the houses of the nobility, as a seventeenth-century botanist noted, that the great maple was most likely to be found. Such trees, thought John Smyth, the Elizabethan steward of the Berkeley family, delighted the beholder with their 'beautiful verdancy'.[40]

In the early modern period planting for ornament and amenity gained momentum, particularly in the towns. John Hammond's map of Cambridge in 1593 shows numerous orchards, like the famous one at

Pembroke, where Nicholas Ridley had learnt St Paul's Epistles ('the sweet smell thereof, I trust, I shall carry with me into Heaven'), and much planting at Peterhouse, Clare Hall, Corpus, Queens', Magdalene and Trinity.[41] By David Loggan's time (1688) a century's further planting by the colleges had resulted in splendid formal walks at the back of Trinity and St John's and an abundance of fine trees in the courts and gardens of every college. The trees in Caius's Tree Court were planted in 1658.[42] In Oxford Christ Church Meadow was encircled by ornamental walks by the beginning of the seventeenth century; and in the late 1660s Dean Fell began the great Broad Walk of elms.[43] Many other cities were almost as well endowed, as can be seen from John Speed's Jacobean town maps. In Norwich trees were so frequent that the place was described in Tudor times as 'either a city in an orchard or an orchard in a city, so equally are houses and trees planted'.[44] In London the citizens had had trees in their gardens since the twelfth century. In the Elizabethan period the Inns of Court and the Livery companies did much planting, while in the reign of James I Moorfields was levelled into walks and lined with trees, making it London's first civic park, and yielding, said a contemporary, 'much delight to the eye'.[45] After the Restoration tree-planting became common, and contemporary illustrations show rows and avenues, not just in Covent Garden, Lincoln's Inn Fields and the royal parks, but in many squares and outside most public buildings. In 1748 a visiting Swede noticed that in London there were trees planted in the gardens of nearly every house, elms in nearly every square, and still more elms on both sides of the road in the villages outside the metropolis.[46]

In the reign of Charles I fashionable society had stayed in their coaches when parading through Hyde Park. But in the Restoration period the desire to take the air and walk became a widespread feature of English social life and the London parks became a much-copied setting for these outdoor fashion parades.[47] At Oxford, for example, in the 1640s, when Charles I's garrison was in the city, it was Trinity Grove which was the rendezvous of the nobility and gentry. After the Restoration, Merton Wall became the fashionable meeting-place. When that became notorious as a place of scandal, high society moved to Magdalen, which in 1723 was said to be crowded with people on every Sunday night in the summer. Twenty years later, it was St John's garden which afforded the 'opportunity of seeing the whole university together almost, as well as the better sort of townsmen and ladies, who seldom fail of making their appearance'.[48] In due course virtually every town of any social pretension became prepared to vote money for a walk or avenue where local beaux and belles might stroll up and down under the trees to display their best clothes and exchange gossip, as a sort of outdoor assembly room.[49] And if they would not pay for it themselves there was always the chance of a local benefactor, like John Kyrle, Pope's 'Man of Ross', who planted elms along the causeway leading to a town, rows of elms on each side of the church, and a public park in which the inhabitants of Ross-on-Wye might walk under the leaves.[50] When Celia Fiennes travelled through England at the end of the seventeenth century she noticed fine rows of trees at Bath, Bristol and Exeter and in the abbey gardens at Shrewsbury, which were kept neatly cut and rolled for polite company to walk in every Wednesday. In the 1720s Defoe wrote of Epsom that 'the town at a distance looks like a great wood full of houses'. No wonder that Thomas Traherne's conception of the perfect place was a city, but a city full of trees.

> Green trees the shaded doors did hide
> . . .
> Beneath the lofty trees,
> I saw, of all degrees,
> Folk calmly sitting in their doors.
> The streets like lanes did seem
> Not pav'd with stones but green.[51]

The idea of a garden city was not invented by Ebenezer Howard in the 1890s; it was formulated in 1661 by John Evelyn, who had been much impressed by the delicious walks and avenues he had seen on his travels in the towns of France, Italy and the Netherlands.[52] In the 1720s Thomas Fairchild urged the creation of more public parks in central London; not just grass plots and gravel walks, but a wilderness of trees and singing-birds, with shade and privacy. If only lime trees could be planted along the Haymarket, Pall Mall, the Strand and Fleet Street, sighed Batty Langley in 1728, 'then might we view a city in a wood'.[53]

Most striking of all was the scale of aristocratic planting. In Tudor and early Stuart times there were innumerable gentlemen who delighted in walks and groves, like the Elizabethan Sir William Hatton, whose 'many young groves newly planted' at Holdenby were commended by John Norden as 'both pleasant and profitable', or the regicide Colonel Hutchinson, who covered the slopes of the Vale of Belvoir with woods.[54]

Such planting began in the gardens of large houses, which by the early seventeenth century were likely to have as one of their features a so-called 'wilderness': a dense plantation of trees, which, despite its name, was laid out in an orderly and geometrical fashion. A wilderness, as a preacher put it, was 'a multitude of thick bushes and trees, affecting an ostentation of solitariness in the midst of worldly pleasures'.[55] Beyond the garden plantations extended outwards into the countryside as trees were used to integrate the house with its surroundings by means of avenues, ridings and shelter belts. In the later seventeenth century it was common to cover the adjacent hills with plantations, intersected with axial vistas, as at Badminton in Gloucestershire, where the lanthorn of the Duke of Beaufort's house formed the centre of a star cut through the woods of the surrounding country, and where the neighbouring gentry deferentially planted their own trees in such a way as to continue the pattern. By reaching outwards so as to shape not only his own ground, but also the surrounding landscape, the duke demonstrated his power to manipulate the lives and environment of lesser mortals and emphasized that all local avenues of power converged upon him.[56] In the same spirit, other magnates planted woods adjacent to their houses so as to make their headquarters appear more beautiful and more impressive. This was what Stephen Switzer in 1718 called 'extensive or forest gardening' and he cited 'the incomparable wood of my Lord Carlisle's at Castle Howard, the wood at New Park belonging to the Right Honourable the Earl of Rochester, the woods at Cashiobury, the design of Bushy Park, etc.' As Horace Walpole would observe, 'a great avenue cut through the woods' had 'a noble air'; it 'announced the habitation of some man of distinction'.[57]

The alternative to cutting a vista through existing trees was to plant a long avenue leading up to the great house. By the early seventeenth century the avenue of lime, elm or horse chestnut had become a recognized aristocratic symbol. At New Hall, Essex, John Tradescant, gardener to the first Duke of Buckingham, planted a mile-long avenue of lime trees in four rows; and there was another of similar length at the royal palace of Theobalds.[58] During the century these great avenues were widely copied. The bird's-eye views engraved by John Kip in the reign of Queen Anne show scores of stately houses encircled by regimented lines of trees, like armies drawn up on the eve of a battle. 'Like a guard on either side,' wrote Andrew Marvell, 'the trees before their lord divide.' At Patshull Park, Staffordshire, reported Celia Fiennes, for a quarter of a mile before the house, 'you ride between fine cut hedges and the nearer the approach the finer still; they are very high and cut smooth.'[59]

Long avenues radiating out from a single centre and stretching relentlessly over the countryside without regard to its natural features were a particularly obvious way of subjecting a whole district to the authority of the great house – what Wordsworth would call putting 'a whole country into a nobleman's livery'.[60] But most forms of aristocratic tree-planting were deliberate assertions of ownership. 'What can be more pleasant,' John Worlidge asked his gentlemen readers in 1669, 'than to have the bounds and limits of your own property preserved and continued from age to age by the testimony of such living and growing witnesses?' In the late eighteenth century William Marshall ruled that the view from the house should be open, if (and only if) all the land in sight belonged to the householder, in which case the vista would gratify his 'love of possession'. But if someone else's property obtruded, then a screen of trees should be erected to conceal it; and, if the boundary was embarrassingly close, then trees should be judiciously planted so as to disguise the exiguousness of the estate. The whole purpose of a belt of trees, explained Uvedale Price, was to conceal the extent of the property if it was small and to display it if it was large.[61] To reveal the full dimensions of the owner's property, agreed Humphry Repton, was 'one of the leading principles' of landscape-gardening; and high among the pleasures which the art afforded was that of 'appropriation'.[62]

Prevailing aesthetic theory thus laid heavy emphasis upon the importance of trees as an essential part of the architectural setting. In the seventeenth century there were still some who thought it unhealthy for a house to be 'choked up with trees', but most gentlemen would have sympathized with Lord Caryll, who in 1699 refused to have the ash trees around his seat cut down, because wood was 'its chiefest ornament'.[63] All buildings needed trees to beautify them, considered Humphry Repton a century later: there should be conifers for classical buildings and round-headed trees for gothic ones.[64] In any fine landscape, trees played an indispensable role. It was by 'frequent plantations' that Addison in the *Spectator* (1712) recommended that a whole estate might be 'thrown into a kind of garden'. Gentlemen of property, thought Sir Alexander Dick in 1762, 'should plant at proper places, and at proper distances, noble clumps of trees of all sorts, to dignify the look of the land.' A place without trees, declared Charles Marshall in 1796, appeared 'disagreeably naked'; and Lord Kames concurred: 'A hill covered with trees appears more beautiful as well as more lofty than when naked.'* 'The great art of improvement,' pronounced Uvedale Price at the end of the eighteenth century, 'consists ... in the arrangement and management of trees.' Planting, agreed Edmund Bartell in 1804 with the needs of the Navy in mind, was 'an almost indispensable duty in every gentleman'.[65]

The motives for aristocratic planting were thus a complex mixture of social assertiveness, aesthetic sense, patriotism and long-term profit. Together, they combined to make planting as much of an aristocratic obsession as dogs and horses. Trees had ceased to be a symbol of barbarism or a mere economic commodity. They had become an indispensable part of the scenery of upper-class life. The depredations wrought in the 1650s upon the estates of the Crown, Church and supporters of Charles I were exaggerated by Royalist propagandists after the Restoration in such a way as to create an association between the wanton felling of trees and republican politics. Evelyn and other contemporaries cleverly represented tree-planting as a way of affirming a gentleman's loyalty to the restored monarchy; they also laid much emphasis upon its supposedly 'heroic' nature.[66] In the years after 1660, accordingly, the pace for planting was set by the Crown itself. In 1664 Charles II had over 6,000 elms planted at Greenwich and in the early 1680s he was responsible for the three great avenues of limes at Hampton Court and the Long Walk stretching three miles from Windsor Castle. Successive monarchs continued the tradition, both at Hampton Court and Windsor and also at St James's, Kensington and Richmond.[67] The royal example was emulated by scores of aristocratic planters, including great magnates like the second Duke of Montagu ('Planter John'), who in the 1730s projected no less than seventy miles of elm avenues at Boughton, the Versailles of the Midlands.[68] A decade earlier Defoe had observed that fine houses 'surrounded with gardens, walks, vistas and avenues' gave 'a kind of character to the island of Great Britain'.[69]

In the mid eighteenth century under the influence of Lancelot

*It is tempting to offer a semi-Freudian explanation for this vehement dislike of bareness and the consequent anxiety to cover with trees the hitherto exposed slopes of hillsides like those in Sussex and Northumberland. But even if the question to bare landscape does not reflect a feeling that nakedness should be concealed by pubic hair, it certainly parallels the fashion for the wig and the reluctance to display bald or shaven heads in public. Trees, as John Smith put it in 1670, were 'to the earth as the hairs of a man's head'; *Englands Improvement Revived* (1670), 77.

('Capability') Brown the style of tree-planting became less formal. Avenues went out of fashion and many were destroyed.[70] But trees were still needed for clumps and shelter belts. Brown himself was responsible for a colossal amount of planting, especially elm, oak, beech and ash, putting in a hundred thousand trees for Lord Donegall at Fisherwick, Staffordshire, alone. 'So much beauty,' he thought, depended 'on the size of the trees and the colour of their leaves.'[71] In due course, his clumps were in turn rejected as 'tasteless deformities', but the Reptonian style was equally dependent on trees. Indeed clumps have remained so prominent a feature of the English landscape as to prove that the reaction against Brown was only partial. 'English park trees,' thought a traveller in 1811, had 'a character of picturesque magnificence, unequalled anywhere else.'[72]

Meanwhile, the planting of forest trees had become a characteristic activity of every 'improving' landlord. It is to the eighteenth century that we owe not just the wooded Chiltern reaches of the Thames, but also the planting of the hitherto treeless Northumberland.[73] Between 1757 and 1835 the Royal Society for the Encouragement of Arts gave gold and silver medals to stimulate large-scale plantations; and the reigns of George III and George IV witnessed enormous forestry projects, as at Hafod, North Cardiganshire, where Thomas Johnes put in almost 5,000,000 trees in the thirty years before 1816, or in Scotland, where the landscape of Perthshire was decisively transformed by three successive dukes of Atholl, who between 1740 and 1830 planted 14,000,000 larches. Timber for the Navy during these years came overwhelmingly from private landowners rather than from the royal forests; and between 1760 and 1835 private landowners are thought to have planted some 50,000,000 timber trees at the very least.[74] Yet, although that was the period of gigantic conifer plantations, there is much evidence earlier in the eighteenth century of larger-scale forestry planting.[75] More research on estate archives is likely to reveal that planting for timber, like planting for ornament, had a longer prehistory than is sometimes appreciated.

Such planting needed not just leisure and a deep purse, but political security and a system of inheritance which gave confidence in the transmission of property. No doubt, this was one of the reasons why it began earlier in England than in troubled Ireland, where, as Sir John Davies reported in 1610, inheritance was so uncertain that the natives never planted orchards, or in Lowland Scotland, where, as Dr Johnson observed with even greater exaggeration, a tree was as much a rarity as a horse in Venice.[76] Even in Hanoverian England landlords were reluctant to plant oak trees because they were so slow growing, taking a century to come to full maturity. (It was a common saying among countrymen that withy and sallow, which could be sold for firewood, would purchase a horse sooner than an oak would buy a saddle.) As a result the more pretentious landowners were reluctant to sow acorns because 'they will not *show* themselves to advantage soon enough'. They preferred conifers or limes or chestnuts, all of which made a show speedily: and they bought young nursery trees to edge their plantations so that their achievement should be seen.[77]

The commercial implications of this immense activity were obvious to everyone. There had been organized sources of supply of quicksets since the fourteenth century. Nurserymen were selling forest trees in Jacobean times; and by 1748 it was said that there were one or more tree-nurserymen in nearly every town and large village in England.[78] They dealt not only in indigenous trees, but also in the exotics which poured in during the early modern period to widen the planters' repertoire.[79] The Tudor period had seen growing interest in the sycamore and white poplar and the introduction of the holm oak. The seventeenth century witnessed the development of the London plane, the arrival of the cedar of Lebanon and the false acacia, and the mounting popularity of the horse chestnut, followed by the lime. In the eighteenth century came the Lombardy poplar and the Weymouth pine. Spruce, silver fir and larch arrived as garden specimens in the Tudor and Stuart periods before being used in the great conifer forestry plantations of the late eighteenth century. There was also an infinity of new garden trees and shrubs: in the sixteenth century, laburnum, philadelphus and lilac; and in the eighteenth hydrangea, buddleia and rhododendron. By 1676 the cypress was so popular that John Rea thought it 'now common in every garden of any note'.[80] At Marden in the parish of Godstone, Surrey, the rich Whig merchant Sir Robert Clayton so changed the face of the landscape that in 1700 a visitor thought it resembled a foreign country with its pines, firs, cypresses, yews, and junipers, 'naturally solitudinarious and exceedingly and pleasantly exotic'. 'I pray God to bless improvements in gardening,' wrote Christopher Smart, 'till London be a city of palm trees.'[81] On one calculation, 89 new species of trees and shrubs came into England in the sixteenth century, 131 in the seventeenth, 445 in the eighteenth, and no less than 699 in the first thirty years of the nineteenth.[82] By 1838 exotics were so much *de rigueur* that J. C. Loudon could pronounce that 'no residence in the modern style can have a claim to be considered as laid out in good taste in which all the trees and shrubs employed are not either foreign ones, or improved varieties of indigenous ones'.[83]

iii. THE WORSHIP OF TREES

So it was that in England trees were not merely domesticated but gradually achieved an almost pet-like status. As the woodlands shrank in area they ceased to terrify and became instead valued sources of pleasure and inspiration. 'The love of woods,' pronounced Addison in 1713, 'seems to be a passion implanted in our natures.' 'I think the beauty of a country consists chiefly in the wood,' said Edwin Lascelles in 1763.[1] As for forests, they were now 'romantic'. 'One of the sublimest objects in natural scenery,' declared Archibald Alison in 1790, 'is an old and deep wood, covering the side of a mountain.' Even the notably urban Samuel Johnson loved the sight of fine forest trees and detested Brighton Downs because they were so bare. 'Walking in a wood when it rained,' recalled Mrs Piozzi, 'was, I think, the only rural image he pleased his fancy with.'[2] Only occasionally did the older, more fearful attitude linger; as in 1734, when the young George Whitefield felt an urge to follow the example of Christ tempted in the wilderness among the wild beasts. The best he could do was to go out after supper into the tree-lined Christ Church Meadow, where he

> continued in silent prayer under one of the trees for near two hours, sometimes lying flat on my face, sometimes kneeling upon my knees ... the night being stormy, it gave me awful thoughts of the day of judgement. I continued ... till the great bell rung for retirement to the College, not without finding some reluctance in the natural man against staying so long in the cold.[3]

Few of Whitefield's contemporaries saw trees in so frightening a light. For many they were, as William Gilpin put it, 'the grandest and most beautiful of all the productions of the earth'. To Alexander Pope a tree was 'a nobler object than a prince in his coronation robes', while for William Shenstone 'a large, branching, aged oak' was 'perhaps the most venerable of all inanimate objects'. The essayist Vicesimus Knox pitied a man who could not fall in love with a tree.[4] In the later eighteenth century many artists began to specialize in portraits of trees and devoted much time to studying their silhouettes. Between 1770 and 1850 books on handsome trees, famous trees, ancient trees and how to draw trees poured off the presses. It was an obsession which would culminate in the writings of John Ruskin.[5] English gentlemen, observed Washington Irving, spent hours discussing the shape and beauty of individual trees, as if they were statues or horses; and in Victorian landscape photography the trees often have greater individuality than the figures standing beside them.[6]

It was these new sensibilities which generated the feeling that, as William Marsden put it in 1783, it was impossible to behold the destruction of an aged tree without a strong sentiment of regret: 'it appears a violation of nature, in the exercise of a too arbitrary right.'[7] Pope's mock petition of 1741 to the Earl of Burlington, imploring him to prevent William Kent from felling a venerable tree at Chiswick, was but one of a long series of poetic manifestos against the fall of old trees. It begins with Drayton's lament for the vanished forests in *Polyolbion*, continues with Anne Countess of Winchilsea's *The Tree* ('To future ages may'st thou stand/Untouched by the rash workman's hand'), and proceeds on through Cowper ('The poplars are felled'), to Wordsworth (whose 'Excursion' includes a 'digression on the fall of beautiful and interesting trees' and who commemorated the yew tree of Lorton Vale – 'Of form and aspect too magnificent/To be destroyed'), Clare (who grieved 'to see the woodman's cruel axe employ'd'), Campbell ('Spare, woodman, spare the beechen tree'), Tennyson ('The woods decay, the woods decay') and Hopkins ('O if we but knew what we do/When we delve or hew – Hack and rack the growing green'); not to mention innumerable lesser works such as the anonymous Welsh poem (probably Elizabethan) on the loss of Glyn Cynon wood or Francis Mundy's protest in 1776 against the felling of Needwood Forest by the Duchy of Lancaster or the bathos of the American George Pope Morris:

> Woodman, spare that tree!
>   Touch not a single bough!
> In youth it sheltered me,
>   And I'll protect it now.[8]

Of course, this writing varied in its preoccupations. Whereas Cowper, for example, lamented the fall of the poplar trees because it reminded him of his own approaching demise, Hopkins was indignant at the felling of the Binsey poplars because it seemed a brutal human onslaught on nature. The poets, moreover, were sometimes ill-informed (the elms to which Cowper referred in 'The Task' were, in fact, poplars[9]). Such literary laments incurred the contempt of many landscape-gardeners, who wanted trees removed if they were infelicitously placed, and of all foresters, for whom felling was the whole point of the business: 'To cut down timber,' said an early-seventeenth-century commentator, was not a fault, but 'a part of good husbandry'; and William Gilpin noted that 'in a cultivated country, woods are considered only as large corn-fields: cut, as soon as ripe'.[10]

Yet there is no doubt that poetic regrets for the felling of old trees reflected a widespread and genuine current of feeling. There was a conflict between economic forestry, which required regular felling, and the needs of ornament, amenity and display. Some English aristocrats refused to cut down trees for timber and many private individuals were bitterly upset when such felling occurred.[11] In the eighteenth century it was generally believed that the English attitude to tree-planting was much less narrowly economic than was that of contemporary France, where all mature trees were marked out by royal surveyors, and where, as Horace Walpole observed, it was a curiosity to see old trees.[12] In England trees were increasingly cherished, not just for their use, not even just for their beauty, but because of their human meaning, what they symbolized to the community in terms of continuity and association.

Part of this feeling might almost be called religious. The English no longer worshipped sacred groves, for early Christian missionaries had always been hostile to so-called 'holy' trees; and in the eleventh century the Church had made it an offence to build a sanctuary around a tree.[13] Yet green branches were carried in procession on May Day or at Midsummer.[14] Churchyards usually contained a yew,* and in popular folklore many trees had a protective significance which made it unlucky to cut them down.† The poets preserved the classical conception of woods as the haunt of sylvan deities:

> Arched walks of twilight groves,
> And shadows brown that Sylvan loves
> Of pine, or monumental oak,
> Where the rude axe with heaved stroke,
> Was never heard the nymphs to daunt,
> Or fright them from their hallowed haunt.[15]

Some of the early Protestants were adamant that prayers could be as effectively said in fields or woods as in churches. In 1429 the Lollard Robert Cavell, a clergyman of Bungay, maintained that no honour was due to images, but that trees were of greater vigour and virtue and fitter to be worshipped than stone or dead wood carved in the shape of a man. In the reign of Edward VI the strongly Protestant curate of St Katherine Cree, London, forsook his pulpit and chose to preach out of a high elm tree in the churchyard.[16] In Jacobean Yorkshire Richard Shanne of Methley, whose 'chiefest delight was planting', carried the Spanish Dominican Luis de Granada's book *Of Prayer and Meditation* (in an English version), avoided company and took 'much pleasure to walk in woods and to be solitary'.[17] In the same way, Mary Rich, Countess

* The reason for this has been the subject over the centuries of much speculation but no certain proof. It has been variously suggested that yews grew in churchyards because Christians originally met in yew groves to worship; or because their branches were used to decorate graves and church interiors on Palm Sunday and Christmas; or because they gave protection against evil spirits; or because they drew off noxious exhalations from graves; or because churchyards afforded a safe enclosure for growing a tree which was indispensable for bow-making, but whose berries were poisonous to cattle; or because they were a wind-break; or simply because the yew had since pre-Christian times been a symbol of immortality. For such speculations see, e.g., Sir Thomas Browne, *Hydriotaphia* (1658), chap. iv; Robert Turner, *The British Physician* (1664), 362–3; Batty Langley, *New Principles of Gardening* (1728), 180; John Evelyn, *Sylva*, with notes by A. Hunter (York, 1776), 381; Gilbert White, *The Antiquities of Selborne*, ed. W. Sidney Scott (1950), 36–7; J[ohn] A[ikin], *The Woodland Companion* (2nd edn, 1815), 85; J. E. Bowman, 'On the Longevity of the Yew ... and on the Origin of its Frequent Occurrence in Churchyards', *Magazine of Natural History*, new ser., 1 (1837), 85–7; J. C. Loudon, *Arboretum et Fruticetum Britannicum* (1838), iv. 2070–72; John Lowe, *The Yew-Trees of Great Britain and Ireland* (1897), chap. vii. The fullest work on the subject is Vaughan Cornish, *The Churchyard Yew and Immortality* (1949), 127.

† Above, p. 75.

of Warwick (1625–78), made it a habit to meditate for two hours a day in the 'wilderness' of her garden. 'We do commonly devise a shadowy walk,' wrote John Beale in 1657, 'from our gardens through our orchards (which is the richest, sweetest and most embellished grove) into our coppice-woods or timber woods. Thus we approach the resemblance of Paradise.' 'The beauty of trees,' remarked Francis Hutcheson in 1725, 'their cool shades, and their aptness to conceal from observation, have made groves and woods the usual retreat to those who love solitude.'[18]

In the eighteenth century the view of woods as a place for privacy and meditation was reinforced by the new conception of nature as a positive religious force. 'Ye woods and wilds,' sang Elizabeth Rowe,

> receive me to your shade!
> These still retreats my contemplation aid:
> From mortals flying to your chaste abode,
> Let me attend th' instructive voice of God![19]

Gothic architecture was widely regarded as an attempt to reproduce in stone the branching of a forest walk.[20] When it returned to fashion in the 1750s, so did the view of woods as primitive churches. Those 'oak priests', the Druids, it was said, had frequented groves because they felt the religious awe which ancient trees engendered. 'It is natural,' thought Alexander Hunter in 1776, 'for men to feel an awful and religious terror when placed in the centre of a thick wood.' To worship a venerable oak was idolatry, admitted William Cowper, but it was 'idolatry with some excuse'.[21] In the Romantic era the analogy between groves and ecclesiastical architecture became commonplace.[22] It was no accident that Coleridge's Christabel went to pray beneath an old huge oak tree or that in *The Rime of the Ancient Mariner* the 'hermit good' lived in a wood. For Wordsworth no moral philosophy could match 'one impulse from a vernal wood', while in nineteenth-century America the Transcendentalists would see the forests as 'God's first temples'. 'In the woods,' thought Emerson, 'we return to reason and faith.' 'If we do not go to church as much as did our fathers,' wrote John Burroughs in 1912, 'we go to the woods much more.'[23]

England had no forests on the North American scale to act as a focus for such feelings. But she did have individual trees which played a crucial part in her social life. From Anglo-Saxon times they had been essential landmarks, demarcating local boundaries or indicating the meeting-place for assemblies.[24] The annual parish perambulation picked its way from one tree to another, pausing to read the scriptures at some 'gospel oak' or 'holy oak'.[25] Such trees were older than any of the inhabitants; and they symbolized the community's continued existence. Few parishes were without their famous trees, while some decrepit pollards and branchless trunks achieved national renown for their age or social importance. There was the Shire Oak at Wentworth Woodhouse, dividing the counties of Yorkshire, Nottingham and Derby; the Greendale Oak at Welbeck, through which a coach could be driven; the elms in a Hampshire parish, which the purchaser in 1648 was not allowed to cut down because they were a landmark guiding ships at sea; the hollow oak at Kidlington in which the magistrate would imprison local malefactors; and the indulgent oak in Staffordshire belonging to Sir Charles Skrymsher, which, it was claimed, bestowed immunity from prosecution upon the parents of any bastard child begotten under its boughs.[26]

Trees provided a link with eternity. In the late eighteenth century the Greendale Oak at Welbeck was over 700 years old, while the great oak in Salcey Forest was said to have been there for 1500 years. As among that African people who will show the visitors the tree under which God created the world, so in 1670 John Smith, a well-known forestry expert, maintained that some living English oaks dated from the first summer after the Flood and that a few went back to the Creation.[27] The desire to associate particular trees with national heroes or historic events was ubiquitous. English travellers to Italy were shown trees supposedly planted by St Francis or St Dominic.[28] But nearer home they could see the oak at Clipston, where Edward I had convened his Parliament in 1289; Herne's oak in Windsor Great Park; Chaucer's oak in Donnington Park, Newbury, supposedly planted by the poet himself; the oak in the New Forest, off which glanced the arrow which shot William Rufus; and the Boscobel oak, which sheltered Charles II after the battle of Worcester. Even today the T.U.C. helps to maintain the Martyrs' Tree at Tolpuddle, while in 1978 *The Times* carried a photograph of Queen Elizabeth's Oak in Hatfield Park, under which Elizabeth I was allegedly sitting when she received news of her accession.[29] *

As social change accelerated, the desire to preserve such visible symbols of continuity grew stronger; and passions became especially intense if the tree in question was identified with the fortunes of a particular family. The analogy between great families and great trees, giving shade and protection to lesser beings, was well established by the seventeenth

* This miserable stump, held together by rope and cement, is now preserved in Hatfield House.

century, when landowners might sometimes plant a tree to mark the birth of a child, and when many poets saw trees as symbols of ancestors. As Andrew Marvell wrote of Lord Fairfax's trees at Bilborough,

> For they ('tis credible) have sense,
> As We, of Love and Reverence,
> And underneath the Courser Rind,
> The Genius of the house do bind.
> Hence they successes seem to know,
> And in their Lord's Advàncement grow.[30]

More than a century's rhetoric preceded Edmund Burke's description of the aristocracy as 'the great oaks that shade a country' and his metaphors of 'roots' and 'stock'. When the fourth Earl of Warwick died in 1673 it seemed natural for the preacher of his funeral sermon to describe him as a 'princely cedar' or 'spreading oak' and lament him as a 'great tree cut down'.[31] In eighteenth-century family portraits the great old tree was often a central feature, symbolizing the continuity of the generations; and, as Mrs Hemans later observed, 'the stately homes of England' stood amidst 'their tall ancestral trees'.[32]

Trees indeed were a kind of family monument, a bid for personal immortality, at a time when gravestones and similar memorials were largely confined to the socially privileged. 'Had not Jo[hn] Pullen of Magdalen Hall in Oxford left a tree of his own planting on Headington Hill,' wrote a contemporary in 1758, 'it had been forgot that such a man had ever existed.'[33] Those unable to plant trees could at least carve their names on them. Of one famous beech tree growing between Oxford and Banbury it was reported in 1657 that so many travellers had cut their names on its bark that there was hardly any space left.[34] For landed families, trees were enduring memorials. At Kedleston, Derbyshire, it was thought that some of the oaks were as old as Lord Scarsdale's family, which had supposedly come over with William the Conqueror.[35]

To fell such a monument was to extinguish the planter's name. Hence the shock in the Lake District when the attainder of the Jacobite Earl of Derwentwater in 1715 was accompanied by the felling and sale of the oaks around his castle; in parts of France and in Scotland it had always been conventional that the finest trees on an estate should be ritually felled if their owner proved a traitor.[36] During the Interregnum no actions aroused more passion than the felling of groves on Royalist estates. The enemies of Root and Branch even planned to cut down the royal walk of elms in St James's Park, 'a living gallery of aged trees', as Edmund Waller described them.[37] In the same way two Restoration fellows of Magdalen College, Oxford, decided to dig up an avenue of elms because it had been 'planted in the fanatic times'. When Londoners demonstrated against the Earl of Clarendon in 1667 they chopped down the trees in front of his house.[38] And there were many parallels to the dismay felt by the Gloucestershire gentleman Christopher Guise, who returned to his family home around 1630 to discover that his grandfather had been persuaded to fell all the old trees, 'an irreparable loss of an honourable symbol of antiquity'. Wordsworth tells of the magnificent tree outside the house of a Lake District yeoman, which a neighbour urged him to fell for profit's sake. 'Fell it?' exclaimed the yeoman, 'I had rather fall on my knees and worship it.'[39]

Trees were thus so personal to the householder and his family that there was nothing paradoxical about Jeremy Bentham's proposal that country gentlemen should embalm their ancestors and place them in avenues, alternating with the trees as what he called 'auto-icons' (having, of course, first varnished their faces to protect them against the weather).[40] Neither was it surprising that rural rioters, like the discontented commoners of Windsor Forest in the early Hanoverian period, sometimes cut down ornamental trees as a means of hitting at their owners. It was natural for those with social grievances against the aristocracy to resent the sight of what the poet John Tutchin called 'a tall and sprightly grove of servile trees'; and such plantations were subject to much vandalism. In the later seventeenth century, for example, the Earl of Essex's gardener, Moses Cook, reported that three trees in the garden had been spoiled 'by some base men or boys'; while a century later the architect Sir William Chambers lamented that youth and the common people would destroy anything: they cut down trees, pelted statues with stones, carved graffiti on buildings and trampled on beds of flowers.[41]

Trees thus provided a visible symbol of human society. Woodmen employed a terminology which was notably anthropomorphic. A fruitful cross was called 'a wife'; one without seed was a 'maiden' or 'widow'. A young shoot of felled coppice was an 'imp' and a tree too old to yield useful timber a 'dotard'.[42] Poets saw in trees a parallel to the social hierarchy among humans. Abraham Cowley contrasted 'old patrician trees' with the 'plebeian under-wood', while Joseph Beaumont wrote of Eden that there

no crook-back'd Tree
Disgrac'd the place, no foolish scrambling Shrub,
No wilde and careless Bush, no clownish Stub.

In the same spirit the Royalist poet, Rowland Watkyns, lamented

That Kingdom is an unhappy case
Where Cedars fall and shrubs possess their place.[43]

Landscape-gardeners thought that trees differed in their sexual attributes. For a hunting-box, ruled William Marshall, the style should be masculine; only the hardiest shrubs – box or holly – were to be planted there. In early Victorian Cheshire the villagers used trees as moral symbols, hanging up branches outside other people's houses on May Day to show how the householders were regarded by their neighbours: oak meant a good woman; birch meant a pretty girl; alder meant a scold.[44]

The oak had been a symbol of strength since at least the sixteenth century. Always the king of trees, it became, with the growth of the Navy, an emblem of the British people and as much a national symbol as roast beef. It represented masculinity, vigour, strength and reliability. It was, thought William Shenstone, in all respects 'the perfect image of the manly character: ... the British one'.[45] 'However low the value of its timber may fall,' declared an early-twentieth-century work on forestry, 'it [the oak] will long be regarded as an emblem of British strength and character.' In the eighteenth century, indeed, shipbuilding suffered from the obstinate prejudice that English oak was a better material than the Baltic or American version.[46]

So close was the relationship of trees to human society that their treatment, like that of horses or children, fluctuated according to changing educational fashion. In the sixteenth and early seventeenth centuries infants were swaddled; and it was widely held that most children would need to be beaten and repressed. Timber trees, correspondingly, were to be pollarded (i.e. beheaded), lopped or shredded (by cutting off the side branches). Hedges had to be regularly laid and trimmed; each county had its own distinctive way of doing so. The trees preserved for ornament were brought severely under human control by gardeners who clipped, pruned and manicured them, even working them into artificial shapes. Yews and privet, particularly in the later seventeenth century, were trimmed into cones, pyramids, birds, animals and human figures. Limes were pleached to form long walls of interlocking branches. In avenues it was customary to lop all the branches off, leaving only a tuft or crown at the top. Fruit trees were splayed out, espalier-fashion, against the garden wall. There were utilitarian reasons for many of these practices, but they were also seen as a kind of moral discipline: 'The luxuriancy and vigour of most healthful trees,' declared John Laurence in 1726, 'is like the extravagant sallies of youth, who are apt to live too fast, if not kept within due bounds and restrained by seasonable corrections.' Regular pruning kept 'all in order, which would otherwise be perfect anarchy and confusion'.[47]

In the eighteenth century, when educational theories became less repressive, the cultivation of trees moved from regimentation to spontaneity. There was a reaction against 'mutilating' trees or carving them into 'unnatural' shapes. Topiary went out of fashion in the reign of Anne.[48] Pollarding was attacked by Moses Cook in 1675: 'I wish there were as strict a law as could be made to punish those that presume to behead an oak, the king of woods, though it be on their own land.' Such mutilation did not merely harm the timber, it was also a distasteful form of violation; and the practice went into decline during the eighteenth century.[49] In the same period the East Anglian habit of shaving trees to leave only a tuft on the top was strongly condemned by Arthur Young, while in 1808 William Mavor denounced the 'vile custom' of lopping or shredding hedgerow elms.[50] In 1790 John Byng attacked what he called the 'savage' Midland practice of barking oaks before they were felled; he compared it to flaying the tree alive. To the aesthete William Gilpin even clipped hedges were unpicturesque. A certain irregularity and wildness of appearance in a hedge, agreed the Scottish poet James Grahame, was more pleasing than a uniform trimness.[51] The tree's free growth symbolized the Englishman's freedom more generally. 'Everyone who has the least pretension to taste,' wrote Alexander Hunter in 1776, 'must always prefer a tree in its natural growth.' In Russia the first action of Catherine II on reading an English book on the 'natural' style of gardening was to forbid any more clipping of trees in the imperial gardens.[52] This was the spirit which would, in due course, lead to the abandonment of swaddling clothes for infants, wigs for men and, for a time, corsets for women, on the grounds that they were unnatural and unspontaneous. In England it even became temporarily unfashionable to remove the bark from felled timber. From the 1750s there was a vogue for the so-called 'rustic' style, with huts, seats and gates made out of undressed branches (like names of modern suburban houses painted on slices of imitation tree-trunk).[53]

Finally, there were people who alleged, as Thomas Tryon reported in 1691, that 'trees suffer pains when cut down, even as the beasts and animals do when they are killed'.[54] There had always been a good deal of anthropomorphic talk in the gardening books about what conditions trees 'loved' or 'hated'; and in fruit-growing areas it was common to wassail trees by singing, firing guns and offering libations. 'Men must learn to discourse with fruit trees, having learned to understand their language,' thought Ralph Austen, a leading seventeenth-century authority on the subject.[55] In 1653 Margaret Cavendish published a dialogue in which an oak complains to the woodman of being tortured: 'You do peel my bark, and flay my skin, chop off my limbs.' When an oak was felled, reported John Aubrey, it gave 'a kind of shriek or groan that may be heard a mile off, as if it were the genius of the oak lamenting. E. Wyld, Esq., hath heard it several times.' And if that be thought typical of the credulous Aubrey, here, over a hundred years later, is John Constable commenting on a drawing of an ash tree:

> Many of my Hampstead friends may remember this *young lady* at the entrance to the village. Her fate was distressing, for it is scarcely too much to say that she died of a broken heart. I made this drawing when she was in full health and beauty; on passing some time afterwards, I saw, to my grief, that a wretched board had been nailed to her side, on which was written in large letters, 'All vagrants and beggars will be dealt with according to law'. The tree seemed to have felt the disgrace, for even then some of the top branches had withered. Two long spike nails had been driven far into her side. In another year one half became paralysed, and not long after the other shared the same fate, and this beautiful creature was cut down into a stump, just high enough to hold the board.[56]

It is not a long step from this to placing conservation orders on trees, in accordance with the view proclaimed by William Morris in 1884 that no one should be allowed 'to cut down for mere profit trees whose loss would spoil a landscape'. Indeed John Ramsay of Ochtertyre, a prominent Scottish landlord (1736–1814), had already declared that 'a noble tree is in some measure a matter of public concern; nor ought its proprietor to be allowed wantonly to strip his country of its fairest ornament.' In 1821 Wordsworth's friend Sir George Beaumont, when travelling in Italy, even bought a pine tree on the skyline at Monte Mario, so as to prevent the local Italians from felling it.[57]

In this now familiar movement to preserve trees, regardless of the economic consequences, we can see many ingredients: planning considerations, a desire for amenity and a feeling that trees were intrinsically beautiful played an obvious part. But people also wanted trees preserved not just for the sake of their appearance, but because of what they stood for. They cherished their associations, their antiquity, their link with the past. A hankering for continuity, a bid for family immortality and a tendency to invest trees with human attributes were all important. Just as men cherished household pets because they were projections of themselves, so they preserved domestic trees, because they represented individuals, families and, in the case of the British oak, the nation itself. Durkheim may have been wrong when he suggested that when men worshipped God they were really worshipping society. But he would have been very near the truth if he had said it about the worship of trees.

iv. FLOWERS

To turn from trees to flowers is to move to an area much less obviously charged with social emotion. But the spectacular growth between the sixteenth and nineteenth centuries of domestic flower-cultivation is a social development that deserves far more attention than it has yet received; and it reflects a shift in sensibilities similar to that reflected in the changing attitudes to trees and to animals. Of course, flowers had been grown in the Middle Ages, and their presence in embroidery, illuminated manuscripts and other forms of artistic decoration shows that they were appreciated for their beauty as well as for their herbal use. In London on the annual feast day at St Paul's it was customary for the dean and chapter to wear garlands of red roses.[1] Since roses, lilies, violets and virtually all other garden flowers were still used for medicinal purposes as well as for perfumes, syrups and distilled waters, it is hard to be sure when they were first cultivated for wholly ornamental reasons. But it is clear that by Tudor times flowers were widely cherished for their appearance and, in those relatively insanitary days, particularly for their scent. Flowers, herbs and even whole branches of trees were brought into the house and kept in pots. Sweet-smelling plants were strewn on the floors; nosegays were placed in bedrooms; in summertime hearths were adorned with boughs, wreaths or flowers; and women wore chaplets and garlands on their person.[2] Fritillaries, as the herbalist John Gerard observed, helped further to beautify 'the bosoms of the beautiful', while feather grass was worn by 'sundry ladies and gentlewomen' instead of a feather. By the Elizabethan period the 'garden of delight', as opposed to the mere vegetable- or herb-garden, was a well-established luxury, the recognized source of what the Jacobean herbalist John Parkinson would call 'flowers of beauty and respect'.[3]

Yet during the next two centuries there occurred an expansion in flower-gardening on a scale so enormous as to justify our adding to all the other revolutions of the early modern period another one: the Gardening Revolution. Mr John Harvey has described the rise of the professional nurseryman and the vast sums spent on plants and seeds. Some sort of organized trade had existed since at least the thirteenth century. But the first recognizable commercial nurseries appeared in Tudor times, multiplied in the early seventeenth century and increased sharply in scale thereafter.[4] In 1546 the keeper of the King's garden was able to buy 3,000 red rose trees at 3*s.* 4*d.* a thousand, while in 1629 John Parkinson felt it necessary to warn his readers against knavish gardeners ('whereof the skirts of our towns are too pitifully pestered'), who gave false names 'to deceive men and make them believe they were the finders-out or great preservers of rarities, of no other purpose but to cheat men of their money'. In the 1690s there were at least fifteen nurseries in the London area. Of these the largest concern, the Brompton Park nursery, contained in 1705 nearly ten million plants – 'perhaps as much as all the nurseries of France put together,' thought a contemporary.[5]

At first most of the trade, like that of other luxuries, centred on the capital. Gentlemen sent to the metropolis for seeds or plants, just as they did for the latest fashion in clothes. When in 1604 seeds and garden shears were needed for a Caernarvonshire garden they had to be ordered from London, 'for there is none to be had in this country'.[6] But by the early eighteenth century there were many provincial firms in existence. Mr Harvey's pioneering compilation is inevitably far from complete, and it is likely that the numbers of seventeenth-century florists and seedsmen were even greater than he suggests.[7] There was also a great deal of trade direct with the continent as the nobility and gentry did much of their buying in France, the Low Countries and elsewhere.[8]

A parallel development was the emergence of the new profession of gardeners and landscape-improvers, in the eighteenth century often Scotsmen, offering their advice to the Crown, aristocracy and gentry, prescribing new fashions like tailors and commanding correspondingly substantial fees. In 1683 the new gardener asked £80 per annum for looking after the grounds at Lyme Hall in Cheshire; he was beaten down to £60, equivalent to the income of a well-to-do clergyman. On Easter Day he appeared in a suit with gold buttons, 'better worth than two of my best suits', wryly commented his employer, Richard Legh.[9] An aristocrat of the profession like George London (died 1713) could combine the post of surveyor of the royal gardens at £200 per annum with the income derived from touring the country, seeing and giving directions once or twice a year 'in most of the noblemen's and gentlemen's gardens in England'. His partner at Brompton, Henry Wise, became gardener to Queen Anne and George I. He was paid £1600 per annum and left over £100,000.[10] In 1764 a visitor to Thomas Mawe, head gardener to the Duke of Leeds, found him 'so bepowdered and so bedaubed with gold lace' that he thought he was in the presence of the Duke himself. In 1760 it was estimated that the country contained ten garden-designers, a hundred and fifty noblemen's gardeners, four hundred gentlemen's gardeners, a hundred nurserymen, a hundred and fifty florists, twenty botanists and two hundred market-gardeners.[11]

Then there was the multiplication of gardening books, recently catalogued by Miss Blanche Henrey. In the sixteenth century there were only about nineteen new titles on botany and horticulture, but in the seventeenth century there were a hundred (of which over eighty came after 1650) and in the eighteenth century no less than six hundred. John Abercrombie's *Every Man His Own Gardener* (1767) went into sixteen editions by 1800, while *The Gardener's Pocket Journal* (1789) sold 2,000 copies a year. In 1787 the appearance of William Curtis's periodical *Botanical Magazine or, Flower-Garden Displayed* (still in existence) heralded the beginning of regular gardening magazines.[12] Many of these works were frankly commercial in intention, advertising the services of particular seedsmen and nurseries and drawing attention to the availability of individual landscape-designers, in some cases the author himself.

A closely associated development was the appearance of luxurious books of flower illustration. Botanical drawing had begun as a practical art, designed to enable herbalists to identify the plants they used for their medicines. But in the seventeenth century there was an increased tendency to draw or paint flowers, not for use, as in the herbals, but simply for the sake of their intrinsic beauty. The Hanoverian period was a great age for flower-painting, whether as illustration for handsome books or for reproduction on calico prints.[13]

Most dramatic of all was the vast influx of exotic flowers and shrubs, first from southern Europe and the Levant, then from America, the West Indies and South Africa, ultimately from all over the world. In Tudor times many new commercial and industrial plants were introduced – hops, saffron, woad, flax, hemp and madder. Other species were brought as fruit and vegetables (potatoes, red peppers, artichokes) or for medicinal purposes.[14] But an infinitely greater number came in for ornament, curiosity and display; even the runner-bean was originally introduced for the sake of its flowers.[15] More flower-gardening had gone on in the Middle Ages than is sometimes appreciated. But the repertoire seems to have been fairly limited, with a heavy concentration on roses, lilies, gillyflowers (pinks), cowslips, marigolds and violets. In 1500 there were perhaps 200 kinds of cultivated plant in England. Yet in 1839 the figure was put at 18,000.[16] Nearly all our garden flowers arrived during the intervening years: in the sixteenth century tulips, hyacinths, anemones, crocuses; in the seventeenth michaelmas daises, lupins, phlox, virginia creeper, and golden rod; in the eighteenth sweet peas, dahlias, chrysanthemums, fuchsia. There were auriculas from the Pyrenees, fritillaries from France, lilies from Turkey, marigolds from Africa, nasturtiums from North America. Nothing did more to make contemporaries aware of the fact of change than this constant influx of new flowers; like new trees, new vegetables and new agricultural crops, they helped to create a landscape vastly different from that which men had previously known.[17]

Exotic gardening was a European phenomenon. Its centres were Italy and Spain (which had close connections with the New World), Vienna (where the Imperial court was well placed to secure flowers from the Levant), France (which had early links with the Mediterranean) and, above all, the Netherlands, which by the 1580s was well established as a primary centre for botany and agriculture. But England participated from the beginning; and in the early stages of plant-collecting many individuals played their part: pharmacists eager to acquire new herbal remedies, like Hugh Morgan, 'a curious conserver of rare simples', or John Rich, in whose London garden William Turner noticed many 'good and strange herbs I never saw anywhere else in all England'; diplomatists like Sir Thomas Smith, who in 1565 sent home hyacinths and asphodels from France for his wife, or Sir Henry Wotton, who, as James I's ambassador at Venice, procured seeds and cuttings, or William Sherard, consul at Smyrna (1702–16), who brought the German scientist J. J. Dillenius to England, founded the Chair of Botany at Oxford and made plant-hunting expeditions in Asia Minor; merchants like Nicholas Leate, who combed the Levant for rare flowers, with which, says John Gerard, he was 'greatly in love';[18] sailors who supplemented their earnings by selling to the plant-collectors on the wharves when the ships came in;[19] gardeners like John Tradescant, who, in the successive service of the Earl of Salisbury, Lord Wotton and the Duke of Buckingham, went in search of shrubs, fruit and flowers to France, the Netherlands, Russia and the Mediterranean, bringing back for his garden at Lambeth the first lilac, the first acacia and the first occidental plane;[20] amateur collectors like Dr Simon Trip, a Winchester physician, who, when in 1582 he heard that Richard Madox was making a voyage to the Moluccas, asked him to be sure to bring him home some good seeds;[21] professional dealers who poured into England with allegedly new species, like Francis Le Veau, whom Parkinson revealingly describes as 'the honestest root gatherer that ever came over to us';[22] botanists and apothecaries who, following continental example, founded the physic gardens in which the new imports were arranged and classified at Oxford (1621), Chelsea (1673), Edinburgh (around 1680), Kew (from about 1759) and Cambridge (1760–62). In the later eighteenth century it was from Kew that the great plant-collecting expeditions were organized.

Until that time the search had depended heavily on aristocratic patrons. In the seventeenth century there was Edward Lord Zouche, whose garden at Hackney was supervised by the botanist L'Obel; the Marquis of Dorchester (1606–80), who had 2,600 plants all arranged in botanical order; Major-General John Lambert, who procured flowers from Algiers and Constantinople for his garden at Wimbledon and was portrayed on a contemporary playing-card holding a tulip; Sir Arthur Rawdon, for whose garden at Moira in Ireland the plant-searcher James Harlow brought back from Jamaica 'a ship almost laden with cases of trees and herbs, planted and growing in earth; ... also a great number of samples of them very well preserved in paper'; Henry Compton, Bishop of London, who had over a thousand exotics in his garden at Fulham and employed the missionary John Banister to find him new species in North America (of whose church he was conveniently the official head); and the first Duchess of Beaufort at Badminton, with her hothouses and conservatories.[23] They had their eighteenth-century successors, like the eighth Lord Petre, who died in 1742 at the age of 29, having already assembled over 219,000 trees, shrubs and plants at Thorndon Hall, Essex, with the help of, among others, Peter Collinson the Quaker merchant, who may have been responsible for introducing more American plants into Europe than any other single person.[24] But ultimately the plant-collecting business fell into the hands of syndicates of professional dealers.

Much has been written about the eighteenth-century landscape garden, in which flowers, of course, played only a subordinate role. It is not so frequently stressed that it was in the early modern period that the taste for small-scale domestic flower-gardening gradually established itself as one of the most characteristic attributes of English life. The beauty of Elizabethan gardens, thought William Harrison, had been 'wonderfully ... increased ... so that in comparison of this present the ancient gardens were but dunghills ... How art also helpeth nature in the daily colouring, doubling and enlarging the proportion of our flowers, it is incredible to report.'[25] In the Tudor period the more ambitious gardens belonged to physicians and apothecaries, concerned to widen their repertoire of herbal remedies. But the printed herbals reveal that there were many keen amateurs – peers, gentlemen, clergy, merchants and the well-to-do – who were connoisseurs of rare flowers and diligent horticultural experimenters, actively swapping plants and seeking new varieties from overseas. William Turner remarked that the best peony he had ever seen was in the garden of a rich clothier at Newbury.[26] At Darfield in Yorkshire in the early 1640s the Rev. Walter Stonehouse had 866 different species in his garden, while in North Wales at the same period Sir Thomas Hanmer was acquainted with 69 varieties of iris alone. Hanmer wrote to Evelyn in 1668: 'We have within these few years in Flintshire and some other shires near to me many gentlemen that have upon my instigation and persuasion fallen to plant both flowers and trees and have pretty handsome little grounds.'[27]

By this time flower-gardening ('this lovely recreation', as John Rea called it) extended far beyond the world of the rich and fashionable. There was 'scarce a cottage in most parts of the southern parts of England,' declared John Worlidge in 1677, 'but hath its proportionable garden, so great a delight do most of men take in it.'[28] A few years earlier William Hughes had produced a book on flower-growing, intended chiefly for 'plain and ordinary countrymen and women'. The first impression sold out in three months.[29] In the reign of Charles II a traveller noticed not just that the innkeeper and his wife at the Talbot in Towcester were 'curious in their house and garden', but that in Gloucester even the gaol had a 'neat garden', created by the gaoler's wife.[30] Sir Thomas Browne observed 'in the country a few miles from Norwich ... a handsome bower of honeysuckles over the door of a cottage of a right good man'. In the later eighteenth century it became common for travellers to comment on the honeysuckle, roses, geraniums and carnations to be found round many cottage doors, contrasting markedly with the gardens of the French peasantry, which contained only apples and cabbages.[31] By the early nineteenth century many cottages had 'a bit of lawn and gravel walk to the front door', in imitation of gentlemen's houses.[32]

During the eighteenth century many provincial towns established flower shows with prizes for the best blooms. In Norwich, where Dutch immigrants had greatly stimulated flower gardening in Elizabethan times, there had been an annual florists' feast since at least 1637. In mid-eighteenth-century Bristol there were two flower shows a year: one in late spring for auriculas, another in high summer for carnations.[33] Particular effort went into the cultivation of the eight so-called 'florists' flowers' (hyacinths, tulips, ranunculi, anemones, auriculas, narcissi, carnations and pinks). These were specially grown with an emphasis on improved varieties, artificial hybrids, double blooms, freakish colours and out-of-season flowering. ('For the art of agriculture is improving,' noted Christopher Smart, 'For this is evident in flowers. For it is more especially manifest in double flowers.'[34]) Such cultivation for the sake of prizes reflected a love of gambling as much as of flowers; and it was often a lower-middle-class activity. 'Small tradesmen, weavers or the like', usually won the prizes at town flower shows, thought a commentator in 1770, because perfect blooms needed constant attention and industrious artisans had the habit of regular application. Auriculas, tulips and pinks were to be found in their highest perfection in the gardens of the manufacturing class, agreed another: one did not see such fine flowers in the gardens of the nobility and gentry because they depended for their gardening upon hired servants, who took less care than those who tended their own flowers for themselves.[35] In the late eighteenth century the best pinks and auriculas were produced by the weavers of Spitalfields, Manchester and Paisley, while one of the first men in England to grow a magnolia from seed was a butcher at Barnes.[36]

'Scarce a person from the peer to the cottager,' it was said in 1779, 'thinks himself tolerably happy without being possessed of a garden.' Gardening, agreed a seedsman in the following year, was now 'cultivated by persons in every rank of life'. Only a very widely based demand could have brought prices down to the remarkably low figures charged by late-eighteenth-century nurserymen for a large range of garden flowers.[37] The Baptist church at Bedford which in 1735 rebuked one of its members for 'carrying about garden things on the Lord's Day' was contending against one of the characteristic recreations of the age. As a later commentator remarked, 'almost every one must have observed the high enjoyment which [gardening] affords to individuals of the lowest class.'[38]

The flowers, moreover, were grown not because they were medicinally useful or symbolically meaningful, but because they were aesthetically pleasing. In the early seventeenth century the herbals were already listing plants which, though useless for medicine or food, were nevertheless widely cultivated, simply for 'the beauty of the flowers'. In the gardening books the emphasis was unequivocally upon the 'delight' which flowers afforded those who grew them.[39]

Not that the emblematic meaning of plants had wholly disappeared. In the sixteenth century roses were sometimes hung over dinner tables to warn the assembled company that what men said to each other in their cups should not be repeated outside; and a Jacobean preacher remarks that 'in some churches I have seen a flower set by the hour glass on the pulpit, to express this conceit that the preaching of the Word is even the flower of the place'. Flowers, thought John Gerard, 'through their beauty, variety of colour, and exquisite form do bring to a liberal and gentlemanly mind the remembrance of honesty, comeliness and all kinds of virtues.'[40] As emblems of purity, beauty and the shortness of life, flowers were extensively used in popular calendar rituals. On St John Baptist Day in early Tudor Morpeth the boys wore garlands, while on May Day Elizabethans would attach nosegays of flowers to their oxen and horses. When a traveller passed through North Staffordshire in 1750 on St Barnabas Day he found everyone wearing white roses. Sweet-smelling flowers grew in such profusion around Norwich, reported Richard Bradley in 1728, that on a certain feast day each year the streets of the city were strewn with them.[41]

Flowers and shrubs were equally conspicuous in the rituals accompanying marriage and burial. At weddings the bridal couple bestowed rosemary (for remembrance) among their friends, and guests carried sprigs of it in their hats.[42] In mourning there was extensive use of evergreens as a symbol of immortality; 'cypress garlands are of great account at funerals among the gentler sort,' noted a herbalist in 1656, 'but rosemary and bays are used by the commons, both at funerals and weddings.' In some parts of the country rosemary continued to be used at funerals until the mid nineteenth century.[43] Flowers were often placed on the hearse, and it was sometimes customary for children to be buried with them in their hands. Roses might be planted on the grave and flowers placed there annually.[44] In funeral sculpture the different species retained an emblematic significance; and by mid-Victorian times there were strict rules as to which flowers were or were not appropriate for use in graveyards. African or French marigolds would be in very bad taste in a cemetery, thought an authority in 1858; what was needed were aconites, rosemary, violets, snowdrops, lily of the valley or other symbols of purity. It was also essential that churchyard flowers should be sweet-scented. Flowers which merely *looked* beautiful should never be admitted.[45] The language of flowers thus continued to have a wide vocabulary and to be extensively spoken. As a commentator remarked in 1824, flowers made a suitable present in contexts where any other form of gift would cause offence. Even today, 'sending flowers' remains an effective medium of exchange when no other currency is acceptable.[46]

These symbolic uses of flowers, however, were largely incidental to the preoccupations of the gardeners. If asked why they grew flowers, most of them would have said that they did so because they enjoyed cultivating them and because they thought the blooms were beautiful and interesting. As Charles Marshall explained in 1796, 'the principal object in the cultivation of flowers' was 'a gay appearance'.[47] Nevertheless, so conspicuous an activity was not without its wider implications; and three features in particular should be emphasized.

The first was the strong ingredient of social emulation. As early as 1629 John Parkinson was writing for 'many gentlewomen and others that would gladly have some fine flowers to furnish their gardens, but know not what the names of those things are that they desire' – in other words were ignorant of flowers, but very anxious to keep up. Parkinson commended particular species because they were popular among 'the better sort of the gentry', stressing that there was 'no lady or gentlewoman of any worth' who did not like tulips, and emphasizing the importance of rarity and novelty (though conceding reluctantly that the violet was 'a choice flower of delight notwithstanding the popularity'). Parkinson's successor John Rea in 1676 dismissed the red lily because it was 'a vulgar flower', common in every countrywoman's garden and therefore 'seldom accepted by any florists'.[48]

Flowers thus followed fashions as much as did clothes. Until the 1620s the most prized were the gillyflowers and carnations. They were overtaken first by tulips and then from the 1680s by auriculas. The tulips which used to be valued, noted Sir Thomas Hanmer in the mid seventeenth century, were only those striped with purple, red and white; but of late any mixture of odd colours had become acceptable. As each flower went out of fashion it lost its commercial value and descended the social scale. Golden rod, remarked John Gerard in 1597, had once sold at 2*s*. 6*d*. an ounce.

> But since it was found in Hampstead Wood, even as it were at our town's end, no man will give half a crown for an hundredweight of it; which plainly setteth forth our inconstancy and sudden mutability, esteeming no longer of anything, how precious so ever it be, than whilst it is strange and rare. This verifieth our English proverb, 'Far fetched and dear bought is best for ladies'.

In the same way Samuel Gilbert in 1682 omitted from his *Florists Vade Mecum* what he called 'obsolete and overdated flowers', while John Rea announced in 1676 that many of Parkinson's flowers had 'by time grown stale and for unworthiness [been] turned out of every good garden'. In 1770 William Hanbury thought that the hollyhock had grown too common and was now beginning to be less admired. The old Turkey ranunculus was also 'now esteemed an antiquated flower'. Gardening, said Hanbury, was much complicated by 'this custom of alternately introducing and expelling of flowers'.[49]

Ornamental trees were equally liable to lose favour if they became too common. The box, which had a disagreeable smell and was thought to have a sterile effect upon the earth, went out of fashion in the reign of Charles II; and the bay gave way to the laurel. The privet was also being replaced in the more 'modish' gardens, noted John Worlidge in 1669.[50] There was no good reason for the disesteem into which the lime had fallen, thought William Boutcher in 1775, save the caprice of fashion: 'Many beautiful plants, as well as other things, have been out and in during my time,' At the beginning of the nineteenth century James Grahame observed that it was wrong to plant firs and pines where deciduous trees were more appropriate, but 'fashion's law' was hard to resist.[51]

Even vegetables and fruit were subject to fashion's law. Parkinson noted in 1629 that the potato had begun as a delicacy for the Queen, but had become so common that even the vulgar despised it. The leek had also been rejected 'by our dainty age' as fit only for the very poor and the Welsh. Garden beans were for the poor and so were pumpkins. White currants, on the other hand, were more desirable than black ones, partly because they had a better taste, but chiefly 'because they are more dainty and less common'. At the beginning of the eighteenth century the most highly esteemed fruit was the pineapple, because it was so difficult and so expensive to grow.[52]

This constant desire to keep ahead of the fashion (or at least to profit by selling to those who wished to keep ahead) was one of the chief stimuli to horticultural innovation. It underlay the preoccupation with rarity, novelty and hybridization. It encouraged the gentry to spend large sums of money on improving new varieties from overseas; and it forced them to install stoves and greenhouses in which tropical plants could be housed and abnormally early or late flowering achieved. (As early as 1608 it was reported that 'Master Jacob of the Glass House' had carnations all the year round.[53]) It also led to the proliferation of an infinite number of new varieties of every fashionable plant. Regional peculiarities, combined with the slow development of a national market or standardized source of supply, meant that fruit and vegetables existed in a great number of different varieties: in 1780, for example, it was possible to buy as many as 320 different kinds of gooseberry.[54] Flowers were equally various; and they became more so with the growth of fashion. In 1629 Parkinson listed approximately 50 varieties of hyacinth, 70 carnations, 70 anemones and 140 tulips. Twenty years later John Evelyn was told by a French florist that there were no fewer than 10,000 different kinds of tulip. No other flower proliferated on such a scale. Even so, Richard Bradley thought in 1728 that there were nearly a thousand different sorts of gillyflower; and in 1777 Richard Weston's seed catalogue offered 208 anemones, 575 hyacinths, over 800 tulips and no less than 1100 ranunculi.[55] The rose, by contrast, though always conventionally regarded as the queen of flowers, does not seem to have become the object of intense experimentation till the end of the eighteenth century. In 1800 there were still fewer than a hundred varieties. But by 1826 there were 1393.[56]

The usual tendency in flower-growing was for a new variety to appear, enjoy a brief moment of glory and then sink into obscurity to be replaced by another. 'All new-raised flowers decrease greatly in price in a few years after they are raised,' explained Weston. But, while its popularity lasted, the fashionable flower could command a very high price. 'Some few years ago,' remarked Richard Bradley in 1717, a single auricula root sold for 20 guineas. In 1777 some kinds of hyacinths were only a few pence each; but 'Princess Gallitzyn' cost fifteen guineas and 'Black Flora' twenty. (It was customary to name new flowers after kings and queens, aristocrats and other notabilities, so that the very nomenclature of these flowers reflected their social pretensions. As that skilled entrepreneur Josiah Wedgwood remarked in another context, a commodity would always sell better when the name of a duchess was attached to it.[57])

Flowers thus varied in social acceptability. Since the seventeenth century it had been recognized that the plants found in cottage gardens were the old-fashioned ones, lagging far behind sophisticated taste. In the 1820s it was taken for granted that pinks, roses and polyanthus went with cottages, whereas villas would aspire to geraniums, dahlias and clematis. The auricula was, like the tulip and the pink, a poor man's flower; and the honeysuckle was 'the poor man's shrub'.

> By stately halls we see thee not
> But find thee near the lowly cot.[58]

Yet even unpretentious flowers were an encouraging sign, for refinement and sensibility were associated with flowers of any kind. By the later eighteenth century flower-gardening had emerged as a means by which humble men could prove their respectability. Gardening, it was believed, had a civilizing effect upon the labouring poor. It attached a man to his home and it spread a taste for neatness and elegance. Honeysuckle around a cottage door was not just picturesque: it was also a sign of the sobriety, industry and cleanliness of the inhabitants within. A greenhouse, thought William Cobbett, had a *moral* value. Hence the landlords' practice of building model cottages which had the whole of their gardens *in front* of the house so that they could be inspected by the passer-by. A well-kept cottage garden was both a pretty sight and a reassuring symbol of social contentment.[59] In the Mendip Hills, wrote the Rev. William Lisle Bowles in 1815, there was

> many a cottage . . .
> With porch of flowers, and bird-cage at the door,
> That seems to say – England, with all thy crimes,
> And smitten as thou art by pauper laws,
> England, thou only art the poor man's home![60]

In the 1860s 'the love of flowers and a taste for gardening' were seriously suggested as a means of reducing the high rate of illegitimacy in Cumbria: to establish a Cottagers' Flower Show would raise the moral tone. It was 'the spread of education amongst the working classes,' thought *The Gardeners Chronicle* in 1894, which had 'helped to cause the growth of the public taste for flowers'; and a Scottish clergyman remarked that on his visits to parishioners he had never had an unfriendly reception in a house which had a flowerpot in the window.[61]

The second feature of flower-gardening was its appeal to townsfolk. 'Citizens have gardens over their houses,' says an early Tudor schoolbook; and in his *Utopia* Sir Thomas More envisaged strife and contention between the inhabitants of rival streets over the trimming, husbanding and furnishing of their gardens.[62] Early maps and surveys reveal that not just London and Norwich, but Nottingham, Southampton, Worcester, York, and most other Tudor towns had their orchards, closes and gardens, in which it is likely that the better-off inhabitants grew not just fruit, herbs and vegetables, but also flowers for delight and ornament.[63] Many of these Tudor gardens and orchards were swallowed up by later building, but the taste for flowers was not easily eradicated. There was 'scarce an ingenious citizen,' wrote Worlidge in 1677, 'that by his confinement to a shop being denied the privilege of having a real garden, but hath his boxes, plots or other receptacles for flowers, plants, etc.' Failing that, he had at least a flower-painting on his wall, 'drawn to satisfy the fancy of those that ... cannot obtain the felicity of enjoying them in reality'. London citizens, noted Thomas Fairchild in 1722, furnished their rooms with basins of flowers and bough-pots, bought from countrywomen, 'rather than not ... have something of the garden before them'. 'Almost everybody whose business requires them to be constantly in town, will have something of a garden' and there was 'scarcely a street or valley without a virginia creeper'.[64] There were always flowers for sale in the London streets, while all the gardening books explained how to grow indoor plants or outdoor flowers which, like the virginia creeper, would resist the smoke of London. Indeed some of the socially more pretentious books regarded flowers as 'fit only for little town gardens' and concentrated on plants which would bloom 'in the winter and that part of spring when persons of distinction are in town'.[65]

This was the townsman's yearning for greenery which Cowper observed:

> Are they not all proofs
> That man immured in cities, still retains
> His inborn inextinguishable thirst
> Of rural scenes, compensating this loss
> By supplemental shifts, the best he may?
> ...
> And they that never pass their brick-wall bounds
> To range the fields and treat their lungs with air
> Yet feel the burning instinct; over-head
> Suspend their crazy boxes, planted thick,
> And watered duly. There the pitcher stands
> A fragment, and the spoutless teapot there;

> Sad witnesses how close-pent man regrets
> The country, with what order he contrives
> A peep at nature, when he can no more.

'Where all is stone around, blank wall and hot pavement,' Charlotte Brontë would write, 'how precious seems one shrub, how lovely an enclosed and planted spot of ground.'[66] By this time even the city streets themselves were coming to be named 'gardens'.

Thirdly, the popularity of the garden had a spiritual dimension. The conception of Paradise, a beautiful plot of earth, ornamented with water, trees and flowers, was an ancient one, with its roots in Greek, Roman and Oriental tradition. In Christian teaching the garden had long possessed a religious significance, as a place for spiritual reverie, a reminder both of Eden and of Christ's agony at Gethsemane. Gardening was the one form of labour which was necessary even before the Fall, for when God put Adam in the Garden of Eden he required him 'to dress it and to keep it' (Genesis, ii. 15). The idea that it would be possible to return to a state of pre-lapsarian grace by the cultivation of the soil was heavily emphasized in the titles of the gardening books: *Paradise Retrieved*; *Paradise Regained*; *Paradisus in Sole*.[67] By 'that lovely, honest and delightful recreation of planting', thought William Hughes, men might regain some of the lost splendour of Eden. Landscapes and fine gardens, felt James Shirley in 1646,

> Shew how art of men
> Can purchase Nature at a price,
> Would stock old Paradise again.[68]

It was common for a select part of a large garden to be known as 'a paradise'. Hampton Court in 1525 was said to be 'more like unto a paradise than any earthly habitation'; and there was still 'a parterre which they call Paradise' when Evelyn visited it in 1662. When John Wesley saw Howell Harris's gardens at Trevecca in 1769, he described them as 'a kind of little paradise'. In the later Middle Ages 'Paradise' had been the term for the pleasure garden of a convent. At Oxford a garden at Grey Friars was known among the inhabitants as 'Paradise', and 'Paradise Street' it has remained to this day. In post-Reformation literature the enclosed garden was a symbol of repose and harmony. Its flowers and trees were emblems of spiritual truths, its walks and arbours a sort of outdoor cloister.[69]

The garden thus became the accepted place for spiritual reflection, in life no less than in literature. For the idea of the Hortulan saint was no mere literary affectation: 'Real people,' it has been rightly said, 'did actually meditate in real gardens.'[70] 'When I was in my earthly garden a-digging with my spade,' said the mystic Roger Crab, 'I saw into the Paradise of God from whence my father Adam was cast forth.' When the young Elizabeth Walker was beset by atheistical thoughts she was saved because her father, a London druggist and a great lover of flowers, kept pots of them all over his shop and in the upstairs parlour window, to which she often went, 'to countermine my temptation, in admiring the curious works of the God of nature. With others there was then in flower a Calcedon Iris, full of the impresses of God's curious workmanship.'[71] Most wise men, thought the Rev. William Turner in 1697, sought a garden as their place for meditation; and the mid-seventeenth-century diarist Ralph Josselin was never more characteristic of his age than when he built a little house in his orchard as a place for retirement and reflection.[72] The professional hermits whom eighteenth-century gentlemen hired by the year to occupy cells in their gardens were a debased and self-consciously Gothic survival of this tradition.[73]

So although some puritanical divines occasionally attacked the vanity and waste of ostentatious gardening,[74] the belief of most clergymen was that gardening was an activity which brought men nearer to God. Gardening, declared John Laurence in 1726, was particularly appropriate for 'clergymen and other studious persons that have a taste for beauty and order'. There were many famous clerical gardeners, and some of the most celebrated horticultural achievements of the day were carried out in the gardens of deans and bishops.[75] Even the Quakers, who disliked the use of colour in clothes and houses and were hostile to many contemporary forms of recreation, permitted their members to 'follow after gardnering'. Although some of their official pronouncements implied that the Friends would do better to stick to vegetables (gardening 'in a lowly mind' and keeping 'to plainness and the serviceable part'), the Quakers in the later seventeenth and eighteenth centuries readily accepted the decorative pleasure of flowers and were celebrated for producing a quite disproportionate number of botanists, plant-collectors and nurserymen.[76]

It was because of the traditional notion of the garden as a spiritual resource that it came increasingly to be seen as the ideal environment for man's final resting-place. Flowers on tombs were an ancient symbol of regeneration and the concept of garden burial had been occasionally canvassed in the seventeenth century: the early Quakers were often buried in their gardens; John Evelyn planned a chapter on the subject; and Christopher Codrington, the benefactor of All Souls College

Oxford, who died at Barbados in 1710, had always hoped to return to England to be buried in his garden.[77] During the eighteenth century the association of graves with rural scenery grew stronger. In an essay on epitaphs written for Coleridge's *The Friend* in 1810 and reprinted as a note to Book 5 of 'The Excursion', Wordsworth maintained that urban cemeteries were inadequate, 'for the want of the soothing influences of nature, and for the absence of the types of renovation and decay which the fields and woods offer to the notice of the serious and contemplative mind'. He cited his now-forgotten contemporary, the poet John Edwards, who had urged that the inhabitants of large towns should be buried in the country, not because urban cemeteries were a danger to health, but because they were spiritually inadequate. Shortly afterwards the garden cemetery movement, which had already begun on the continent, spread to England; and in the 1820s what proved to be the first of innumerable garden cemeteries was opened at Liverpool.[78]

Finally, the garden was an infinite source of personal satisfaction. 'From heavy hearts and doleful dumps, the garden chaseth quite,' sang an Elizabethan poet.[79] This was partly because it provided the sedentary with much-needed exercise: 'Gentlewomen,' observed a gardener in 1657, 'if the ground be not too much wet, may do themselves much good by kneeling upon a cushion and weeding.' But it was also because the garden was an escape, a source of renewed vitality, a private domain which the gardener, however beaten down by the world, could order, arrange and manipulate without fear of contradiction. 'In a garden,' wrote John Laurence in 1716, 'a man is lord of all, the sole despotic governor of every living thing.'[80] This was one of the reasons why gardening so appealed to women, to whom other spheres of activity were closed. Long before the 'old wives that gather herbs'[81] of Tudor times, women and gardening had been closely associated, since food and medicine were both largely female responsibilities. William Lawson's *Country Housewife's Garden* (1617) was the earliest gardening-book aimed specifically at the female sex, but it had many successors. In his herbal John Parkinson often addressed himself to gentlewomen alone, though adding that their husbands might also share the delights of flower-gardening 'as much as is fit'. Flowers indeed were often seen as the woman's particular domain. Their care, thought Sir William Temple in 1685, was 'more the ladies' part than the men's'.[82] Flowers were conceded to women, partly because of the association in men's minds between the ephemeral beauty of women and flowers, partly because flower-gardening was a useless but decorative pursuit appropriate for the growing number of leisured, well-to-do females. In the early nineteenth century botany was generally regarded as particularly suitable for young girls; and J. C. Loudon could declare that 'there is scarcely such a thing to be found as a lady who is not fond of flowers'. The upper-class ladies who were so important in the history of late-nineteenth-century gardening had many Stuart predecessors: like Lady Bacon of Audley End, who in 1629 was buying tulips from Holland, 'the rarest that can be had', or the female connoisseurs of rare plants cited by Parkinson in his herbal, such as Mistress Thomasin Tunstall of Bullbank near Hornby Castle, Yorkshire, a 'great lover of these delights, ... who hath often sent me up ... roots to London'.[83]

It was the scope which gardening afforded for individual self-expression which explains why it could become such an obsession. 'If [a man's] heart be on his garden,' said a Stuart preacher, 'Oh how neatly it is kept! All the rare roots and slips that can be got for love or money shall be sought for it.'[84] This is why gardening appealed to so many different sorts of people, from Dr Pinke, the early-seventeenth-century warden of New College, Oxford, who at the age of nearly eighty would get up early and dig for an hour or two every morning, or that great autocrat Sir Isaac Newton, who was 'very curious in his garden, which was never out of order, in which he would at some seldom times take a short walk or two, not enduring to see a weed in it',[85] down to today's suburban gardener, with his hedge demarcating his property, his prize roses and his neatly mown lawn. The latter was a direct development from the Tudor bowling alley, which was not an indoor affair but a green surrounded by trees. By the mid seventeenth century the bowling green had become one of the most distinctive features of the English landscape. 'Nothing is more pleasant to the eye than green grass kept finely shorn,' thought Bacon.[86] The invention of the lawn-mower in 1830 would put this pleasure within the reach of many.[87]

By the beginning of the nineteenth century there was no country in which flower-gardening had as socially wide an appeal as in England. The lower-class English taste for trees and flowers was exported to Australia, whose earliest towns were indeed garden cities.[88] But it was largely absent from Wales, where in the 1790s John Byng claimed to have found no real gardens: 'I never saw a flower.'[89] It was much less advanced on the continent of Europe: in Lisbon in the 1830s, for example, there were no flower shops or nurseries whatsoever.[90] It also failed to take root in the United States. When William Cobbett went there in 1817, he found no counterpart to the English working-man's cottage with roses round the door:

> we see here the labourer content with a shell of boards, while all around him is as barren as the sea-beach ... This want of attention in such cases is hereditary from the first settlers. They found land so plenty, that they treated small spots with contempt. Besides the *example* of neatness was wanting. There were no gentlemen's gardens, kept as clean as drawing-rooms, with grass as even as a carpet.[91]

The difference is still visible today. In England, however, working-class gardening was encouraged by land-shortage, social imitation and a developed sense of private property. Like pets and trees, gardens became a means of strengthening their owner's sense of identity and adding to his self-esteem.* ('Most of the so-called love of flowers,' D. H. Lawrence would write, 'is merely this reaching out of possession and egoism: something I've got: something that embellishes *me*.'[92]) The cultivation of flowers is an historical phenomenon of great importance to anyone concerned to know how the working classes would use their leisure and direct their emotional energies. It explains why large-scale tenements have seldom been built in England, for they would have deprived working men of the gardens which they regarded as a necessity; and it accounts for the growth of the allotment movement in the nineteenth century. The preoccupation with gardening, like that with pets, fishing and other hobbies, even helps to explain the relative lack of radical and political impulses among the British proletariat.[93]

It is also important as an indication of that non-utilitarian attitude to the natural world whose emergence has been the theme of this book. The vegetable garden and the flower garden represented two fundamentally opposed ways of using the soil. In the one, men used nature as a means of subsistence; its products were to be eaten. In the other, they sought to create order and aesthetic satisfaction and they showed a respect for the welfare of the species they cultivated. The contrast must not be overstated, for agriculture and vegetable-cultivation were not without their aesthetic dimensions.† But the new attitude to trees and flowers closely paralleled the more sentimental view of animals which was emerging during the same period.

Yet even flowers were grown for human purposes. Like animal pets,

* Their role in this respect was delightfully conveyed in the light-hearted but revealing programme *Front Garden*, first shown on BBC2 television on 25 December 1979.

† See below, pp. 256–7.

they were raised in artificial conditions and they were wholly subservient to man's whims. It was an inevitable development that the next step would be for many persons to look for their emotional satisfaction beyond the garden fence, in wild nature itself.

VI
THE HUMAN DILEMMA

The creation of the mental realm of phantasy finds a perfect parallel in the establishment of 'reservations' or 'nature-reserves' in places where the requirements of agriculture, communications and industry threaten to bring about changes in the original face of the earth which will quickly make it unrecognizable. A nature reserve preserves its original state which everywhere else has to our regret been sacrificed to necessity. Everything, including what is useless and even what is noxious, can grow and proliferate there as it pleases.

Sigmund Freud, *Introductory Lectures on Psycho-Analysis (part iii)* (Standard Edn of the Complete Psychological Works of Sigmund Freud, trans. James Strachey *et al.*, xvi (1963), 372).

Yet, if we wield the sword of extermination as we advance, we have no reason to repine at the havoc committed, nor to fancy, with the Scotch poet, that 'we violate the social union of nature' ... We have only to reflect, that in thus obtaining possession of the earth by conquest, and defending our acquisitions by force, we exercise no exclusive prerogative. Every species which has spread itself from a small point over a wide area, must, in like manner, have marked its progress by the diminution, or the entire extirpation, of some other, and must maintain its ground by a successful struggle against the encroachments of other plants and animals ... The most insignificant and diminutive species, whether in the animal or vegetable kingdom, have each slaughtered their thousands, as they disseminated themselves over the globe, as well as the lion, when first it spread itself over the tropical regions of Africa.

Charles Lyell, *Principles of Geology* (1830–33), ii. 156.

It's so horrible, things having to be killed for us to eat them – it feels so wicked. Yet we have to do it – or die ourselves.

Kate Greenaway to Violet Dickinson, 14 June 1897; M. H. Spielmann and G. S. Layard, *Kate Greenaway* (1905), 190.

At the beginning of this book it was suggested that, at the start of the early modern period, man's ascendancy over the natural world was the unquestioned object of human endeavour. By 1800 it was still the aim of most people and one, moreover, which at least seemed firmly within reach. But by this time the objective was no longer unquestioned. Doubts and hesitations had arisen about man's place in nature and his relationship to other species. The detached study of natural history had discredited many of the earlier man-centred perceptions. A closer sense of affinity with the animal creation had weakened old assumptions about human uniqueness. A new concern for the sufferings of animals had arisen; and, instead of continuing to destroy the forests and uproot all plants lacking practical value, an increasing number of people had begun to plant trees and to cultivate flowers for emotional satisfaction.

These developments were but aspects of a much wider reversal in the relationship of the English to the natural world. They were part of a whole complex of changes which, by the later eighteenth century, had helped to overthrow many established assumptions and to create new sensibilities of a kind which have gained in intensity ever since. It is these wider changes which this last chapter will attempt briefly to evoke.

i. TOWN OR COUNTRY?

The first great change was one which G. M. Trevelyan never ceased to lament: the growth of cities and an intensification of what he called 'the harsh distinction between rural and urban life'. In 1700 over three-quarters of the British population still lived in the countryside; only 13 per cent, it has been estimated, dwelt in towns with over 5,000 inhabitants. But by 1800 the urban proportion had risen to 25 per cent and by 1851 the inhabitants of the towns were in a majority. Moreover, these nineteenth-century towns were more sharply differentiated from the country than their early modern predecessors had been. Before the end of the eighteenth century England had become, with the exception of the Netherlands, easily to most urbanized country in Europe.[1]

In Renaissance times the city had been synonymous with civility, the country with rusticity and boorishness. To bring men out of the forests and to contain them in a city was to civilize them. As an Elizabethan dialogue put it, a gentleman brought up in the town would be more 'civil' than one reared in the country.[2] The town was the home of learning, manners, taste and sophistication. It was the arena of human fulfilment. Adam had been placed in a garden, and Paradise was associated with flowers and fountains. But when men thought of heaven they usually envisaged it as a city, a new Jerusalem.[3] For centuries town walls had symbolized security and human achievement; and to the traveller their sight was always reassuring. In his tour in the 1530s John Leland often commented on the visual pleasures of the townscape: the 'pretty market' at Leeds; the 'fair streets' of Exeter; the radiance of Bewdley glittering 'as it were of gold' at sunrise; the 'beauty' of Birmingham. Rice Merrick, the Tudor historian of Glamorgan, thought Cardiff 'beautified with many fair houses and large streets', while in the 1690s Celia Fiennes could readily take pleasure in the sight of a 'neat town'.[4] In the eighteenth century there was much satisfaction expressed at the beauty of the London squares and the new buildings in Bath or Edinburgh New Town; and we know that in 1802 Wordsworth thought that earth had nothing fairer to show than the sleeping city of London seen from Westminster Bridge.

Yet, long before 1802, it had become a commonplace to maintain that the countryside was more beautiful than the town. 'No one,' wrote William Shenstone in 1748, 'will prefer the beauty of a street to the beauty of a lawn or grove; and indeed the poets would have found no very tempting an Elysium, had they made a *town* of it.'[5] It was partly the physical deterioration in the urban environment which encouraged this view. There had been complaints about London air since the thirteenth century.[6] By Elizabethan times the increasing use of coal for industrial as well as domestic purposes had created a major pollution problem. Queen Elizabeth stayed away from the capital city in 1578 because of the 'noisome smells'; and for centuries the first sight of the capital caught by approaching travellers was the overhanging pall of smog. Margaret Cavendish records the emotion felt by her husband, the royalist Marquis of Newcastle, when, returning from enforced exile in 1660, he caught sight once more of 'the smoke of London, which he had not seen in a long time'. An early-eighteenth-century poet wrote:

> While thus retir'd, I on the city look,
> A group of buildings in a cloud of smoke.[7]

The coal which was burned in the early modern period contained twice as much sulphur as that commonly used today; and its effects were correspondingly lethal. The smoke darkened the air, dirtied clothes, ruined curtains, killed flowers and trees and corroded buildings. By the mid eighteenth century the statues in London of some of the Stuart kings were so black that they looked like chimney sweeps or Africans in royal costume.[8] In 1700, Timothy Nourse wrote:

> 'Twere endless to reckon up all the mischiefs which houses suffer hereby, in their furniture, their plate, their brass and pewter, their glass . . . A bed of fourscore or one hundred pounds price, after a dozen years or so, must be laid aside as sullied by the smoke . . . The vast number of coal-dust carts trotting up and down the town, perpetually scatter very liberally of their precious cargo in the streets . . . from whence it is, that the complexions of men, and women too if they do not wash and daub, are soon tarnished and become sooty.

Dirt in the air meant dirt in the street; and in the summer the clouds of dust raised by the wheels of the traffic choked passers-by and made it difficult to walk with one's eyes open.[9]

Equally noxious was the pollution caused by the fumes and the waste matter generated by brewing, dyeing, starch-making, brick manufacture, and all the other industries which were carried on in the middle of the city. Since the reign of Richard II there had been intermittent legislation against the pollution of the Thames, and in the early seventeenth century there were many conflicts over the noisome effects of urban industry. James I issued a series of proclamations against the pollution caused by London starch-makers; the inhabitants of St Botolph's, Aldgate, complained in 1627 that the fumes from the alum factory at St Katherine's by the Tower were poisoning the inhabitants and that the waste matter was killing the fish in the Thames; a few years later Archbishop Laud harried a Westminster brewer for polluting the London air.[10] In 1657 there was a parliamentary debate on the smell given off by the London brick kilns. In the reign of George II the Duke of Chandos, staying in his new house in Cavendish Square, found himself 'poisoned with the brick kilns and the other abominate smells which infect these parts'.[11]

Overcrowding made London notoriously unhealthy, but many other towns were little better. In 1608 visitors to Sheffield expected to be 'half choked with town smoke', while in 1725 a traveller to Newcastle found that 'the perpetual clouds of smoke hovering in the air make every thing look black as at London'.[12] Even in Oxford the air was so bad that an eighteenth-century antiquary estimated that the Arundel Marbles had 'suffered more in seventy or eighty years there than in perhaps two thousand in the countries from whence they were first fetched'.[13] Inevitably there was more plague in the towns than in the country and a higher level of mortality.[14]

'Immers'd in smoke, stunn'd with perpetual noise',[15] it is no wonder that town-dwellers came to pine for the imagined delights of rural life.

Visitors to London soon found themselves beginning to cough; and many chronic invalids, like John Locke, chose to avoid the city altogether for the sake of their lungs. Sir William Temple was 'sensible extremely to good air and good smells, which gave him so great an aversion to the Town that he once passed five years at Sheen without seeing it'. Even King William III chose to live at Kensington for reasons of health.

> Who, that has reason, and his smell,
> Would not among roses and jasmine dwell,
> Rather than all his spirits choke
> With exhalations of dirt and smoke?[16]

asked Abraham Cowley. It was prolonged exposure to what Drayton called 'the loathsome airs of smoky, citied towns', which enhanced the desire for the sunlight and 'fresh air' of the countryside.[17]

Yet in contemporary thought the objection to urban life was less to the physical environment of the city than to the moral behaviour of its inhabitants. John Norris wrote:

> Their manners are polluted like the air,
> From both unwholesome vapours rise
> And blacken with ungrateful steams the neighbouring skies.

As a speaker remarked in Thomas Starkey's early Tudor *Dialogue between Pole and Lupset*, there was most vice in the towns and most virtue in the countryside.[18] The classical convention that country-dwellers were not just healthier, but morally more admirable than those who lived in the city, was a conspicuous literary theme in English literature of the seventeenth and eighteenth centuries. It was exemplified both by the innocent shepherd of arcadian pastoral and by the sturdy husbandman of Horace's second Epode, living a blameless and independent life in contented obscurity.[19] It had little justification in social fact, for agriculture was the most ruthlessly developed sector of the economy, small husbandmen were declining in number, wage-labour was universal, and the vices of avarice, oppression and hypocrisy were at least as prominent in the countryside as in the town. But since it was in the city that the rural profits were consumed, it was there that one found the most sophisticated society, the latest fashions and the most expensive vices. In the countryside, by contrast, clothes were simpler; and powder and paint were not worn. Moreover, life in the country lacked that anonymity which made the city a better setting for clandestine intrigue. There were 'more frequent fornications and adulteries' in

London than in the country, thought the political economist Charles Davenant.[20] In 1692 several country M.P.s even opposed a parliamentary bill against hawkers, on the grounds that if pedlars no longer came to the door, then a country gentleman's servants would have to be sent to do the shopping in the town, where they would be sure to learn debauchery.[21]

In part, therefore, the appeal of the country was negative. It offered an escape from urban vices and affectations, a rest from the strains of business, and a refuge from the dirt, smoke and noise of the city. Yet most of the gentry had more positive reasons for living in the countryside. For it was on their agricultural estates that their wealth and prestige had always rested. In the reign of Henry VIII Thomas Starkey lamented that it was impossible to persuade them to make their chief residence in town and deplored the 'great rudeness and barbarous custom' of dwelling continuously in the country. Similarly in 1579 another author observed that, whereas in some foreign countries gentlemen inhabited 'the cities and chief towns', 'our English manner' was for them 'to make most abode in their country houses'.[22] In fact, the aristocracy by Elizabethan times was tending to spend much of the year in London or in the larger provincial towns. The introduction of private coaches in the later sixteenth century made it easier to travel to the city; and under the early Stuarts many of the upper gentry habitually wintered in the town, despite attempts by the government to make them return home to their locality.[23] But they always retreated to the country in summer. The duration of the London season would vary considerably over the next two centuries, but fashionable society was never in town for the whole of the year.[24] The habit of living in the country, claimed an essayist in 1620, 'hath been more familiar with us than other nations; so that we have in a kind appropriated it to ourselves'.[25] It is true that the country houses to which the aristocracy retreated were not rural cottages, but splendid mansions, designed to bring urban civilization to country surroundings. Nevertheless, they provided a base for a distinctly 'countrified' style of life, intermingled with a certain amount of politics and administration. 'In this island,' wrote William Blane in 1788, 'from the nature of our government, no man can be of consequence without spending a large portion of his time in the country.'[26]

Meanwhile, for other town-dwellers the countryside was coming increasingly to be seen as the place for relaxation and refreshment. Even in the twelfth century it had been customary for the rich citizens of large towns to hold rural property nearby; and in the later Middle Ages the idea of a 'summer house' in the country became increasingly familiar to prosperous town-dwellers.[27] By the reign of James I half the aldermanic bench of Gloucester possessed houses in the adjoining countryside, while in Norwich it was exceptional for urban magistrates not to own property outside the city. In Tudor London the building of 'summer houses' or garden pavilions in the rural suburbs and adjacent villages became popular among the well-to-do.[28] In due course, many of the richer London merchants even chose in summertime to commute from a country retreat. ''Tis very frequent for [tradesmen] to place their families here,' wrote Defoe of Epsom, 'and take their horses every morning to London, to the Exchange, to the Alley or to the warehouse and be at Epsom again at night.' Merchants in Bristol, Hull and other large towns followed a similar pattern.[29] A rural haven was healthier and quieter; and it afforded more room for gardens and orchards.

Even those who lived over the shop in the week might avail themselves of that early modern invention, the country weekend. In 1667 Samuel Pepys and his wife anticipated many later domestic discussions when they decided not to buy a house in the country, because it would involve extra responsibility and tie them down to a particular spot; instead they would buy a coach and go to different places each weekend. Ultimately, they joined with another couple and rented a country villa at Parson's Green. One of the reasons the Fire of London did so much damage in 1666 was that it broke out in the early hours of a Sunday morning, when most of the chief merchants were away for the weekend.[30] In 1748 a Swedish visitor noted that among the market gardens between Fulham and Chelsea were scattered large brick houses belonging to London gentlemen who went there on Saturday afternoons for the sake of some country air; and in 1754 an essay in the *Connoisseur* made fun of the little weekend villas in Turnham Green or Kentish Town to which London tradesmen would retire with their families for 'the end and the beginning of every week', even though it took most of Saturday to pack up food and clothes for the journey and most of Monday was spent 'unpinning, uncording, locking up foul linen, and replacing empty bottles in the cellar'. 'Even citizens who breathe the smoke of London five days in the week throughout the year,' observed Arthur Young in 1770, 'are farmers the other two.'[31]

By this time town-dwellers had started to idealize the country cottage, with its thatched roof, curling smoke and roses round the door: what Uvedale Price described as 'one of the most tranquil and soothing of all rural objects'. In 1772 Queen Charlotte built her cottage in the woods at Kew; and by the end of the eighteenth century many 'people of fortune' would condescend to pass an occasional weekend in an 'ornamental cottage', usually built for the purpose and equipped to a degree of luxury wholly unfamiliar to the ordinary country-dweller.[32]

Those too poor to afford the weekend cottage still looked to the country for occasional refreshment. John Stow described how on May Day Elizabethan Londoners 'would walk into the sweet meadows and green woods, there to rejoice their spirits with the beauty and savour of sweet flowers, and with the harmony of birds'. It was partly the desire to keep adjacent fields for recreation which underlay the repeated attempts to prevent fresh building in the London suburbs.[33] Country jaunts or 'rambles' were a common form of relaxation in the seventeenth century; and there were many town ladies like Mrs Turner, who in July 1667 went out with friends on to Epsom Downs and, reports Pepys, 'did gather one of the prettiest nosegays that ever I saw in my life'.[34] By the mid eighteenth century the keepers of the inns, beer-shops and lodging-houses in Hampstead, Chelsea and other villages on the fringe of London could sustain a thriving business by catering for the swarms of weekend trippers from the city.[35]

Even religion played its part in forming this new taste for country life. 'After the 1640s,' writes a literary historian, 'rural retirement no more was a mere defence mechanism against a corrupt world; it was an open gateway to paradise before the Fall.' The countryside was portrayed as a holier place than the town; and much of the devotional literature of the ensuing century exhibited what the poet John Clare would call 'the religion of the fields'. When on a country walk, declared the young poet Henry Needler, 'my thoughts naturally take a solemn and religious turn'. Fields and groves, agreed the Platonist Peter Sterry, naturally awakened a sense of the divine. Even the matter-of-fact Bulstrode Whitelocke cited 'him who the Popish authors call St Francis' in support of the view that every leaf, plant and briar was a book of God declaring his power and goodness.[36]

Long before William Cowper and with as little justification,* many seventeenth-century writers said explicitly that God made the country, man made the town. The inhabitants of cities and towns, thought a Jacobean preacher, 'see for the most part but the works of men ...

* 'The landscape of Olney,' it has been justly remarked, 'like the English countryside in general, is as artificial as any urban scene'; H. C. Darby, 'On the Relations of Geography and History', *Trans. Institute of British Geographers*, 19 (1953), 6.

whereas such as are conversant in the fields and woods continually contemplate the works of God'. The Quaker William Penn preferred the country life, 'for there we see the works of God; but in cities little else but the works of men'. 'The country is so lovely,' D. H. Lawrence would write in 1928, 'the man-made England is so vile.'[37]

It was, of course, the growth of a sharp division between town and country, sharper than anything to be found in the Middle Ages, which encouraged this sentimental longing for rural pleasures and the idealization of the spiritual and aesthetic charms of the countryside. Those who were most entranced by rural scenes were sophisticated city-dwellers, like Queen Henrietta Maria, who dallied at Wellingborough in 1628 because she liked the countryside and was amused by the dances of the peasantry; or Samuel Pepys, who in 1667 recorded his fascination on meeting an authentic country shepherd and his boy on the downs near Epsom, 'in his woollen knit stockings of two colours mixed, and ... his shoes shod with iron shoes, both at the toe and heels, and with great nails in the soles of his feet, which was mighty pretty'. As the eighteenth-century critic Hugh Blair would observe, a taste for pastoral depended upon the prior growth of towns, for men did not pine for the countryside so long as they lived on terms of daily familiarity with it.[38] It was no accident that it was in Renaissance Italy that the taste first emerged for *villeggiatura*, retirement to an elegant country villa during the summer season,[39] for it was there that town life was earliest developed. In early modern England the yearning for the countryside was intensified by the enormous growth of London. But it also drew strength from what has been called the 'de-ruralization' of the towns:[40] the shrinking of gardens and orchards, the disappearance of trees and flowers and the increasing density of building in response to the mounting pressure of population.

Of course, the growing tendency to disparage urban life and to look to the countryside as a symbol of innocence rested on a series of illusions. It involved that wholly false view of rural social relationships which underlies all pastoral. The idealized shepherds of the literary idylls so popular in the early seventeenth century bore no relationship to the wage-labourers of Stuart England. As the fashionable dramatist John Fletcher explained, his pastoral play, *The Faithful Shepherdess* (pre 1611; acted at Court in 1633), was not about 'country-hired shepherds, in grey cloaks, with cur-tailed dogs'; on the contrary, the shepherds of pastoral were 'such as all the ancient poets, and modern, of understanding, have received them: that is, the owners of flocks, and not hirelings'.[41] The social inequality of the English countryside meant that arcadia had vanished (if it had ever existed). Even Horace's ideal of the self-sufficient husbandman was wholly unrealistic: as John Evelyn observed, he could never remain 'in that desirable state, without the active lives of others to protect him from rapine, feed and supply him with bread, clothes and decent necessities'. Country-dwellers were not more innocent than townsmen. They were not even more religious; for, as John Beale remarked in 1657,

> whereas the rural life should in all reason be the most humble, tame and innocent: yet daily experience showeth that where any trade of manufacture is driven on, there the word of God bears a price: where trade thrives not, there the word of God is at the best but as a pleasant song: if sometimes they hear it, yet seldom they obey it.[42]

The poets and artists who fed the new rural longings preferred to conceal such harsh realities. Most of them depicted the countryside as free from social tension; they ignored the gentry's economic reasons for being there and they manifested an extreme reluctance to mention the practical aspects of rural life.[43] *

The cult of the countryside was, therefore, in many ways a mystification and an evasion of reality. It did not even necessarily indicate a genuine desire to live in the country, for, as has been justly said, much of the writing about Horace's happy husbandman was 'the conventional patter of poets who would not have been caught an hour away from town, unless a patron's invitation to his country estate made that absence politic, or a creditor's dunning made it imperative.'[44] It certainly did not stop more and more people from moving to the towns. The pleasures, the vitality and the economic opportunities of metropolitan life were irresistible. Indeed it was because the Georgian upper classes were so addicted to London and Bath that they wrote so much about the virtues of the country. They seldom sang the praise of cities, for it was unnecessary to do so.[45]

Much celebration of the countryside, moreover, emanated from those

* A recent critic comments on the ideal rural landscape depicted in English poetry of the mid seventeenth century that 'there is virtually no mention of land-clearance, tree-felling, pruning, chopping, digging, hoeing, weeding, branding, gelding, slaughtering, salting, tanning, brewing, boiling, smelting, forging, milling, thatching, fencing and hurdle-making, hedging, road-mending and haulage. Almost everything which anybody *does* in the countryside is taboo'; James Turner, *The Politics of Landscape* (Oxford, 1979), 165.

whom political failure had driven from the city against their will. This explains the enormous vogue in the 1650s of Walton's *Compleat Angler* and similar literature, for the defeated Royalist gentry sensibly made a virtue of necessity by extolling the merits of country life. 'In the twenty-four years of the reign of Charles I,' it seems, 'more poems were written about the happiness of a retired country life than in the sixty-seven years of Elizabeth and James I put together'; and the peak was attained between 1645 and 1655.[46] With the return of Charles II in 1660, the attractions of a rural existence became less compelling for Royalists; but they still appealed to other unsuccessful politicians and disappointed careerists. Many of the best-known rural idylls of the seventeenth century were compensatory myths, composed by or for the sake of such disconsolate figures: Thomas, Lord Fairfax, whose self-imposed exile from politics in the 1650s at Nun Appleton inspired Andrew Marvell; or Bulstrode Whitelocke, who, after escaping punishment at the Restoration, retired to Chilton Park in Wiltshire, where he wrote reflections on the superiority of rural life; or Sir William Temple, who retreated to Moor Park, Surrey, after being struck off the list of privy councillors in 1681 and wrote his essay 'Upon the Gardens of Epicurus'; or the poetess Anne Finch, Countess of Winchilsea, who, having retired with her husband to the countryside because they were unable to reconcile themselves to the Revolution of 1688, composed poems celebrating the virtue of contented obscurity.[47] As Shelley would write, 'in solitude, or that deserted state when we are surrounded by human beings and yet they sympathise not with us, we love the flowers, the grass, the waters and the sky'.[48]

Yet those who went to the country of their own accord often found that a weekend was long enough. 'That brutal state called a country life', as the third Earl of Shaftesbury termed it,[49] was too boring for urban sophisticates. 'Those of the best condition, who have been constantly used to much converse,' observed the fourth Lord North, soon found the solitude of the country abhorrent. When the young John Locke returned from Oxford to his Somerset home, he was speedily disillusioned: 'I am in the midst of a company of mortals that know nothing but the price of corn and sheep, that can entertain discourse of nothing but fattening of beasts and digging of ground and never thank God for anything but a fruitful year and fat bacon.' There were many others who found that time dragged painfully in the country or who, like the antiquary William Stukeley, gave up living there altogether because they missed the literary conversation of London.[50]

Yet, despite all its falsities, the growing rural sentiment reflected an authentic longing which would steadily increase, both in volume and intensity, with the spread of cities and the growth of industry. This longing was expressed in an unprecedented volume of writing about nature and the countryside. Since its first publication in 1653 *The Compleat Angler* has gone into nearly four hundred editions or separate reissues, while *The Natural History of Selborne* remains one of the steadiest sellers of all time.[51] In their wake follows that long sequence of works, like *Lark Rise to Candleford* or *A Shepherd's Life*, which continue to fuel the rural nostalgia of the town-dweller. As William Hazlitt observed in a memorable essay, 'On the Love of the Country' (1814), one essential ingredient of this nostalgia is that natural objects – trees, flowers, farm animals and birds – are valued for their early associations: they bring back memories of childhood in a way which is more vivid and immediate than any human being ever can; the natural objects, unlike humans, are perceived as classes, not individuals; and a primrose can be instantly recognized as the same primrose one saw as a child, whereas a person cannot.[52]

This nostalgia also drew strength from increasing distaste for the physical appearance of the town. In the seventeenth century those who loved the country did not necessarily hate the town. 'My joys were meadows, fields and towns,' sang Thomas Traherne; and John Ray believed that God took equal delight in the beauties of nature and the work of man 'in adorning the earth with beautiful cities'.[53] But it was increasingly common to maintain that the most beautiful city was the one which had the most rural appearance; and the de-ruralization of the towns led to mounting dissatisfaction with the urban environment. Ebenezer Howard drew on a long tradition when he proclaimed in the 1890s that 'town and country must be married'.[54] The ideals of the garden city and the green belt have proved enduring; indeed the problem of how to combine the social and economic opportunities of the town with the physical environment of the country remains a dominant theme in urban planning.

By the eighteenth century, therefore, a combination of literary fashion and social facts had created a genuine tension between the relentless progress of urbanization and the rural longing to which an increasing number of people were subject. These longings provided a clear indication that there were many who felt that, although the natural world should be tamed, it ought not to be completely dominated and suppressed. This ancient pastoral ideal has survived into the modern industrial world. It can be seen in the rural imagery so often employed to advertise consumer goods; and in the vague desire of so many people to end their days in a country cottage. Sentimental though they are, such feelings reflect the unease generated by the progress of human civilization; and a reluctance to accept the urban and industrial facts of modern life.

ii. CULTIVATION OR WILDERNESS?

The second indication of changing sensibilities was a growing reaction against the relentless advance of cultivation. This represented a striking departure from previous attitudes. To the agricultural propagandists of the sixteenth and seventeenth centuries, untilled heaths, mountains and fens had been a standing reproach. They wanted the bracken, gorse and broom removed; and they cherished the ground which had been painfully 'stubbed or won from wood, bushes, broom or furze'.[1] The old rushy pasture lands should be ploughed up and drained; deer parks were wasteful and there were far too many chases and forests. Hampstead Heath, thought John Houghton in 1681, was a 'barren wilderness', in urgent need of cultivation.[2] In the sixteenth century the emphasis was on tillage: the common law, ruled Sir Edward Coke, gave arable land 'preeminency and precedency before meadows, pastures, woods, mines and all other grounds whatsoever'. In the seventeenth century there was greater appreciation of the value of pastoral farming.[3] In either case agricultural improvement and exploitation were not just economically desirable; they were moral imperatives. God had created land, declared the Elizabethan Sir George Peckham, 'to the end that it should by culture and husbandry yield things necessary for man's life'. He had 'committed the earth to man to be by him cultivated and polished,' agreed Edward Hyde, Earl of Clarendon. The cultivation of soil was a symbol of civilization, whereas 'wild and vacant lands', 'encumbered with bushes [and] briars', were 'like a deformed chaos'. An uncultivated common, thought Timothy Nourse in 1700, was 'the very abstract of degenerated nature'.[4]

All through the early modern period the work went on, pushing cultivation up the hills, reclaiming marshes, draining the fens, converting the heaths into arable. Yet at the end of the seventeenth century Gregory King calculated that out of 39 million acres of land in the country (really 37.3) there were still 10 million acres of heaths, mountains and barren land, plus another 3 million of forests, parks and commons.[5] It was an over-estimate, but it shows how an intelligent contemporary thought that the battle between man and nature was still going on; and it explains why in the eighteenth century the ideology of improvement was so widely dispersed, not just among professional agriculturalists, like Arthur Young, who wanted to bring 'the waste lands of the kingdom into culture' and to 'cover them with turnips, corn and clover, instead of ling, whins and fern,' but also among urban observers, like the writer Mrs Elizabeth Carter, who thought in 1769 that the country was 'disgraced by ... tracts of uncultivated land'.[6]

Those who urged on this activity sometimes appeared indifferent to the aesthetic disadvantages of economic progress. 'Suppose coal in Northampton, Buckingham and Oxfordshire,' cried Walter Blith in 1649, 'what a great benefit to those countries it would be!' Arthur Young similarly lamented the 'monstrous proportion' of waste land in the United Kingdom and thought it a 'scandal to the national policy' that Otmoor (today a haven for bird and plant life) should remain unenclosed.[7] But such men did not normally put utility above beauty. To them a tamed, inhabited and productive landscape *was* beautiful. Theirs was the ancient classical ideal which associated beauty with fertility. In the sixteenth and seventeenth centuries it was always the fruitful and cultivated scenery which travellers admired. Like John Leland, they liked 'marvellous fair meadows', 'good corn ground' and 'goodly gardens, orchards and ponds'.[8] Improvement meant more food and more employment, but its advantages were not exclusively material. 'Beside the excessive profit you will reap,' wrote Sir Richard Weston in 1645, 'imagine what a pleasure it will be to your eyes and scent to see the russet heath turn'd into greenest grass, which doth produce most sweet and pleasant honeysuckles.' By labour and investment, thought Timothy Nourse, men could remove the curse of thorns and briars which had come with the Fall and restore barren heaths to their primitive fertility and beauty; in lush fields of corn, flowering meadows, groaning fruit trees and 'curious groves and walks' would be reflected the 'Restauration of Nature'. Human labour, agreed Thomas Traherne, could restore 'the beauty and order of Eden'. John Norden reported that, at Harrow, 'towards the time of harvest, the men may behold the fields round about so sweetly to address themselves to the sickle and scythe, with such comfortable abundance of all kind of grain, that the husbandman which waiteth for the fruits of his labours cannot but clap his hands for joy to see this vale so to laugh and sing'.[9]

This landscape of cultivation was distinguished by increasingly regular forms. Ploughing had always been a symmetrical business; and all sixteenth-century farmers would have understood William Cobbett's delight in the visual pleasure of a furrow a quarter of a mile long and as straight as if it had been laid with a level.[10] The practice of planting corn or vegetables in straight lines was not just an efficient way of using limited space; it was also a pleasing means of imposing human order on the otherwise disorderly natural world. In the reign of Henry VIII Richard Harris was reported to have planted over a hundred acres with fruit trees at Teynham, Kent, 'so beautifully as they not only stand in most right lines, but seem to be of one sort, shape and fashion, as if they had been drawn through one mould or wrought by one and the same pattern'. A topographer subsequently found that in Kent the cherry orchards and gardens were 'beautifully disposed in straight lines'.[11] Wood plantations, ruled Walter Blith in 1653, could be square, triangular, rectangular, oval or circular; but they should not be made 'rudely and confusedly'; a straight hedgerow was 'a thing of delight'. Herbs and flowers, agreed Stephen Blake in 1664, should be placed 'in uniform ranks'. Symmetry and regularity were essential features of good husbandry and no formation was more admired than that of the quincunx: 'It is a great pleasure and delight,' says the husbandman in Ralph Austen's dialogue on fruit trees (1676), 'to walk among you, so many beautiful fruit trees; seeing ye grow so handsomely and uniform; ye grow in order, in straight lines every way (look which way a man will).'[12]

Neatness, symmetry and formal patterns had always been the distinctively human way of indicating the separation between culture and nature. But the trend towards uniform planting seems, if anything, to have increased in the early modern period. Certainly John Parkinson in 1629 thought that the orchards of his day were planted in more formal patterns than had been common earlier.[13] The movement paralleled the shift in architectural taste from the Gothic to the Classical. For the neo-classical theorists of the later seventeenth century, it was axiomatic that geometrical figures were intrinsically more beautiful than irregular ones. There was no one, save those 'as stupid as the basest of beasts', said Henry More, who would not agree that a cube, a tetrahedron or an icosahedron had more beauty in them 'than any rude broken stone lying in the field or highways'.[14] Roger North declared that order was the essence of beauty, 'as in trees, planting which is done usually in equal spaces, and straight ranges'. John Laurence said the same: 'Beauty requires that the hedges should be in straight lines ... and what will be thus more pleasing to the eye will be cheapest and most convenient; straight lines are the shortest, and Gothic buildings are vastly more expensive than the majestic simplicity of Grecian architecture.'[15] The long, straight, quickset hedges of the eighteenth century contrasted markedly with the straggling irregularity of earlier field patterns. Indeed modern field archaeologists assume that, if the edge of a boundary of a wood is perfectly straight, then it is very likely to be post-1700 in date.[16]

Throughout the eighteenth century and beyond, the improvers continued to praise this regular landscape of opulence and productivity and to deplore the uncultivated waste.[17] 'What painter,' asked John Laurence in 1716, 'can draw a landskip more charming and beautiful to the eye than an old Newington peach-tree laden with fruit in August?' William Cobbett detested the 'rascally heaths' near Marlborough. 'I have,' he wrote, 'no idea of picturesque beauty separate from fertility of soil.' The gardener Samuel Collins spoke for many contemporaries when he said in 1717 that the best of all flowers was a cauliflower; and the affectations of the landscape-gardeners received short shrift from Dr Johnson, who hated talk about prospects and views. 'That was the best garden (he said) which produced most roots and fruits; and that water was most to be prized which contained most fish.'[18] 'The generality of people,' observed William Gilpin in 1791, found wild country in its natural state totally unpleasing: 'there are few who do not prefer the busy scenes of cultivation to the greatest of nature's rough productions. In general indeed, when we meet with a description of a pleasing country, we hear of haycocks, or waving cornfields or labourers at their plough.' Wordsworth agreed: 'In the eye of thousands and tens of thousands,' he lamented, 'a rich meadow, with fat cattle grazing upon it, or the sight of what they would call a heavy crop of corn, is worth all ... the Alps and Pyrenees in their utmost grandeur and beauty.'[19]

In keeping with this attitude, unproductive mountains had traditionally been regarded as physically unattractive. They were supposedly the home of uncivilized people, like the Zapoletes of More's *Utopia*, 'hideous, savage and fierce, dwelling in wild woods and high mountains', or the wild Welshmen of Elizabethan Pembrokeshire, whom, it was reported, 'the rest call the mountain men'.[20] Early modern travellers usually found mountainous country unpleasant and dangerous. William Camden thought Radnor 'hideous after a sort to behold, by reason of the turning and crooked by-ways and craggy mountains' and described Craven as 'rough all over and unpleasant to see to, with craggy stones, hanging rocks and rugged ways'. Celia Fiennes hated the Pennines and was glad to descend from the mountain rain to sunshine and the singing of birds. She thought the Lake District 'desert and barren' and its mountains 'very terrible'.[21] In the 1670s Chief Justice North observed the 'hideous mountains' on his way from Carlisle to Appleby, while in 1697 Ralph Thoresby found both the Border country and the Lake District full of horrors: dreadful fells, hideous wastes, horrid waterfalls, terrible rocks and ghastly precipices. In the same spirit Dr Johnson wrote of the Scottish Highlands that 'an eye accustomed to flowery pastures and waving harvests is astonished and repelled by this wide extent of hopeless sterility'. Infinitely preferable was the softer and more fertile landscape of a county like Northamptonshire, of which John Morton boasted in 1712 that 'here are no naked and craggy rocks, no rugged and unsightly mountains, or vast solitary woods to damp and intercept the view'.[22]

It would be wrong to exaggerate the extent to which people were actually afraid of mountains. The supposed horrors of the Welsh hills did not prevent seventeenth-century botanists from clambering over Snowdonia and Cader Idris in search of specimens. The real objection to mountains like the Alps, thought James Howell in 1621, was less that they were 'high and hideous' than that they were useless, unlike 'our mountains in Wales', which 'bear always something useful to man or beast, some grass at least'.[23] But there is no denying that before 1700 most sophisticated contemporaries found hilly country distasteful and infinitely preferred the tamed and fertile landscape over which man had asserted his control.

Yet, before the end of the eighteenth century, taste had changed dramatically. In place of the clipped and manicured formal garden which had been the old horticultural ideal there had developed a distinctively English style of landscape garden, so informal as at times to be barely distinguishable from an uncultivated field; and, even more remarkably, wild, barren landscape had ceased to be an object of detestation and become instead a source of spiritual renewal. 'What are the scenes of nature that elevate the mind in the highest degree, and produce the sublime sensation?' asked Hugh Blair, lecturing in Edinburgh in the 1760s. 'Not the gay landscape, the flowery field, or the flourishing city; but the hoary mountain, and the solitary lake; the aged forest, and the torrent falling over the rock.'[24] The wilder the scene, the greater its power to inspire emotion. The mountains which in the mid seventeenth century were hated as barren 'deformities', 'warts', 'boils', 'monstrous excrescences', 'rubbish of the earth', 'Nature's *pudenda*', had a century or so later become objects of the highest aesthetic admiration.[25]

This new attitude to wild nature had first become apparent during the course of theological controversy. In his anxiety to refute the view that the earth had degenerated since the Creation, the clergyman George Hakewill was led in 1635 to defend mountains on the pragmatic grounds of their utility and 'pleasing variety'. This defence was developed by his theological successors, anxious to prove that all God's works served a purpose. Mountains, wrote Henry More in 1653, might 'seem but so many wens and unnatural protuberances upon the face of the earth', but when one remembered that without them there would be no rivers, one could hardly deny their utility. In 1681 Thomas Burnet in his *Sacred Theory of the Earth* restated the view that the earth had originally been as smooth as an egg until broken up and 'deformed' by the Deluge. In the ensuing controversy, his opponents denied that the earth was imperfect in its design, pointing out with donnish ingenuity that mountains served sundry indispensable purposes, whether in creating rivers, providing natural frontiers or offering a congenial home for goats.[26]

These justifications of God's design took on an increasingly aesthetic dimension. The 'natural pulchritude' of the earth, declared Burnet's most determined opponent, a Suffolk clergyman, Erasmus Warren, in 1690, was

> made up of such things as Art would call rudenesses; and consists in asymmetries and a wild variety ... That roughness, brokenness and multiform confusion in the surface of the Earth, which to the inadvertent may seem to be nothing but inelegancies or frightful disfigurements, to thinking men would appear to be as the turnings and carvings and ornamental sculptures that make up the lineaments of nature, not to say her braveries.

For a deist like the third Earl of Shaftesbury, not just mountains, but even deserts, had 'their peculiar beauties'. 'The Wildness pleases,' he declared in 1709, 'we seem to live alone with Nature. We view her in her inmost recesses.'[27]

In the later seventeenth century the growth of nature mysticism among the theologians and philosophers was paralleled by the feeling of an increasing minority of contemporaries that mountainous country was enjoyable because it offered the most wholesome air and the best views. The later taste for the sublime was anticipated by a traveller in 1682 who confessed that he felt 'a kind of pleasant horror' at the sight of the Wrekin and the Malvern Hills; and the growing association of mountains with religion is revealed by the Welshman who declared in 1686 that Snowdon was a 'Paradise': 'I am sure 'tis one of the nearest places to Heaven that is in this world.'[28]

During the course of the eighteenth century the rage for mountain scenery took firm possession of the holiday-seeking public. By the 1760s visitors were pouring into the Lake District, the Wye Valley, Snowdonia and the Scottish Highlands, in search of exciting scenic effects. When John Byng climbed Cader Idris in 1784, he was accompanied by a guide who had been taking up tourists for the previous forty years; by 1800 Coleridge could complain that the Lakes were alive and swarming with tourists for a third of the year.[29] The more adventurous went further afield, to Savoy or Switzerland. Those who stayed at home could buy the pictures and prints of mountain scenery which from the middle of the century were produced in increasing numbers.

By the later eighteenth century the appreciation of nature, and particularly wild nature, had been converted into a sort of religious act. Nature was not only beautiful; it was morally healing. The value of the wilderness was not just negative; it did not merely provide a place of privacy, an opportunity for self-examination and private reverie (which was an ancient idea); it had a more positive role, exercising a beneficent spiritual power over man. 'All the noblest convictions and confidences of religion,' declared Archibald Alison, 'may be acquired in the simple school of nature.'[30] The feeling of awe, terror and exultation, once reserved for God, was gradually transposed to the expanded cosmos revealed by the astronomers and to the loftiest objects discovered by explorers on earth: mountains, oceans, deserts and tropical forests. The inhabitants of mountain areas ceased to be universally despised for their barbarism; instead they were praised for their innocence and simplicity. The mountains themselves were no longer repugnant; they had become the highest form of natural beauty and a reminder of God's sublimity.

'The farther I ascend from animated nature, from men and cattle and the common birds of the woods and fields,' wrote Coleridge in 1803 after climbing the Kirkstone Pass in a storm, 'the greater becomes in me the intensity of the feeling of life ... "God is everywhere," I ... exclaimed.' The more spectacular Alpine scenes, thought Sir Richard Colt Hoare in 1786, would 'awe even an atheist into belief'.[31]

This semi-religious devotion to wild landscape was, of course, a European phenomenon, whose prophets included Rousseau and Alexander von Humboldt. But it was the English who went furthest towards what has been called 'the divinisation of nature'.[32] Certainly it was they who created the greatest mystique about mountain-climbing, representing it as a semi-religious activity. In the early nineteenth century it was not the French or the Spaniards who were found in the Pyrenees botanizing and searching for the sublime, but the English; while in Switzerland, after the foundation of the Alpine Club in 1857, it was, in the words of the *Alpine Journal*, notorious that 'if you met a man in the Alps, it was ten to one that he was a University man, eight to one (say) that he was a Cambridge man, and about even betting that he was a fellow of his college.'[33]

Explanations for the rise of the new taste for wild nature have tended to concentrate on the eighteenth-century improvements in communications which made mountains more accessible to town-dwellers and somewhat less dangerous when they got there. Just as improved navigational techniques robbed Scylla and Charybdis of their terrors for eighteenth-century sailors, so easier travel made mountains less forbidding to tourists. It has even been suggested that the appreciation of sublime scenery 'increased in direct ratio to the number of turnpike acts'. Better roads, better horses, more maps and signposts both account for and reflect the growth of tourism. Visitors to the Lakes multiplied after 1763, when the first coach travelled over Shap Fell; and from 1773 there was a regular coach service from London to Carlisle.[34]

But improvements in transport do not really explain the taste for wilderness as such, any more than do new techniques of aquatinting and lithography explain the demand for books with picturesque views. After all, climbing in the Alps did not cease to be dangerous once the Swiss had built railways. Greater comfort in ordinary life made occasional bouts of hardship positively attractive to the middle classes when they were on holiday; a certain degree of danger was part of the appeal.

A much more likely explanation, both for the new taste for wild landscape and for the growth of more informal gardening styles, is provided by the progress of English agriculture. During the eighteenth century a further 2 million acres of land were brought into regular cultivation as arable or pasture; and between 1760 and 1820 alone 2½ million acres of the ground already farmed were carved up by parliamentary enclosure into regular fields. In 'The Excursion' (published 1814), Wordsworth mused:

> Wheresoe'er the traveller turns his steps,
> He sees the barren wilderness erased,
> Or disappearing.[35]

To the agricultural improvers these changes were pure gain. Cobbett, for example, thought the old common fields 'very ugly things' and he praised the new, 'neat' landscape of quickset enclosure. But to the lovers of the picturesque, 'all the formalities of hedgerow trees and square divisions of property' were, in William Gilpin's words, 'disgusting in a high degree'.[36] It was in self-conscious reaction to this new agricultural pattern that, from the beginning of the eighteenth century onwards, those who set the fashion in landscape-gardening opted for increasingly natural forms: curves rather than straight lines, and, by the 1740s, a subtle merging into the surrounding countryside rather than a sharp distinction between the cultivated and the wild.[37] It was no coincidence that it was England which became famous for this 'natural' style, or that landscape-gardening became one of the country's most distinctive cultural achievements. For it was in England that the ordinary countryside most closely approximated to the effect produced by the geometrical gardens of the past; and it was there, accordingly, that the opposite quality of informality made its greatest aesthetic appeal. As a visitor remarked in 1810, the ordinary English countryside was 'too much chequered with enclosures for picturesqueness'.[38]

Of course, the new taste in landscape was shaped by continental models: the gardens of Italy, the poetry of Horace and Virgil, the paintings of Claude, Poussin and Salvator Rosa. But it was English agricultural progress which made these models so seductive. As the pioneer historian of gardening J. C. Loudon observed in the 1830s, 'the modern style of gardening' was 'unsuitable to countries not generally under cultivation'. 'What delight or distinction,' he asked, 'can be produced by the English style in Poland, for example, where the whole country is one forest, and the cultivated spots only so many open glades, with the most irregular and picturesque sylvan boundaries?' His explanation for the emergence of the informal English style was clear-cut:

> As the lands devoted to agriculture in England were, sooner than in any other country in Europe, generally enclosed with hedges and hedgerow trees, so the face of the country in England, sooner than in any other part of Europe, produced an appearance which bore a closer resemblance to country seats laid out in the geometrical style; and, for this reason, an attempt to imitate the irregularity of nature in laying out pleasure-grounds was made in England, with some trifling exceptions, sooner than in any other part of the world; and hence the style became generally known as 'English gardening'.[39]

In 1783 William Marsden had drawn on his years of service in Sumatra with the East India Company to make the same point:

> In highly cultivated countries, such as England, where landed property is all lined out, and bounded and intersected with walls and hedges, we endeavour to give our gardens ... the charm of variety and novelty, by imitating the wildnesses of nature in studied irregularities. Winding walks, hanging woods, craggy rocks, falls of water, are all looked upon as improvements; and the stately avenues, the canals, and lawns of our ancestors, which afforded the beauty of contrast, in ruder times, are now exploded. These different tastes are not merely the effect of caprice, ... but result from the change of circumstances. A man who should attempt to exhibit on Sumatra the modern or irregular style of laying-out grounds would attract but little attention, as the unimproved scenes, adjoining on every side, would probably eclipse his labours. Could he, on the contrary, raise up, amidst these magnificent wilds, one of the antiquated parterres, with its canals and fountains, whose symmetry he has learnt to despise, his work would produce admiration and delight. A pepper garden cultivated in England would not, in point of external appearance, be considered as an object of extraordinary beauty; and would be particularly found fault with for its uniformity; yet, in Sumatra, I never entered one after travelling many miles, as is usually the case, through the woods, that I did not find myself affected with a strong sensation of pleasure.[40]

Formal gardens thus became less fashionable as more of the country was brought under rigorous and symmetrical cultivation; and the taste for wild or mountainous landscape was enhanced by the same development. 'Give me wild barren, haggard mountain scenes in preference to all these close, well-cultivated messuages and manors,' wrote the poet George Darley in 1846. 'Above all plenty of *grey rock*!'[41]

This ability to derive pleasure from scenes of relative desolation represented a major change in human perception. Inevitably, it was more likely to be found among those who by virtue of their social and economic position could contemplate with equanimity the prospect of leaving land uncultivated which might otherwise have produced food. Only when the threat of starvation receded could such an attitude prevail. It is not surprising that in a poor country like Scotland the inhabitants were said, when it came to landscape-gardening, to be in 1790 'at least half a century behind the English'.[42] * Even in England, the taste for the wild and the irregular was much more likely to seduce the well-to-do than the poor, struggling for subsistence, or the agriculturalists, still battling with the land. As Archibald Alison observed in 1790, the common people universally followed the older, more formal style of gardening, while 'even the men of the best taste,' when cultivating waste or neglected lands, still enclosed them in uniform lines and regular divisions, 'as more immediately signifying what they wish should be signified, their industry or spirit in their improvement'.[43]

Those still engaged in the struggle to wrest a living from the land were also understandably reluctant to adopt a mystical attitude towards wild, uncultivated scenery. The man who lived permanently in 'romantic' countryside, noted the same observer, tended to regard it in a very different light from that in which it was seen by the cultivated tourist who came only on a brief visit. To the latter, the streams were known 'only by their gentleness or their majesty, the woods by their solemnity, the rocks by their awfulness or terror'. But to the former

> they serve as distinctions of different properties, or of different divisions of the country. They become boundaries or landmarks, by which his knowledge of the neighbourhood is ascertained . . . Even a circumstance so trifling as the assignation of particular names contributes in a great degree to produce this effect, because the use of such names in marking the particular situation or place of such objects naturally leads him to consider the objects themselves in no other light than that of their place or situation. It is with very different feelings that he must now regard the objects that were once so full of beauty. They now occur to his mind, only as topographical distinctions, and are beheld with the indifference such qualities naturally

* This was something of an exaggeration; but, as a recent authority remarks, 'the early informal gardens in Scotland were decisively shaped by English tastes and men'; A. A. Tait, *The Landscape Garden in Scotland 1735–1835* (Edinburgh, 1980), 3.

produce. Their majesty, their solemnity, their terror, etc., are gradually obscured ... and ... he must be content at last to pass his life without any perception of their beauty.[44]

The new taste for wild nature was therefore not an intuitive affair. Just as the appreciation of the English landscape garden in the early eighteenth century required a classical education and some knowledge of history and literature, necessary to catch all the references to Horace and Virgil or the allusions to Poussin and Claude, so the attraction to unimproved nature was initially a sophisticated business, reflecting the highly literary and intellectual inspiration of the new sensibilities. Of course, most people, however uneducated, had from time immemorial been spontaneously attracted by large views and open prospects. John Constable was surely right when he said that 'there has never been an age, however rude or uncultivated, in which the love of landscape has not in some way been manifested.'[45] But the self-conscious appreciation of rural scenery which developed so spectacularly during the eighteenth century was a different matter, for it depended upon prior acquaintance with the tradition of European painting. The initial appeal of rural scenery was that it reminded the spectator of landscape pictures. Indeed the scene was only called a 'landscape' because it was reminiscent of a painted 'landskip'; it was 'picturesque' because it looked like a picture. The circulation of topographical art in which human figures were absent or unimportant thus preceded the appreciation of rural landscapes and determined the form it took.[46] When Edward Waterhouse praised the English countryside in 1663 he said that it had 'pleasantness of situation in the landskip of it, having woods, rivers, springs, meadows interwoven'; and when Celia Fiennes visited Epsom thirty years later she remarked that the prospect showed 'the country like a landskip, [with] woods, plains, enclosures and great ponds'. Others admired British mountain scenery because they saw in it a vague approximation to the bizarre rock backgrounds of late medieval painting or to the wild landscapes of Salvator Rosa. It could hardly be denied that mountains were 'pleasant objects to behold,' thought John Ray, when 'the very images of them, their draughts and lanskips are so much esteemed'.[47]

Since at least the 1680s there had been an established market in prints of 'landskips' to hang on the walls of middle-class houses. At first most of them were Netherlandish or Italian, but during the course of the eighteenth century it was English scenery which became the object of increasing artistic attention; and the reign of George III saw an unparalleled level of achievement in English landscape art. By the 1780s there was a torrent of published tours and guides to the beauties of England, embellished by aquatints of picturesque views from 1775 and steel engravings from 1810.[48] These artistic representations, whether English or foreign, shaped the taste of the educated classes. It was the reproduction of Paul Sandby's water-colours of 1747–52 which would make the Falls of Clyde a popular tourist attraction in the 1790s; and the initial appeal of the Lake District was only faintly parodied by Thomas West's guide of 1778, which led the tourist from the 'delicate touches of Claude' at Coniston, past 'the noble scenes of Poussin' at Windermere, to the 'stupendous romantic ideas of Salvator Rosa' at Derwent Water.[49] Even Gilbert White, whose direct, unstylized appreciation of the natural world is so remarkable, could not free himself from the influence of previous artistic models. In rural Hampshire he could be moved by 'Italian skies' or by 'a lovely picturesque scape' or a scene 'worthy the pencil of a Rubens'.[50]

By the early nineteenth century the taste for wild nature had far transcended this earlier dependence on prior artistic models, just as it had exceeded the bounds of the most 'informal' landscape garden. For the Romantics, nature 'improved' was nature destroyed. 'A gentleman's park,' wrote Constable in 1822, 'is my aversion. It is not beauty because it is not nature.'[51] 'Picturesque travel' was equally suspect. Just as the landscape-gardeners sought to collect together all natural beauties and to shut out everything unpleasing or inharmonious, so the picturesque travellers looked to nature only for conformity to a preconceived pattern or accepted model of aesthetic harmony. Usually they were disappointed, for, as Gilpin remarked, it was very seldom that 'a purely natural scene' was 'correctly picturesque'. There was always a 'rudeness' in the works of nature; she never produced 'a polished gem'.[52] Even Gainsborough confessed that English landscapes usually failed to measure up to artistic ideals: 'With regard to *real views* from Nature in this country, he has never seen any place that affords a subject equal to the poorest imitations of Gasper or Claude.'[53] Gilpin had no hesitation in pronouncing that nearly all actual mountains, lakes or waterfalls exhibited 'deformities' which 'a practised eye would wish to correct'.[54]* As Wordsworth would remark, the habit of comparison thus served only to obscure 'the spirit of the place'. For him, as for

* For that matter, the ruins of Tintern Abbey could do with some improvement: 'a mallet judiciously used (but who durst use it?) ...'; William Gilpin, *Observations on the River Wye* (2nd edn, 1789), 47.

Constable or Clare, there could be no 'improvers'; nature had no deformities and was impossible to improve. It was the unchecked spread of human cultivation which was the real threat.[55]

Yet, as Wordsworth recognized, many people were prevented by the defects of their education or social situation from viewing nature as he did. Much of his opposition in 1844 to the proposed Kendal–Windermere railway, which, in his view, threatened to flood the Lake District with what he called 'the whole of Lancashire and no small part of Yorkshire', reflected the same assumption that there were social differences in perception. A feeling for romantic scenery was not inherent in mankind, he urged. It took a long course of aesthetic education to instil a taste for barren rocks and mountains. The urban lower classes could derive no good from immediate access to the Lakes. What they needed was a preparatory course, starting with Sunday excursions into nearby fields.[56]

By the end of the eighteenth century, therefore, the old preference for a cultivated and man-dominated landscape had been decisively challenged. Encouraged by the ease of travel and by immunity from direct involvement in the agricultural process, the educated classes had come to attach an unprecedented importance to the contemplation of landscape and the appreciation of rural scenery. 'Within the last thirty years,' wrote Southey in 1807, 'a taste for the picturesque has sprung up; and a course of summer travelling is now looked upon to be ... essential ... While one of the flocks of fashion migrates to the sea-coast, another flies off to the mountains of Wales, to the lakes in the northern provinces, or to Scotland; some to mineralogize, some to botanize, some to take views of the country, – all to study the picturesque, a new science for which a new language has been formed, and for which the English have discovered a new sense in themselves, which assuredly was not possessed by their fathers.'[57] What was notable about this new taste was that the scenery which was most particularly admired was no longer the fertile, productive landscape, but the wild and romantic one. Henceforth there would be a growing concern to preserve uncultivated nature as an indispensable spiritual resource.

That concern had many ingredients: an aesthetic reaction against the regularity and uniformity of English agriculture; a dislike for the artificialities of the gardening movement; a feeling that wilderness, by its very contrast with cultivation, was necessary to give meaning and definition to the human enterprise; a preoccupation with the freedom of open spaces as a symbol of human freedom ('A wilderness is rich with liberty,' thought Wordsworth); and an element of alienation or lack of sympathy for the dominant trends of the age; for whether we think of the early Christian hermits or the medieval Cistercians or of Jean-Jacques Rousseau,* the pull of wild nature can always be recognized as an essentially anti-social emotion.[58]

Perhaps the growth of population helped to foster that anti-social feeling. For in previous, less populated centuries, it had been conventional to regard loneliness as a human misfortune: 'To man by nature,' Thomas Hobbes would write, 'solitude is an enemy.'[59] Only the religious contemplative sought the desert. But in the Elizabethan age the humanist cult of the individual fostered the idea that temporary withdrawal from society could be positively pleasurable. In addition the spiritual desirability of periodically being alone was urged by many Protestant divines in the century after the Reformation. It became an increasingly conspicuous poetic theme from the mid seventeenth century onwards; and in the late eighteenth century it was given wide circulation by the writings of Rousseau and of the German author J. G. Zimmermann, whose meditations on *Solitude* enjoyed an enormous vogue in translation in the England of the 1790s.[60] By the nineteenth century wild scenery was cherished because it provided an escape from the increasing bustle of the cities and the factories. It is revealing that when Queen Victoria was deeply moved by natural scenery, it was always the *solitude* of the place which she singled out as its greatest feature.[61] In 1848 John Stuart Mill would ground his case for a stationary population on the importance of preserving at least some areas where men could still be by themselves. 'Solitude, in the sense of being often alone,' he declared, was indispensable for human fulfilment. It was

> essential to any depth of meditation or of character ... Solitude in the presence of natural beauty and grandeur is the cradle of thoughts and aspirations which are not only good for the individual, but which society could ill do without.

'Nor,' he added

> is there much satisfaction in contemplating the world with nothing left to the spontaneous activity of nature; with every foot of land brought into cultivation, which is capable of growing food for human beings; every flowery waste or natural pasture ploughed up, all quadrupeds or birds which are not domesticated for man's use

* Or indeed of G. M. Trevelyan.

exterminated as his rivals for food, every hedgerow or superfluous tree rooted out, and scarcely a place left where a wild shrub or flower could grow without being eradicated as a weed in the name of improved agriculture.[62]

It was this recurring urge of town-dwellers to turn to the wild for spiritual regeneration which lay behind the subsequent movement to preserve mountain scenery or tracts of waste and moorland before they were all swallowed up by human progress. Reservations of scenery, thought the American Charles Eliot in 1896, had become 'the cathedrals of the modern world'.[63]

iii. CONQUEST OR CONSERVATION?

Mill's reference to wild flowers brings us to another instance of the changing sensibilities which were leading to a revaluation of the natural world. The gardeners had always made a sharp distinction between cultivated blooms, which they cherished, and 'wild' flowers, which they despised. The herbalist John Parkinson, for example, ruled in 1629 that the scabious was not a flower 'of beauty or respect' and should therefore be left in the fields. Single marsh marigolds belonged only in ditches, but double ones could be brought into the garden. The primrose, which grew under every hedge, should be left to wild habitation, 'being not so fit for a garden'. Sometimes, individual species were downgraded and transferred from one category to another: borage and bugloss, explained Parkinson, had once been grown in 'gardens of pleasure' and their flowers copied in ladies' needlework; but now they belonged only to the kitchen garden.[1]

The farmers had drawn equally rigid distinctions between 'crops', which were to be cultivated, and 'weeds', to be ruthlessly exterminated. To the agriculturalist a weed was an obscenity, the vegetable equivalent of vermin. In the Gloucester dialect, for example, the word 'filthy' applied equally to a man with lice on his body and a field full of weeds. In forestry a 'weed' tree was an undesirable survivor of natural woodland.[2] For the Elizabethans, 'the darnel, hemlock and rank fumitory' were 'savagery' which the plough should deracinate; 'hateful docks, rough thistles, kecksies, burrs' had neither beauty nor utility. Later agricultural improvers, like Walter Blith, hated gorse, ferns, rushes, bracken, broom and all other 'such filth'.[3] In the eighteenth century the agricultural writer William Ellis denounced not just charlock, wild sorrel ('this ugly weed'), darnel ('a rampant weed'), coltsfoot ('most pernicious'), black bennet, crow needle, thistle, hemlock and cow garlic ('the devil of a weed'), but also wild marigold, wild iris, honeysuckle and water lilies. 'Weeds,' ruled the late-seventeenth-century aesthete Roger North, 'have no beauty.'[4] Even today there are few farmers who are cheered by the sight of poppies in the corn.*

But town-dwellers, with the encouragement of artists, naturalists and poets, were coming to regard many of these despised or hated plants as beautiful. Seventeenth-century Londoners sought willow herb, foxgloves and poppies to decorate their houses and constantly scrutinized the wild for plants worth importing to town gardens. In 1657 a herbalist noted that some gardeners loved 'to feast themselves even with the varieties of those things which the vulgar call *weeds*; and indeed [he added] there is a great deal of prettiness in every one of them if they be narrowly observed'. In rural Northamptonshire, goodwife Cantrey's garden in the 1650s included scabious, campion and larkspur.[5] To later Stuart naturalists, like Robert Sharrock in 1660, even the great horsetail, a con-

* This hierarchy of plants closely paralleled that of human society. In 1700 Timothy Nourse thought that the common people should be 'looked upon as trashy weeds or nettles'; while in 1838 the gardening expert J. C. Loudon explained that 'to compare plants with men, we consider aboriginal species as mere savages, and botanical species ... as civilised beings'; Timothy Nourse, *Campania Foelix* (1700), 16; J. C. Loudon, *Arboretum et Fruticetum Britannicum* (1838), i. 216.

temptible plant found in bogs and ditches, was beautiful in construction. Seventeenth-century artists, like Henry Peacham and Richard Waller, spent time making careful water-colours of wild flowers and grasses.[6]

Apothecaries had always believed that many neglected wild plants were valuable medically ('Weeds or grass,' lamented William Turner, were the names which the ignorant gave to 'precious herbs').[7] From the sixteenth century onwards, botanists began to record the location of wild flowers. The first published local flora was the catalogue made by Thomas Johnson in 1632 of the plants growing in Kent and on Hampstead Heath; he followed it with an account of his plant-hunting expeditions in the south of England. In 1650 William How's *Phytologia Britannica* was the first attempt at a complete British flora. Even in Elizabethan times there had been many gentlemen and apothecaries who had noted where wild plants were to be found. William Mount had worked on Kentish flora in the early 1580s and Richard Shanne of Methley (fl. 1577–1617) had studied the distribution of plants in the north of England. Thomas Johnson's edition in 1633 of Gerard's herbal reveals the existence of many apothecaries and amateur botanists who engaged in the search for rare plants. By the end of the seventeenth century there was an informal botanical club which met at the Temple Coffee House in London.[8]

The motives for the early plant-hunting expeditions had been practical: the aim was to record herbs of medicinal utility and bring them back to be grown in physic gardens. But by the later seventeenth century naturalists had developed an interest in plants for their own sake. In the Hanoverian age botany was established as a familiar pastime of middle-class ladies and gentlemen. Equipped with a pocket guide to the Linnaean classification and a portable press for drying the plants, they roamed the fields and woods in search of new discoveries. In the later eighteenth century local floras multiplied: by 1800 there were at least four separate floras of Cambridgeshire alone; and by the 1850s most parts of England had had their wild flowers carefully listed by some devoted local naturalist. In 1788–9 James Bolton, a self-taught weaver, devoted three whole volumes to the fungi growing around Halifax, 'the result of more than twenty years' observation'. 'Botanists,' commented Samuel Pegge in 1796, 'allow nothing to be weeds.'[9] Meanwhile the expansion of British colonial influence had been accompanied by enormous interest in tropical plants, which were dried, sent home and assembled in private herbaria. When Sir Hans Sloane amassed his vast accumulation in the early eighteenth century (now known as the Sloane Herbarium and preserved in the Natural History Department of the

British Museum), he did it by buying up the collections of scores of individuals: apothecaries, merchants, sea captains, ships' surgeons, missionaries and foreign naturalists.[10]

To these new sensibilities even the so-called 'weeds' were beautiful. Gorse was the enemy of every improver, but the story was that when Linnaeus (others say Dillenius) came to England and saw gorse for the first time he fell on his knees to give thanks for so beautiful a plant.[11] The eighteenth-century gardening expert William Hanbury thought heather very elegant and looked kindly upon meadow-sweet and even thistles. The agricultural writer William Marshall described blackberry flowers as 'beautiful beyond expression'; and the royal gardens at Richmond were noted for their 'rude, cultivated' tract of furze and broom. William Cowper celebrated the beauty of a common overgrown with fern and gorse, while John Clare devoted many poems to the beauty of plants which the farmers hated: ragwort, yarrow, rushes, spear thistle, poppies in the corn.[12] In the 1830s J. C. Loudon thought that the briar, sloe thorn, fern and bramble 'would, if introduced into the picturesque grounds of a residence, have a most enchanting effect'. Ferns were a reliable sign of poor land, but James Bolton in 1785 considered no 'tribe of plants so singular and beautiful'. Middle-class Victorians found them delightful and filled every nook and cranny of their houses with them during the great fern craze, which reached its peak in the mid 1850s. 'What effect,' asked one propagandist, 'could be more pleasing ... to wearied town dwellers than the sight of graceful fern-fronds *everywhere* surrounding them in and about their houses?'[13]

Just as Shakespeare's Perdita had dismissed 'streak'd gillyvors' as 'nature's bastards', so, in reaction to the competitive cultivation of prize blooms, the Romantics preferred the common wild flowers, which in Ruskin's words had never been 'provoked to glare into any gigantic impudence at a flower show'. To Ruskin a flower-garden was

> an ugly thing, even when best managed: it is an assembly of unfortunate beings, pampered and bloated above their natural size, stewed and heated into diseased growth; corrupted by evil communication into speckled and inharmonious colours; torn from the soil which they loved, and of which they were the spirit and the glory, to glare away their term of tormented life among the mixed and incongruous essences of each other, in earth that they know not and in air that is poison to them.[14] *

* To the reformer H. S. Salt a garden was merely 'a zoo with the cruelty omitted'; *The Call of the Wildflower* (1922), 9.

In the late 1830s the young Tennyson had expressed a similar dislike for hothouse plants, echoing his predecessors Gray and Wordsworth:

> Better to me the meanest weed
> That blows upon its mountain,
> The vilest herb that runs to seed
> Beside its native fountain.

'Long live the weeds,' wrote Gerard Manley Hopkins. From 1888 onwards local councils began to bring in by-laws for the protection of wild plants.[15]

If weeds now had their friends, the same was true of the wild animals and birds against whom earlier generations had battled in their struggle for subsistence. Here, too, new security was the essential precondition for greater tolerance. Already at the beginning of the early modern period England was distinctive among European countries because she had no wolves.* This was a matter of some importance and the occasion of much self-congratulation. It made English sheep-farming less labour-intensive, for shepherds no longer had to guard their flocks by night, as in the days of Aelfric or of Walter of Henley, or lock them up in stone sheepcotes; and it explains why in post-medieval England a shepherd usually drove his sheep in front of him, whereas in France or Italy, where the wolf survived until the nineteenth century, the sheep followed behind and the shepherd, with a mastiff or wolfhound rather than a sheepdog, went in front as their protector.[16] Wolves also lingered in Ireland. When in the late seventeenth century John Dunton spent a night in County Galway, he was 'strangely surprised to hear the cows and sheep all coming into my bed-chamber. I enquired the meaning and was told it was to preserve them from the wolf, which every night was rambling about for prey.'[17]

In England, however, the wolves receded into legend, along with the loathly 'worms' and 'serpents' slain by twelfth-century North-countrymen,[18] and the 'very many' lions which the Elizabethan William Harrison believed had once stalked Scotland (and against which Bede had warned English shepherds to beware).[19] But other predators remained. 'So noisome and offensive are some animals to human kind,' observed a seventeenth-century clergyman, 'that it concerns all mankind to get quit of the annoyance, with as speedy a riddance and despatch as may be, by any lawful means.'[20] In early Tudor times the campaign

* The wolf seems to have survived on the North Yorkshire Moors and other high parts of England until the fifteenth century. It lasted in Scotland until the late seventeenth century or, according to some traditions, the 1740s.

was placed on a statutory basis. An Act of Parliament in 1533 required parishes to equip themselves with nets in which to catch rooks, choughs and crows. In 1566 another authorized churchwardens to raise funds to pay so much a head to all those who brought in corpses of foxes, polecats, weasels, stoats, otters, hedgehogs, rats, mice, moles, hawks, buzzards, ospreys, jays, ravens, even kingfishers. Many parishes continued to make payments under these and later acts until the nineteenth century, the persecution shifting from one species to another according to prevailing agricultural needs.[21] In the sixteenth century the main attack was directed against crows preying on the corn. In later Stuart times the campaign turned against kites and ravens because they were a menace to poultry and agriculture; hitherto they had been protected as indispensable scavengers, but they became more vulnerable when urban authorities took to cleaning the streets and selling the manure to farmers. Also harried were the jays and bullfinches who nipped the buds off the fruit trees. In the eighteenth century there was a new onslaught on the rats who ate the corn in granaries. In the early nineteenth century the focus shifted again and there was a proliferation of suburban sparrow clubs whose members competed to see who could shoot the largest number.

As surviving parochial records show, the destruction effected under these Acts of Parliament was colossal, particularly from the later seventeenth century, when guns were increasingly used to shoot birds on the wing. At Tenterden, Kent, for example, they killed over 2,000 jays in the 1680s. At Deeping St James, Lincolnshire, in 1779 they killed 4,152 sparrows. At Prestbury, Cheshire, in 1732 they killed 5,480 moles. At Northill, Bedfordshire, between 1666 and 1812 the toll included 95 foxes, 130 badgers, 917 hedgehogs and 1,018 polecats. As for sparrows, the same parish between 1764 and 1774 alone saw the destruction of nearly 14,000, plus 3,500 eggs. Frequently, these trophies were displayed in the churchyard or nailed up in the barn – what Gilbert White called 'the countryman's museum'.[22]

It is easy now to forget just how much human effort went into warring against species which competed with man for the earth's resources. Most parishes seem to have had at least one individual who made his living by catching snakes, moles, hedgehogs and rats; and even the King had his official rat-catcher, who in the eighteenth century wore a special uniform of scarlet and yellow worsted, on which were embroidered figures of mice devouring wheatsheaves.[23] Every gardener destroyed smaller pests, and it was usual for the gardening-books to contain a calendar like the one drawn up by John Worlidge in

1668: 'January: set traps to destroy vermin. February: pick up all the snails you can find, and destroy frogs and their spawn. March: the principal time of the year for the destruction of moles. April: gather up worms and snails. May: kill ivy. June: destroy ants. July: kill ... wasps, flies.' And so on throughout the year.[24]

Yet pleasure rather than necessity accounted for the slaughter of many wild species. It was sheer bravado which led Fulke Greville in Ireland in 1580 to climb a crag 'to fetch an eagle from its nest' or induced dwellers on the sea coasts to let boys down in baskets to raid the birds' nests on the cliffs.[25] The voyages of Hawkins and other Elizabethans showed that the first reaction of English seamen when confronted by penguins and other sea-birds which had not yet learnt to avoid human beings was to slaughter them indiscriminately.[26] At home the countryside teemed with wild life which no one felt inhibited about shooting. In 1605 James I was said to have 'fallen into a great humour of catching larks'; he took 'as much delight in it or more than hunting' deer. When Lord Spencer gave a banquet at Althorp for Charles I in 1634 the menu included ruffs, reeves, redshanks, dotterels, godwits, curlews, swans, bitterns, mallards, peewits, herons, storks and dozens of other wild birds, some of them now gone for ever.[27] Generations of country boys were encouraged to make what William Ellis called a 'pleasant sport' of robbing nests and destroying both eggs and birds.[28] Hurling stones at kingfishers seems also to have been a popular activity.[29] In the eighteenth century the first impulse of many naturalists on seeing a rare bird was to shoot it.[30] Enormous depredations were also made to satisfy the growing craze for collections of eggs and stuffed birds.

By 1800 many species were disappearing which had been common a few centuries earlier. Who today has ever seen a kite in England? Yet in sixteenth-century towns kites were so common that they would swoop down and snatch the food out of children's hands.[31] Eagles, bustards, goshawks, marsh harriers, hen harriers, cranes, ospreys, ravens and buzzards have similarly diminished.* So have the martens and polecats which were regularly slaughtered by gamekeepers. Of course, changing land-use was a more important cause of extinction than deliberate persecution. The felling of the forests and draining of the marshes eliminated some species, just as the planting of the hedges and growth of human habitation increased others.[32] Long before the coming of pesticides and chemical fertilizers, pollution of the rivers killed the barbel,

* It has been suggested that there was more change in British bird life in the seventeenth and eighteenth centuries than in any other comparable period; *Book of British Birds* (Reader's Digest and A. A., 1969), 9.

trout, bream, dace, gudgeon, flounders and other fish which had in Elizabethan times swum in the London Thames; just as it would reduce the thirty different kinds of fish to be found in the Trent in the later Stuart period.[33] The overall effect of human action, whether deliberate or inadvertent, was to bring about a dramatic reduction in the wild life with which England had once teemed.

The need for artificial measures to preserve those wild species on which men depended for food or sport had long been appreciated. Since medieval times royal forests and private parks had protected beasts of the chase. As these animals grew rarer they had to be managed like domesticated beasts. Since the thirteenth century there had been numerous attempts by statute, proclamation or forest law to prescribe a close season and to protect red and fallow deer, otters, hares, salmon, hawks and wild fowl during the breeding period.[34]* Pheasant-breeding developed in early Tudor times and had by the mid eighteenth century generated strict rules for the preservation of the young birds. By 1773 the shooting seasons for pheasants, grouse and partridges had assumed their modern form.[35] In the Elizabethan period the fox (as has already been seen)† also joined the ranks of socially necessary species, to be artificially protected. In the seventeenth century the projectors who embarked on the draining of the Fens had to meet objections from those who contended that it would lead to the 'decay of fish and fowl'. In the eighteenth century General Howe even tried to breed wild boar in Wolmer Forest until the enraged local inhabitants rose and destroyed them.[36]

More notable than the preservation of animals to be hunted were the first stirrings of the view that wild creatures should be preserved even if they had no utility. Already some species were protected for curiosity or prestige, like the wild white cattle kept at Chillingham, Holdenby, and other private parks in the sixteenth century[37] and the swans 'preserved for their beauty' at Abbotsbury, Dorset, since medieval times.[38] The privilege of owning swans was a mark of high social status carefully controlled by the Crown, and those who possessed it would go to great trouble to safeguard their property. At Leconfield in the

* The earliest use of the term 'conservation' (originally 'conservacy') seems to have been in connection with the river Thames. The Lord Mayor and Aldermen of London were 'conservators' of the statutes made in the later Middle Ages for the upkeep of the river and thus came to be entrusted with its 'conservacie'. 'The word "conservacie",' explained a later commentator, 'doth extend itself to the preservation of the stream, and the banks of the river, as also the fish and fry within the same'; John Stow, *A Survey of the Cities of London and Westminster*, enlarged by John Strype (1720), i. 38.

† Above, p. 164.

East Riding the villagers' animals were excluded from the Fens in 1570 because they disturbed the breeding of the wild swans prized by the Earl of Northumberland.[39]

Exotic animals had always been prized possessions and an appropriate gift for one ruler to bestow on another. Since the twelfth century the kings of England had collected lions, leopards and other ferocious beasts; and their menagerie in the Tower lasted until 1834. 'There is an elephant given to the King,' notes an early Tudor schoolmaster, 'but none can guide him but they that came with the present.'[40] The royal menagerie symbolized its owner's triumph over the natural world; some medieval rulers even demonstrated their valour by fighting against their captive beasts. Later the zoo became a symbol of colonial conquest as well as of wealth and status. But it also provided aesthetic satisfaction; one of the leopards in the Tower Zoo, thought John Strype in 1720, was 'very beautiful and lovely to look upon; lying and playing, and turning her back wantonly, when I saw her'; and the lions there were described as 'the darlings, the delight of the people'.[41]

With the growth of European exploration and discovery, the import of rare species from every part of the world for private menageries assumed unprecedented dimensions. The Elizabethan William Harrison wrote of 'our costly and curious aviaries'; in the seventeenth century a collection of colourful birds became a standard feature of every aristocratic garden, and there were many specialist dealers in exotic species.[42] The illustrators of the lavish eighteenth-century books on tropical animals and birds sometimes copied dead specimens brought home from abroad, but, without leaving the Home Counties, they could see living examples of monkeys, lizards, turtles, buffalo, goldfish and macaws, whether in gentlemen's parks, domestic households or London inns and coffee-houses.[43] Scores of Hanoverian aristocrats owned large menageries of rare beasts and birds. The Duke of Cumberland kept ostriches in Windsor Great Park, while the zebra painted by Stubbs belonged to George III.[44]

Among the general public there was a keen interest in viewing unfamiliar animals and there were commercially-minded individuals ready to exploit it. People made long journeys to the metropolis, noted the Elizabethan Thomas Muffett, to have a chance of buying seats at a display of elephants, lions or rhinoceros. In 1560 the schoolboys at Eton gave money to see 'a camel in the College', while in 1653 Daniel Fleming on a visit from Westmorland to London paid fourpence 'for the sight of the dromedary'. In 1623 Sir Simonds D'Ewes in London saw an elephant 'out of Spain'. Later in the century Lord Keeper Guilford was rumoured to have so forgotten his dignity as to seize the chance of riding upon a rhinoceros which a merchant was exhibiting for profit. 'Shows of strange creatures,' conceded the godly Richard Baxter, were 'desirable and laudable.'[45] In provincial towns a touring menagerie was always certain of a good audience, though there were often risks for the animals. In Dublin in 1682 few gained admission to see an elephant, 'by reason of the great rates put upon the sight of him'; and when the beast was killed in an accidental fire, its proprietor had to have a file of musketeers to guard the carcase until the skeleton was ready for exhibition. In London in 1720 another elephant died after being publicly exhibited, its disorders being 'heightened by the great quantity of ale the spectators continually gave it'.[46]

There was therefore nothing new about the artificial preservation of ornamental or unfamiliar creatures or the cherishing of exotic birds and animals for amusement and display. More novel, however, was the growth of inhibitions about eliminating any wild animal, whether ornamental or not. 'We dispute in [the] schools,' wrote John Bulwer in 1653, 'whether, if it were possible for man to do so, it were lawful for him to destroy any one species of God's creatures, though it were but the species of toads and spiders, because this were taking away one link of God's chain, one note of his harmony.'[47] The continuation of every species was surely part of the divine plan.

The modern idea of the balance of nature thus had a theological basis before it gained a scientific one. It was belief in the perfection of God's design which preceded and underpinned the concept of the ecological chain, any link of which it would be dangerous to remove. The argument for design contained a strong conservationist implication, for it taught that even the most apparently noxious species served some indispensable human purpose. In the eighteenth century most scientists and theologians accordingly maintained that all created species had a necessary part to play in the economy of nature.[48] At the same time some of them had become increasingly aware that man's persecution really could eliminate individual species, a possibility which earlier generations had always denied.[49]

A mixture of theology and utility thus lay behind the increasing feeling that wild creatures ought, within limits, to be preserved. When the movement to protect wild birds gathered force in the nineteenth century it would lay much emphasis on the indispensable functions (eating grubs and keeping down insects and other vermin) performed even by those species thought most pernicious. Jays, magpies, bullfinches and ants were all useful in their different ways and it was therefore wrong to kill them. As the Somersetshire adage had it, 'If it were not for the Robin-Riddick and the Cutty-Wren, a spider would overcome a man.' In keeping with this view Lord Erskine wrote in his poem of 1818, *The Farmer's Vision*:

> Instant this solemn oath I took
> No hand shall rise against a rook.

When sea-birds gained legislative protection in 1869 it was argued that they were necessary to guide sailors and to show the fishermen where the herrings were.[50]

But from the seventeenth century onwards less utilitarian arguments for the preservation of wild species had also been advanced. Sir Matthew Hale urged mercy and compassion towards all wild creatures, in view of 'the admirable powers of life and sense ... in the birds and beasts ... All the men in the world could not give the like being to anything, nor restore that life and sense which is once taken from them.' John Locke thought it wrong to waste any food which would sustain a wild creature, even the birds of the air; and in the eighteenth century it became a mark of human sensibility to throw crumbs to wild birds in winter.[51] The bird-fanciers continued to catch and sell every kind of wild species, but this activity encountered increasing opposition. It was a platitude among seventeenth-century writers that every cage-bird would prefer the hardships of freedom to captivity, however mild;[52] and in the Hanoverian period the cruelty of trapping wild birds, clipping their wings, slitting their tongues and confining them in cages became a common theme of poetic lament. By 1735 it was necessary for the author of *The Bird-Fancier's Recreation* to refute the 'common objection, which some austere men (pretending to more humanity than the rest of their neighbours) make against the confining of songbirds in cages'. Two years later a 'lover of birds' protested against the practice of blinding chaffinches in preparation for captivity. By the end of the century moralists and aesthetes alike agreed that the song of a bird in a cage could give no pleasure.[53] Wild birds were a symbol of the Englishman's freedom and even aviaries were objectionable. As Lord John Russell told the Commons in the 1820s: 'It was not from the bars of a prison that the notes of English liberty could ever be heard; to have anything of grace and sweetness they must have something of ... wildness in their composition.'[54] Similar attacks were made on bird-nesting and on shooting wild birds for sport. 'Blessed be the name of the Lord Jesus against the destruction of small birds,' exclaimed Christopher Smart.[55]

Much of this eighteenth-century writing against cruelty to wild birds had a distinctly anthropomorphic character. The poets lamented the anguish of the mother bird whose eggs were stolen by marauding schoolboys or whose offspring were shot by heartless sportsmen.

> Again the slaughtering gun is heard,
> And wildly screams the parent bird.
> All night she mourns her lessen'd brood ...[56]

The poets also tended to favour some wild species more than others. The sparrow attracted less sympathy than did the robin; it was no accident that it was a robin redbreast in a cage which put Blake's Heaven 'in a rage'. Yet the volume of poetic attacks on cruelty to wild birds of every kind increased steadily from the mid seventeenth century; and it had an incalculable effect upon middle-class sensibilities. Many moralists now taught that it was only self-defence which could justify the destruction of wild species. God required goodness towards every living creature and men had no right to shoot the eagle on the mountain-top.[57] The killing of the albatross would bring retribution to Coleridge's Ancient Mariner. Much later sentiment was anticipated by such works as Margaret Cavendish's *Dialogue of Birds* (1653) or Thomas Tryon's *The Complaints of the Birds and Fowls of Heaven to their Creator for the Oppressions and Violences Most Nations on the Earth do Offer Them* (1683).[58] Many individuals who had shot birds in their youth or raided their nests suffered subsequent torments of remorse. The conversion of the Quaker John Woolman dated from the time when, as a child, he threw a stone and killed a female robin, whereupon he realized with horror that if he did not kill the young robins as well they would starve to death. As a boy, Thomas Bewick threw stones at bullfinches, until he killed one, whereupon he never threw them again. Byron shot an eaglet, which subsequently died; he never repeated the act. In the same way John Wilkes was sickened by the Italian habit of shooting small birds to protect their vines.[59] Although the nineteenth century would be the age of the great *battues*, the practice of killing birds for sport had become a controversial affair. 'A bittern was shot and eaten at Keswick by a young Cantab,' wrote Robert Southey, 'for which shooting I vituperate him in spirit whenever I think of it.' The ornithologists began to curb their urge to shoot a rare specimen on sight, discarding their guns and moving instead to spyglasses and ultimately cameras.[60]

It was from those who studied birds for recreation or curiosity that the pressure for conservation would arise; and it was the naturalists who pushed through a series of Acts which, from 1869 onwards, gave an increasing degree of statutory protection to wild birds.[61] This was the culmination of several hundred years of mounting interest in the natural world. Since the seventeenth century the study of birds, shells, fungi, butterflies, seaweed, fossils, flowers and wild animals had become firmly established as middle-class recreations. Gentlemen, clergymen and townsfolk (their wives included) had turned in increasing numbers to the natural world for pleasure, curiosity and emotional satisfaction. The movement was not peculiar to England, for, though notoriously slow to develop among the middle classes of Italy and Spain,* the taste for natural history was widespread in France and Germany; indeed it was a visiting German who in 1738 recommended botany to English country gentlemen as a healthy alternative to books and the bottle.[62] But nowhere did natural history become more popular than in England. As a modern authority remarks, 'The flora and fauna of the British Islands have probably been more exhaustively studied than those of any comparable area of country.'[63]

The achievements of the great English naturalists during these centuries are well known. What is not so often stressed is that, from the Tudor period onwards, almost every one of these pioneer scientists was assisted by dozens of now-forgotten amateur helpers and correspondents. Their names can be found in the Tudor and Stuart herbals or the county natural histories of the later seventeenth century. In his *Natural History of Northamptonshire* (1712), for example, the Rev. John Morton reveals his dependence at every stage upon information supplied by local gentry and clergy. He consults the cabinet of 'the ingenious Sir Matthew Dudley' at Clapton, the stuffed birds of Captain Saunders of Brixworth, 'the pleasant aviary of the ingenious Mr Mansell of Cosgrave' and the botanical learning of many other inhabitants of the county.[64] Later in the eighteenth century the works of Thomas Pennant reveal the names of scores of amateur students of natural history. Clergymen were particularly well placed to carry on such studies. John Ray's clerical helpers and correspondents included Lewis Stevens, an expert on seaweed and Cornish plants; Matthew Dodsworth, Rector of Sessay, Yorkshire, who worked on ferns; William Stonestreet, Rector

* In 1701 Jezreel Jones, clerk to the Royal Society, reported that in Cadiz he had been 'suspected for one that studies witchcraft, necromancy and a mad-man by some who observed me following butterflies, picking of herbs and other lawful exercises'. In 1788 the Earl of Bristol found that the natural history of Spain was 'as yet *vierge – parfaitement Pucelle*'; and fifty years later Richard Ford confirmed that Spain was still 'little better than a *terra incognita* to naturalists, geologists, and all other branches of ists and ologists'. See *The Sloane Herbarium*, compiled by James Britten, ed. J. E. Dandy (1958), 144; William S. Childe-Pemberton, *The Earl Bishop* (n.d.), ii. 407; Richard Ford, *Gatherings from Spain* (1846), 268.

of St Stephen Walbrook, who had a great collection of shells; Samuel Langley, the incumbent of Tamworth, who provided information on smelts; the Rev. Adam Buddle, who was an authority on mosses; and William Derham, the vicar of Upminster, Essex, who collected insects and frogs, dissected worms and fish and carefully observed birds building their nests.[65] The special sensibilities and literary skills of the Hampshire curate Gilbert White have assured his immortality, but it is worth remembering that in the eighteenth century he was but one of innumerable country clergy who occupied their time recording the events of nature and the passing of the seasons.

Middle-class women, who enjoyed a comparable degree of leisure, were also much involved. Dr Robert Plot cited Madame Offley, 'a lady that has an excellent artifice in preserving birds'; and John Ray was helped in botanical matters by Mrs Ward, 'an ingenious gentlewoman' of Guisborough in Cleveland. In 1750 a quarter of the subscribers to Benjamin Wilkes's *English Moths and Butterflies* were women. The Earl of Bute's *Botanical Tables* (1785?) were composed 'solely for the amusement of the fair sex'. By the end of the eighteenth century many women had themselves published works on botany.[66]

Of course some of this activity had a utilitarian bent. Physicians retained their interest in the medicinal uses of plants; entomologists studied insects in order to learn how to destroy tiresome pests. But it was not a concern with utility which bred what Lord Chesterfield in 1748 disparagingly called 'the numerous and frivolous tribes of insect-mongers, shell-mongers, and pursuers and driers of butterflies, etc.'.[67] The scope of natural history in the early modern period far transcended practical needs, and derived from a combination of religious impulse, intellectual curiosity and aesthetic pleasure. It was religion which taught that the natural world was God's book and that its study was a direct route to understanding the divine wisdom. For Henry Power, author of *Experimental Philosophy* (1664), the contemplation of the natural world was a moral duty, a form of homage due to the creator: ''Tis a tribute we ought to pay him for being men.' A century later Thomas Pennant agreed that the principal end of natural history was 'to exalt our veneration towards the Almighty',[68] while George Edwards boldly dedicated his *Natural History of Birds* (1743–51) to God, *tout court*.

To this moral impulse was added the more insistent pressure of fashion and pleasure. The eighteenth century saw the popularization of natural history by authors who wrote in the vernacular rather in Latin and who aimed to entertain as well as to instruct. In France library catalogues reveal that the Abbé Pluche's *Spectacle de la Nature* (1732) and the Comte de Buffon's *Histoire Naturelle* (1749–1804) had a greater vogue than even the writings of Voltaire. In England popular writers of natural history like John Hill, Oliver Goldsmith, Thomas Pennant and William Bingley enjoyed similar success; Peter Collinson thought in 1747 that works on natural history 'sell the best of any books in England'.[69] This tradition of popular writing on botanical and zoological subjects would have a long subsequent history. In Victorian England the Rev. J. G. Wood's *Common Objects of the Country* (1858) sold 100,000 copies in a week, while Eliza Brightwen, author of such works as *Glimpses into Plant Life* (1898), *Wild Nature Won by Kindness* (1890) and *Inmates of My House and Garden* (1895), attributed her popularity to 'that love of animated nature which is engrained in English hearts'.[70] Eighteenth-century England also had artists who specialized in natural history subjects, like Eleazar Albin, Moses Harris, William Lewin and Thomas Bewick. Their meticulous depictions of butterflies and spiders, birds and birds' eggs, and flora and fauna of every kind appeared in expensive colour plates for a luxury market, as well as in black-and-white engravings for more general consumption.[71]

The demand for such commodities revealed that natural history had become a highly fashionable affair. 'We have great numbers of nobility and gentry that know plants very well,' wrote Peter Collinson in 1755.[72] The vogue was enhanced by the accession of George III, whom a contemporary later described as 'one of the most scientific botanists in Europe' and whose minister, the Earl of Bute, was certainly more successful as a botanist than as a politician. Natural history, declared *The Critical Review* in 1763, had become 'the favourite study of the times'. In the following year a guidebook reported that multitudes of people were going to Margate to collect pebbles, shells and seaweed, while at Freshwater, Isle of Wight, the cliffs were visited by sightseers who wanted to see the large number of exotic birds which nested there each year. By 1776 the botanist William Curtis could boast that 'men from the other end of the town call on him in their coaches to desire private lectures' on his subject.[73] It was the eighteenth century which saw the beginning of the clubs and societies for the study of natural history and field botany which were to become so characteristic a feature of provincial life in Victorian England. By 1800 all but five of the known English butterflies had been discovered and recorded.[74]

Much of this activity was distinctly acquisitive in character. Ladies vied with each other to emulate the great collections of shells, plants and insects amassed by aristocrats like the Duchess of Beaufort and the Duchess of Portland. Some even bought caterpillars from the poor and bred them to enlarge their collections of butterflies.[75] In the 1730s and 40s the shell craze generated innumerable private collections: Thomas Martyn's *The Universal Conchologist* (1784) gives a fine list of private collections from that of the Countess of Bute downwards. These private collections often lacked the classified rigour of the didactic ones formed by serious scientists, but they showed how fashionable an interest in natural history had become. In 1739 a visitor to Charmouth in Dorset discovered that even a local labourer there had assembled a large collection of fossils.[76]

In such ways England became a mecca for foreign naturalists and botanical illustrators. It supplied Linnaeus with a greater following than he ever enjoyed in Sweden and it provided a home for his fellow-countrymen Daniel Solander (1736–82), who became Keeper of the Natural History Department in the British Museum, and James Dryander (1748–1810), who was made Librarian of the Royal Society. The German J. J. Dillenius (1687–1747) came to Oxford and became its first Professor of Botany. When J. J. Audubon produced his great *Birds of America* (1826–38) it was not in the United States but in Edinburgh and London that the work was published; of the 180 subscribers he listed in 1831 all but 18 were British, 29 coming from the Manchester area alone.[77]

It was in these years, when natural history had not yet been professionalized but was still an amateur hobby, that the feelings were engendered which would ultimately produce the legislation in the late nineteenth and twentieth centuries for nature conservation and the protection of wild creatures. For those aspects of the natural world which it was now fashionable to cherish were precisely those which early generations had despised or even sought to eliminate. As a contemporary noted in 1704, the virtuoso carefully preserved 'those creatures which others industriously destroy',[78] just as he sedulously cultivated those plants which others rooted up as weeds. Naturalists kept private menageries of live species which they could observe, and scientifically-minded gentlemen created reservations for animals which others regarded as useless or even harmful. At Dalkeith the Duchess of Buccleuch introduced the red squirrel about 1772, at a time when it seemed on the verge of extinction in Scotland. In Virginia, Thomas Jefferson wanted his garden to be an asylum for every kind of wild animal. The Evangelical minister Rowland Hill in the 1790s thought it cruelty to destroy the toads. 'In my country abode I even attempted to make them a place of retirement and called it a *toadery*.'[79] In the same decade John Byng lamented the slaughter of rooks, squirrels and singing-birds. If the persecution continued, he thought, the whole race of birds would be extinguished by farmers and gardeners: 'the country is stript of a chief beauty; and the contemplative man misses a prime satisfaction.' At Blenheim, at least, the Duke of Marlborough had forbidden his servants to disturb birds who nested in the shrubbery, though they were allowed to shoot them if they hopped over the wall into the kitchen garden.[80]

This, in microcosm, was the problem which would face every later conservationist. How could one preserve wild nature and yet keep it out of the kitchen garden? For by the later eighteenth century the competing demands of utility and beauty, production and consumption, body and spirit, seemed more irreconcilable than ever. Many sensitive persons no longer took pleasure in the growth of cities, the destruction of the forests, the spread of agriculture, the elimination of wild predators, the slaughter of the birds, the uprooting of the weeds. In 'The Excursion' Wordsworth wrote:

> I grieve, when on the darker side
> Of this great change I look; and there behold
> Such outrage done to nature as compels
> The indignant power to justify herself;
> Yea, to avenge her violated rights ...[81]

Early in the eighteenth century Joseph Addison had noted that 'the materials of a fine landscape' were 'not always the most profitable to the wonder of them'.[82] By the end of the century most aesthetes had come to regard the classical ideal of a union of beauty and utility as increasingly unattainable. 'Wherever man appears with his tools,' wrote William Gilpin, the self-appointed authority on the picturesque, 'deformity follows his steps. His spade and his plough, his hedge and his furrow, make shocking encroachments on the simplicity and elegance of landscape.' England, he thought, would 'be more beautiful in a state of nature than in a state of cultivation ... The regularity of cornfields disgusts and the colour of corn, especially at harvest, is out of tone with everything else.' Morally, cultivation was pleasing. Picturesquely, it aroused disgust.[83] His contemporary Archibald Alison agreed: landscapes were disfigured by the works of man whether 'traces of manufactures', 'the regularity of enclosures' or 'attempts towards improvement'. The sad truth was, as William Mitford put it in 1824, that 'the cultivation of the soil, necessary for supplying the wants of mankind, is highly adverse to the beauty of landscape'.[84]

Even the new kind of farm animals was distressingly unaesthetic. To breed sheep in the manner of Robert Bakewell, thought Uvedale Price, was to think only of 'their disposition to produce fat on the most profitable parts' – 'a very grazier-like and material idea of beauty'. The painter's or poet's idea of a beautiful bull or pig was very different from the farmer's.[85] As for trees, 'the picturesque eye,' ruled Gilpin, 'scorns the narrow conceptions of a timber-merchant.' Forestry plantations of fast-growing conifers were disliked by all connoisseurs of landscape. Uvedale Price thought larches monotonous and criticized large plantations which did not harmonize with their surroundings. Wordsworth attacked the Scotch firs and larches which disfigured the Lake District: a wretched 'vegetable manufactory'.[86] The uniform, rectangular hedges of parliamentary enclosure incurred similar strictures. Though convenient for use and aesthetically similar to the regularity of Georgian architecture, they seemed tiresomely inhuman and monotonous. For Humphry Repton, straight fences, pollarded trees and confined animals were 'objects of profit not of beauty'.[87]

Inevitably there were some desperate attempts to reunite utility with beauty. Frequent plantations could turn a whole estate 'into a kind of garden', thought Addison. 'Why [not] throw all your haystacks into the form of pyramids,' asked the exquisite William Shenstone, 'and choose out places where they may look agreeably?' But the *ferme ornée* and similar mid-eighteenth-century experiments in aesthetic farming invariably proved business failures.[88] As William Marshall wrote in 1796, 'The man of business and the man of taste are rarely united in the same person.' Ornament and profit, agreed Repton, were 'incompatible'.[89]

The early modern period had in fact engendered that split sensibility from which we still suffer. What was useful and productive was most likely to be ugly and distasteful. This attitude had a long pre-history: for poets and artists had always been selective about the human activities they chose to portray as beautiful or ennobling; and, as we have seen, protests against industrial pollution were not new.* But there was no real precedent for the volume of late-eighteenth-century complaint about the disfiguring effects of new buildings, roads, canals, tourism and industry. Such complaints were not universal, for some contemporaries thought mills and furnaces were sublime and inspiring.[90] But they became increasingly widespread and they have never ceased. Modern

* G. M. Trevelyan wrote that under the early Stuarts 'what paid best was beautiful'; only in modern times did beauty and economics conflict; *England under the Stuarts* (20th edn, 1947), 30; above, p. 14. The reaction of Jacobean Londoners to the fumes of the alum factories, the smells of the brick kilns and the pollution of the Thames by tanners (above, p. 245) suggests that Trevelyan's view of the early Stuart period was unduly optimistic.

writers and artists have still not succeeded in creating a new aesthetic model of an ideal industrial landscape, both pleasing and productive, to replace the old image of the *paysage riant*.[91]

Yet the irony was that the educated tastes of the aesthetes had themselves been paid for by the developments which they affected to deplore. The Hoare family who made Stourhead so exquisite were London bankers who must have helped to finance many new developments. The Dudleys could afford the beauty of Himley, with its trees, park and lakes, because their ironworks had despoiled the Staffordshire landscape further east.[92] The aesthete Richard Payne Knight was himself the grandson of a Shropshire ironmaster. Such men seldom allowed their aesthetic sensibilities to get in the way of the productive process. In the ensuing century and a half these private sensibilities would have to be gratified by the creation of special reservations, landscape gardens, green belts and animal sanctuaries: artificial oases or peepshows into an idealized world, whose very existence underlined their essential opposition to the fundamental values of ordinary society.

iv. MEAT OR MERCY?

It was not only aesthetic sensibilities which were now offended by the human conquest of nature. There were moral objections as well, particularly to the subjugation of the animals. As the threat from wild beasts receded, so man's right to eliminate wild creatures from whom he had nothing to fear was increasingly disputed. It had always been a feature of the Christian millenarian ideal that wild animals would one day lose their ferocity and live once more, as in Eden, on peaceable terms with man.[1] At fairs it was common to display booths foreshadowing this golden age to come. Thus in 1654 a lamb and a lion living on friendly terms with each other were put on public show in London; and in 1831 on one of the London bridges a showman exhibited animals in a state of reconciliation: cats, rats and mice in one cage, hawks and small birds in another.[2] In their utopian novel *Millennium Hall* (1762) Elizabeth Montagu and Sarah Scott depicted a sanctuary where man was no longer 'a merciless destroyer' and where animals moved unmolested.* The construction by Hanoverian gentlemen of private sanctuaries for the preservation of animals and birds was followed in the late nineteenth century by the foundation of fauna-protection societies who successfully agitated for the statutory protection of many hitherto-despised forms of wild life. In modern times even the dangerous Indian tiger would be protected.[3]

Human authority over the domestic animals was also challenged. Just as many eighteenth-century thinkers, like Rousseau, believed that civilization had corrupted natural man, so many naturalists followed Buffon in believing that domestication, far from improving animals, had merely degraded them. Oliver Goldsmith wrote:

> In all countries, as man is civilised and improved, the lower ranks of animals are oppressed and degraded. Either reduced to servitude or treated as rebels, all their societies are dissolved and all their united talents rendered ineffectual. Their feeble arts quickly disappear and nothing remains but their solitary instincts or those foreign habit[ude]s which they receive from human education.

Tamed animals were like once-proud aborigines, demoralized by their European conquerors.[4] In the nineteenth and twentieth centuries many defenders of animal rights would urge that zoos and menageries were offensive to the natural dignity of the inmates. Even pet-keeping was degrading and should be forbidden.[5]

Yet the objection to the domestication of animals went deeper than this. For, once it was accepted that animals should be treated with kindness, it inevitably seemed increasingly repugnant to kill them for meat. The tradition that man was originally vegetarian is ancient and worldwide. It may reflect the actual practice of our remote ancestors,

* The animals, however, were only native ones. The lady proprietors of the sanctuary considered it cruel and unprofitable to imprison lions or tigers in a menagerie far removed from their native element.

for apes are largely vegetarian and it was probably only with the rise of a hunting economy that the change to meat-eating occurred.[6] Its expression in much Greek and Roman literature ensured its transmission to early modern England. 'On roots, not beasts, they fed,' sang the seventeenth-century poetess Katherine Philips of the Golden Age.[7]

Vegetarianism was also encouraged by Christian teaching, for all theologians agreed that man had not originally been carnivorous. In Eden, wrote Alexander Pope,

> Man walk'd with beast, joint tenant of the shade;
> . . .
> No murder cloth'd him, and no murder fed.[8]

Many biblical commentators maintained that it was only after the Flood that humans became meat-eaters; in the period of disorientation following the Fall they had remained herbivorous.[9] Others, noting that Abel was a herdsman, suggested that it was the Fall which had inaugurated the carnivorous era and that the liberty of eating flesh which God gave Noah was merely the renewal of an earlier permission.[10] Commentators argued as to whether meat-eating had been permitted because man's physical constitution had degenerated and therefore required new forms of nutriment, or because the cultivation of the soil to which he was condemned required a more robust food, or because the fruits and herbs on which he had fed in Eden had lost their former goodness.[11] But everyone agreed that meat-eating symbolized man's fallen condition. 'God allow[s] us to take away the lives of our fellow creatures and to eat their flesh,' wrote Richard Baxter in 1691, 'to show what sin hath brought on the world.'[12] The death of brute animals to supply the wants of sinful man could even be made a paradigm of Christ's Atonement.[13]

Meanwhile, the permission to eat meat was regarded as a concession to human weakness, not a command. For the pagan writers Seneca and Porphyry, voluntary abstinence from flesh had symbolized the triumph of the spirit over the body; many austere medieval Christians deliberately renounced meat for the same reason (fish remained acceptable, partly because they were bloodless, partly because they were not produced by sexual congress).[14] In seventeenth-century England there were still a few such ascetics who gave up meat to vanquish the flesh, like the future mining projector Thomas Bushell, who in the 1620s lived for three years in a hut on a diet of herbs, oil, mustard and honey, or Mrs Traske, wife of John Traske, the Judaist, who abstained from meat and drank only water for seven years when imprisoned in the reign of Charles I, or the Ranter John Robins, who in the early 1650s required his disciples to abstain from 'meat and drinks'.[15] In the eighteenth and early nineteenth centuries there were sectarians, influenced by the German mystic Boehme and by William Law's *Serious Call* (1738), who, along with some Southcottians and Swedenborgians, followed a similarly austere regime of abstinence from animal food.[16]

A good deal of anxiety was also generated by the Old Testament prohibition (Genesis, ix. 4) on eating blood. This ban could not be dismissed as part of the now obsolete Jewish ceremonial law, for the prohibition had been repeated in the Acts of the Apostles (xv. 20; xxi. 25). It lingered among the early Christians and survived in the Eastern Church. Some early modern commentators tried to explain it away allegorically or to interpret it as an injunction against unnecessary cruelty or the consumption of living animals.[17] But others took it literally, and as late as the 1730s the issue occasioned a brisk clerical controversy.[18] In Stuart England there were numerous 'tender and curious persons', particularly among the Civil War sects, who refused on conscientious grounds to eat gravy or blood-soaked black puddings. The Scots were also said to have religious objections to black puddings.[19] Not everyone shared these scruples, for Samuel Pepys noted in 1667 that Mr Andrews, a timber-merchant, liked his meat raw and ate it 'with no pleasure unless the blood run about his chops'; and in the eighteenth century the English were notorious among foreign visitors for serving their beef underdone.[20] But tastes were changing. Blood, thought Nathaniel Lardner in 1762, was 'filthy and highly disagreeable ... it is never brought, neither alone, nor mixed with other things, to the tables of polite people.' By early Victorian times the roles had been reversed. It was now the English travellers abroad who recoiled from the half-cooked meat served in Continental restaurants. 'Unless our appetites are very keen,' observed a mid-nineteenth-century guide to dining-room etiquette, 'the sight of much meat reeking in its gravy is sufficient to destroy them entirely.' The Evangelical Zachary Macaulay, father of the historian, regarded a taste for underdone meat as a deadly sin, comparable to smoking or lying in bed in the morning.[21]

More notable than this enduring inhibition about blood, however, was the appearance in mid-seventeenth-century England of individuals who rejected meat, not because it was undercooked or for ascetic reasons, but because they believed it wrong to kill animals at all. One of the sectarian 'errors' listed by Thomas Edwards in 1646 was the doctrine that it was forbidden to kill any lawful creature; he cited the case of a Hackney bricklayer named Marshall, a follower of the

Familist Giles Randall, who taught that it was 'unlawful to kill any creature that had life, because it came from God'.[22] At Ickenham near Uxbridge, the mystic Roger Crab, formerly a Chesham hatter, had from about 1641 held it sinful to eat flesh, both because it strengthened human lusts and because it was produced by 'bloody butchers', who destroyed 'their fellow-creatures'. He had a disciple, Captain Norwood, who died from attempting to follow his frugal regime; it may possibly have been another follower who appeared in Yorkshire in 1674, dressed in white and claiming to have drunk only water and eaten only roots for the past fourteen years.[23] In 1691 the Waterford landowner, Robert Cook, who had lived for a time in England, published a paper in defence of the 'Pythagorean' regime, which he followed on grounds of conscience, refusing any raiment or food which came from animals.[24]

The most notable vegetarian in this sectarian tradition was the Behmenist Thomas Tryon, whose views on animals we have already encountered.* In 1657 Tryon gave up meat and fish and refused to wear leather. He rejected flesh-eating partly because he thought it introduced an animal element into the body, giving man a 'wolfish, doggish nature', partly because he thought it unhealthy, but mainly because he opposed 'killing and oppressing his fellow creatures'. Animals, he declared, bore the image of their creator and were entitled to be treated according to the golden rule ('do unto all creatures as they would be done unto'). He developed his views in a remarkable series of tracts which, though published in the last two decades of the seventeenth century, carried on the authentic radical tradition of the Interregnum. He attacked not just cruelty to animals, but also negro slavery, war-games, the criminal code, the harsh treatment of the insane, and even the practice of making all persons behave as if they were naturally right-handed. He advised his readers to be sparing in their consumption of meat, rather than to give it up altogether, but his own practice was unambiguous and he made a number of converts both in his lifetime and posthumously (Aphra Behn and Benjamin Franklin among them).[25] What is notable about his arguments is that they reveal that the long-established habit of praising red meat because it supposedly made men virile and courageous had produced the inevitable reaction. For Tryon, the adoption of animal food after the Fall was associated with the beginning of fighting and quarrelling among men. It was important 'to prevent the growth of all fierceness, wrath and violence, even in the bud'. Vegetarianism was for him a means of curbing aggression, of van-

* Above, pp. 155, 170.

quishing 'a tumultuous, envious spirit'.[26] This argument would have a long currency, since it was generally accepted that food affected the character. As an eighteenth-century naturalist put it: 'Vulgar and uninformed men, when pampered with a variety of animal food, are much more choleric, fierce and cruel in their tempers than those who live chiefly on vegetables.'[27]

By the later seventeenth century man's right to kill animals for food was being widely debated. The vegetarian teachings of Plutarch and Porphyry were well known to the educated, while Pythagoras's moral objections to meat-eating (based on his belief in the kinship of all animate nature) gained wide currency through successive translations of Ovid's *Metamorphoses*. In his version of 1700, Dryden interpolated the ringing lines

> Take not away the life you cannot give:
> For all things have an equal right to live.[28]

Along with these conscientious objections to meat-eating went more practical considerations. Later-seventeenth-century scientists like Walter Charleton, John Ray and John Wallis were much impressed by the suggestion that human anatomy, particularly the teeth and intestines, showed that man had not originally been intended to be carnivorous.[29] This argument subsequently provided additional support for the view that meat-eating was 'unnatural'. Many scientific writers also felt, very reasonably, that the heavy meat diet which was every Englishman's ideal was distinctly unhealthy. As one of them put it in 1721, 'It is that dreadful mixture of the souls ... of so many thousand animals, destroyed to pamper one, that raises that terrible war in the blood which has made it a prey to such distempers as have baffled the skill of the most learned physicians.' A simple diet would keep the blood free of 'noxious juices' and conduce to a longer life.[30] John Evelyn wrote a tract to prove that it was possible 'to live on wholesome vegetables, both long and happily', while the naturalist Edward Bancroft agreed that 'not only humanity, but self-interest, conspire to engage us at least to abridge the quantity of animal food which at present we devour with so much avidity.' In 1780 the philosopher Adam Ferguson was restored to health by following a 'Pythagorean course of diet'.[31] The notion that meat was unhealthy thus became central to much later vegetarian teaching. Beef tea, it would be claimed, had killed more people than had Napoleon; and no vegetarian had bad breath.[32]

In the 1730s and 1740s the health argument was strongly propagated by the influential physician George Cheyne, who at one point had himself weighed 32 stone and whose idea of going on a diet was to 'scarce ever eat animal food, above once a day,' and drink 'very little above a pint of wine, or at most, not a quart one day with another'. Cheyne did not urge that meat be renounced altogether: such a change would be 'unnatural; out of the order of providence, and in some degree, immoral'. But he confessed to being unable 'to find any great difference on the foot of natural reason and equity ... between feeding on human flesh and feeding on brute animal flesh, except by custom and example.' 'To see the convulsions, agonies and tortures of a poor fellow-creature ... dying to gratify luxury ... must require a rocky heart and a great deal of cruelty and ferocity.'[33]

But many hearts were far from rocky. In 1548 John Foxe, the future martyrologist, had written that 'such is my disposition that I can scarce pass the shambles where beasts are slaughtered, but that my mind recoils with a feeling of pain.' A century later Sir Matthew Hale confessed that the sight of sheep grazing always made him feel that God must have intended 'a more innocent kind of food to man'. 'I am convinced that to eat flesh is lawful,' wrote the aged Richard Baxter, 'yet all my days it hath gone, as against my nature, with some regret.' John Ray agreed that a vegetable diet was preferable to 'butchery and slaughter of animals', while Sir Isaac Newton was said to have found 'a frightful contradiction' between accepting that animals could feel and making them suffer. 'He yielded only reluctantly to our barbarous usage of feeding on the blood and flesh of beings like ourselves.' 'Today we killed a swine,' wrote the Rev. Robert Meeke in his diary in 1692, 'I heard his cry into my study – many creatures die for us, but sinful man deserveth death most of all.'[34]

In Margaret Cavendish's poems of the 1650s the cooking of meat became a symbol of death and cruelty, as in her description of a battle, where

> beasts and men both in their blood lay masht,
> As if that a French cook had them minc'd, so hasht,
> Or with their blood a jelly boil
> To make a bouillon of the spoil.

Slaughtering creatures for food, she suggested, was blatantly unjust:

> As if that God made creatures for man's meat,
> To give them life and sense for man to eat;
> ...
> And that all creatures for his sake alone
> Was made for him to tyrannise upon.[35]

No doubt these sentiments did not prevent the duchess from enjoying roast beef, any more than they impeded her poetic successor James Thomson, who in 1728 included a section recommending a vegetable diet in *The Seasons*.[36] But even if such passages indicated poetic sentiment rather than real conviction, they nevertheless betrayed the existence of a distinct uneasiness. Margaret Cavendish's arguments reappeared in 1721, when an anonymous medical writer denounced the whole practice of animal slaughter as a tyranny over God's creatures: 'It is true man is the lord of the creation; so is a master of his family: but what lord devours his own subjects? Or what father feasts upon his own children and servants?'[37] In the eighteenth century meat-eating was often described as a 'dreadful, disgusting' practice which only long usage had made familiar. One could hardly condemn the cannibals of Guiana, said Edward Bancroft in 1769, since they were merely victims of habit similar to that which enabled the English 'to survey without an involuntary horror the mangled carcasses of inoffensive animals, exposed in a London market'. Richard Ford would say the same of the Spaniards and their bullfights: 'They are reconciled by habit, as we are to the bleeding butchers' shops which disfigure our gay streets and which if seen for the first time would be inexpressibly disgusting.'[38]

Butchers, accordingly, were regarded with suspicion, not just because of the noise, smell, blood and pollution which their activities involved, but also because of a widespread aversion to the act of slaughter itself. In More's *Utopia* the bondsmen did all the slaughtering; freemen were not even allowed to witness it lest their human clemency be eroded.[39] In medieval and early modern times civic authorities tried to prevent the slaughter of animals in public places. They regarded the shambles as an offensive nuisance and frequently tried to drive them outside town walls altogether.[40] The butchers themselves became the object of prejudices not unlike those directed against the public executioner. Their trade was 'odious', thought William Vaughan in 1608. They handled raw flesh, which, it was said, everyone else felt it too repugnant even to touch.[41] In a poetical dictionary of 1657 they were described as 'greasy, bloody, slaughtering, merciless, pitiless, cruel, rude, grim, harsh, stern, . . . surly'; and the epithets frequently recur. Butchers got 'a greasy living by killing beasts', thought a later Stuart preacher.[42] In 1716 John Gay urged pedestrians in the London streets

> To shun the surly butcher's greasy tray,
> Butchers, whose hands are dy'd with blood's foul stain,
> And always foremost in the hangman's train.

'Taking away the lives of animals, in order to convert them into food,' thought the philosopher David Hartley in 1748, 'does great violence to the principles of benevolence and compassion. This appears from the frequent hard-heartedness and cruelty found amongst those persons whose occupations engaged them in destroying animal life, as well as from the uneasiness which others feel in beholding the butchery of animals.' 'The trade of a butcher,' agreed Adam Smith,' is a brutal and an odious business.'[43] In Victorian times the slaughtermen were frequently said by social investigators to be the most demoralized class of all.[44] No wonder that it was widely believed in the early modern period that butchers were ineligible for jury service in capital cases, owing to their cruel inclinations. There seems to have been no legal authority for this notion, but it was held throughout the seventeenth and eighteenth centuries by scores of commentators who should have known better.[45]

By the beginning of the eighteenth century, therefore, all the arguments which were to sustain modern vegetarianism were in circulation: not only did the slaughter of animals have a brutalizing effect upon the human character, but the consumption of meat was bad for health; it was physiologically unnatural; it made men cruel and ferocious; and it inflicted untold suffering upon man's fellow-creatures. By the end of the century these arguments had been supplemented by an economic one: stock-breeding was a wasteful form of agriculture compared with arable farming, which produced far more food per acre.[46]

At first vegetarianism made only a trickle of temporary converts, such as James Boswell, who was convinced at the age of sixteen by the Scottish Pythagorean John Williamson of Moffat, or the future Lord Chesterfield, who, as an undergraduate at Trinity Hall in 1714, gave up meat for a period after being much affected by reading Pythagoras's speech in Ovid's *Metamorphoses*.[47] The Essex Quaker Benjamin Lay, who emigrated to Philadelphia in 1731, had so tender a conscience that he would eat no food nor wear any garment which had been procured at the expense of animal life (or indeed at the cost of slave labour). The future Scottish Minister James Gillies held the conventional doctrine that it was unlawful to kill animals save out of necessity, but as a student at Aberdeen in the 1770s he took it to its logical conclusion: discovering that he could live without animal food, he gave it up altogether.[48]

From about 1790 there developed a highly articulate vegetarian movement. Its most prominent representatives included the antiquarian Joseph Ritson (1752–1803), who had been converted about 1772 by reading Bernard Mandeville's reflections on animal slaughter in *The Fable of the Bees* (1714) and who later published *An Essay on*

Abstinence from Animal Food as a Moral Duty (1802); the radical Scot John Oswald, author of *The Cry of Nature* (1791), who had learnt his vegetarianism from the Hindus when serving with a Highland regiment in India; the Yorkshire printer George Nicholson (1760–1825); and the London physician William Lambe (1765–1847), who had become a vegetarian by 1807 and in turn converted his patient John Frank Newton, whose subsequent book defending a vegetable regime, *The Return to Nature* (1811), provided much of the basis for the poet Shelley's *Vindication of Natural Diet* (1812).[49] * At Salford in 1809 the Bible Christians were founded as a schismatic branch of the Swedenborgians by William Cowherd (1763–1816), who, influenced by arguments which combined humanitarianism with a concern for physical health, a search for gnostic religion, and a social distaste for wining and dining, made vegetarianism a condition of entry and secured 300 members.[50] Among them would be the free-trader and parliamentary reformer Joseph Brotherton (1783–1857), whose wife was the author of *Vegetable Cookery* (1821). It was Cowherd who converted William Metcalfe (1788–1862), who led a branch of the church to Philadelphia, where he in turn in 1830 converted Sylvester Graham (1794–1851), who was to become the apostle of dietary reform in nineteenth-century America and who drew heavily on the works of William Lambe and other English writers of this period.[51]

In the 1790s vegetarianism had markedly radical overtones. Ritson liked to be known as 'Citizen Ritson', while Oswald died fighting for the Jacobins against the Vendée. Richard Phillips (1767–1840), who had given up meat on humanitarian grounds around 1780, was a Republican who founded the *Leicester Herald* to uphold the rights of man. He was gaoled in 1793 for selling Tom Paine's book and, though knighted in 1808, retained his radical sympathies, dedicating one of his books in 1826 to 'Simon Bolivar the Liberator'.[52] Vegetarianism at this period had a millennial flavour. By promising to remove the ferocity from human nature, it struck 'at the root of all evil,' thought Shelley; and William Lambe asserted that if men gave up eating meat there might be no more wars. To the converts meat-eating was 'horrible', 'barbarous' and 'unnatural'; and they were convinced the time would come when unregenerate flesh-eaters would see the error of their ways.[53] But much of

* Some doubt is thrown upon the seriousness of Shelley's conversion to the 'Pythagorean system' in 1812 by the tone of his wife's invitation to a friend: 'Mrs Shelley's comp[liment]s to Mrs Nugent and expects the pleasure of her company to dinner, 5 o'clock, as a murdered chicken has been prepared for her repast'; *The Letters of Percy Bysshe Shelley*, ed. Roger Ingpen, i (1909), 284n.

the impulse subsided after the period of revolutionary ferment. In 1847 the Vegetarian Society of Great Britain was founded, but fifty years later its membership was still only about 5000.[54]

The early vegetarians thus made little appeal to the masses. Their inspiration was often literary, many claiming to have been converted by reading the arguments of Pythagoras or Plutarch. They wrote at a time when meat was still a precious luxury for many people and consequently a matter of status. In attacking roast beef they were hurling themselves against a cherished national symbol as well as against the weight of medical opinion, which continued to insist that some flesh intake was necessary to human health. Their cause was also hindered by its association with unfashionable dissenting groups. It is true that many kinds of religion were represented among the early vegetarians: Thomas Forster (1789–1860) was a Catholic, Lewis Gompertz a Jew; and Joseph Ritson was an atheist. But heterodox sects like the Quakers, the Bible Christians, the Swedenborgians and the Behmenists (later Theosophists) were disproportionately prominent. Inevitably the 'Pythagoreans' tended to be regarded as cranks and eccentrics by their contemporaries.

Nevertheless, they offered a notable challenge to conventional practice, and one to which official thought no longer had a ready answer. In Elizabethan times it had been easy for Thomas Muffett to refute the classical vegetarians by invoking the anthropocentric teachings of the day:

> Whereas Plutarch objecteth how loathsome a thing it is to see butchers and cooks sprinkled with blood in killing and dressing flesh, I answer him that the sight is not so loathsome to nature, but to niceness and conceit. For what God permits to be eaten, nature permits to dress and kill; neither rebelleth she more at the death of an ox than at the cutting down of hay or corn. Nay, furthermore, sith all was made for man's use, ... she giveth us liberty to kill all things that may make for the maintenance of our life or preservation and restoring of our health.[55]

By the end of the eighteenth century this claim that animals were made only for man's use was still being advanced, but it no longer carried general assent. Very soon, it would disappear almost altogether.

Undaunted, many utilitarian thinkers continued to argue that to kill animals for human food was wholly consistent with benevolence and virtue, so long as the beasts were carefully looked after during their lifetime and slaughtered with the minimum of cruelty. Animals, they urged, could not anticipate their deaths and they felt no terror. If beasts were not killed for food, maintained John Lawrence in 1798, they would overstock the earth; so 'in numberless cases' it was 'an act of mercy to take their lives'. Moreover, as the Nonconformist Philip Doddridge pointed out, many people subsisted by breeding and selling cattle; what would happen to their livings if the custom of eating flesh were suddenly laid aside?[56]

But it was no longer enough to say that cows and sheep would never have been bred in the first place if they were not to be slaughtered, for, as Dr Johnson observed in 1776, 'the question is whether animals who endure such sufferings of various kinds, for the service and entertainment of man, would accept of existence upon the terms on which they have it.'[57] In the eighteenth century defenders of meat-eating found themselves increasingly forced back upon the mandate of the Old Testament. The right to slaughter for food, remarked Francis Hutcheson, was 'so opposite to the natural compassion of the human heart that one cannot think an express grant of it by revelation was superfluous'. Without the explicit authority of scripture, thought William Paley, man's right to kill beasts for meat would be difficult if not impossible to justify. Ever since the 1680s, when Thomas Tryon questioned the legitimacy of meat-eating, it had been biblical precedent upon which the defenders of the status quo had based their case.[58] But in a secular world arguments founded on scripture alone would prove increasingly unimpressive. As Hutcheson shrewdly observed, if there was force to the humanitarian argument against meat-eating, then any grant by revelation of the right to slaughter would appear that much more incredible.[59]

All that remained was the Hobbesian view, justifying the human species in doing anything it felt necessary for its survival. The rights which brutes had over us, declared Spinoza, we had over them. The objection to killing animals was 'based upon an empty superstition and womanish tenderness, rather than upon sound reason'. Civilization would be impossible if humanity acted justly towards nature; man could not survive without being a predator.[60] This was the argument which overcame Lord Chesterfield's scruples about meat-eating. 'Upon serious reflection I became convinced of its legality, from the general order of nature, who has instituted the universal preying upon the weaker as one of her first principles.' 'Philosophy,' remarked David Hartley, 'has of late discovered such numberless orders of small animals in parts of diet formerly esteemed to be void of life, and such an extension of life into the vegetable kingdom, that we seem under the perpetual necessity, either of destroying the lives of some of the creatures, or of perishing ourselves.' The whole of nature, agreed Erasmus Darwin, was 'one great slaughter-house'. Anyway, man was a superior species and his interests should come first.[61]

The brutal realism of this view conflicted sharply with the principles of benevolence and good nature to which it was now customary to pay lip-service. As a contemporary wrote of Joseph Ritson, 'to follow his plan of abstinence were absurd, and nearly impossible; yet it is surely a disagreeable necessity which drives us to form part of a system where ... the powerful exist by preying on the weak.'[62] John Tweddell (1769–99), a Cambridge classical scholar who gave up eating flesh on conscientious grounds, declared himself 'persuaded we have no other right, than the right of the strongest, to sacrifice to our monstrous appetites the bodies of living things, of whose qualities and relations we are ignorant'.[63] No wonder the vegetarians were so confident that future ages would come to share their view of meat-eating as a hideous barbarity.

Meanwhile they were contemptuous of the sentimentalists, able to eat, yet unable to kill, particularly when previously acquainted with the animal concerned. As Mandeville observed in 1714, there were now many meat-eaters who would themselves have been reluctant to wring a chicken's neck. When the second Duke of Montagu was talking to a visitor at Boughton, a flock of sheep went past. 'The duke admired the prettiness, the simplicity, the innocence of the animals,' but confessed that 'when by chance he saw 'em killing one, he turned away his head and could not bear to look.' This was the humbug denounced by the poet Nathaniel Bloomfield:

> Well might he who eats the flesh of lambs
> ...
> Boast his humanity, and say 'My hand
> Ne'er slew a lamb;' and censure as a crime
> The butcher's cruel, necessary trade.[64]

In 1756 Gilbert White planted four lime trees at Selborne between his house and the butcher's yard opposite, 'to hide the sight of blood and filth'. His action symbolized a growing effort, not to abolish slaughter-houses, but to hide them from the public gaze. Dr Johnson, who had 'a kind of horror of butchering,' said 'he was afraid there were slaughter-houses in more streets of London than one supposes'. Even in Elizabethan England there were people who were too 'squeamish' to see animals killed. By 1714 Mandeville could write of the growing aversion to animal slaughter that 'in this behaviour methinks there appears something like a consciousness of guilt'.[65] In the past it had been customary to serve pigs, calves, hares and rabbits at the table with their heads attached, but around the end of the eighteenth century there seems to have been a growing tendency to conceal the slaughtered creature's more recognizable features. 'Animals that are made use of as food,' wrote William Hazlitt in 1826, 'should either be so small as to be imperceptible, or else we should ... not leave the form standing to reproach us with our gluttony and cruelty. I hate to see a rabbit trussed, or a hare brought to the table in the form which it occupied while living.'[66] Killing animals for food was now an activity about which an increasing number of people felt furtive or uneasy. The concealment of slaughter-houses from the public eye had become a necessary device to avoid too blatant a clash between material facts and private sensibilities.

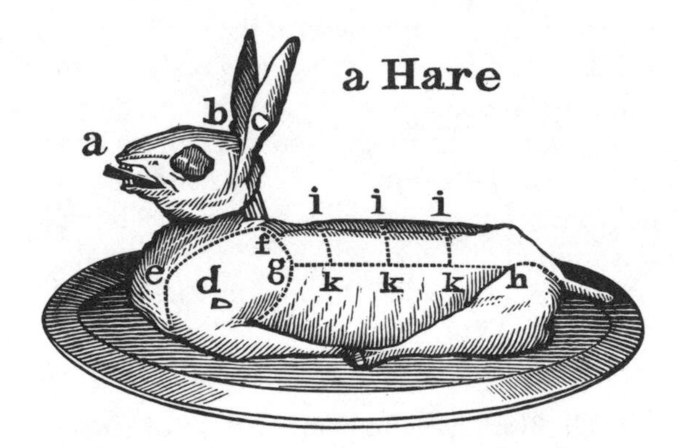

V. CONCLUSION

The embarrassment about meat-eating thus provides a final example of the way in which, by the end of the eighteenth century, a growing number of people had come to find man's ascendancy over nature increasingly abhorrent to their moral and aesthetic sensibilities. This was the human dilemma: how to reconcile the physical requirements of civilization with the new feelings and values which that same civilization had generated. It is too often assumed that sensibilities and morals are mere ideology: a convenient rationalization of the world as it is. But in the early modern period the truth was almost the reverse, for, by an inexorable logic, there had gradually emerged attitudes to the natural world which were essentially incompatible with the direction in which English society was moving. The growth of towns had led to a new longing for the countryside. The progress of cultivation had fostered a taste for weeds, mountains and unsubdued nature. The new-found security from wild animals had generated an increasing concern to protect birds and preserve wild creatures in their natural state. Economic independence of animal power and urban isolation from animal farming had nourished emotional attitudes which were hard, if not impossible, to reconcile with the exploitation of animals by which most people lived. Henceforth an increasingly sentimental view of animals as pets and objects of contemplation would jostle uneasily alongside the harsh facts of a world in which the elimination of 'pests' and the breeding of animals for slaughter grew every day more efficient. Oliver Goldsmith wrote of his contemporaries that 'they pity and they eat the objects of their compassion'.[1] The same might be said of the children of today who, nourished by a meat diet and protected by a medicine developed by animal experiments, nevertheless take toy animals to bed and lavish their affection on lambs and ponies. For adults, nature parks and conservation areas serve a function not unlike that which toy animals have for children; they are fantasies which enshrine the values by which society as a whole cannot afford to live.

By 1800 the confident anthropocentrism of Tudor England had given way to an altogether more confused state of mind. The world could no longer be regarded as having been made for man alone, and the rigid barriers between humanity and other forms of life had been much weakened. During the religious upheavals of the 1640s and 1650s contemporaries had been shocked to hear sectaries like the Ranter Jacob Bauthumley asserting that 'God is in all creatures, man and beast, fish and fowl, and every green thing.'[2] But, in a secularized form, this kind of pantheism was to become very general in the eighteenth century, when it was widely urged that all parts of creation had a right to live; and that nature itself had an intrinsic spiritual value. Not everyone now believed that mankind was uniquely sacred. Some Romantics preferred the once-condemned mystical view that 'each shrub is sacred, and each weed divine'. As William Blake put it, 'Every thing that lives is Holy.'[3] *

Of course most people in practice, like G. M. Trevelyan himself, retained their faith in the primacy of human interests, even if they lamented the effect of material progress on the natural world.

> Whether trees, or animals, ought to be preserved 'for their own sakes' [wrote Trevelyan] is an interesting question on which different opinions might be held. But the argument for the preservation of natural scenery and the wild life of English fauna and flora may be based on motives that regard the welfare of human beings alone, and it is those arguments alone that I wish here to put forward. To preserve the bird life of the country is required in the spiritual interests of the human race, more particularly of the English section of it, who find such joy in seeing and hearing birds.[4]

As Trevelyan implied, it was not for the sake of the creatures themselves, but for the sake of men, that birds and animals would be protected in sanctuaries and wild-life parks. In 1969 the United Nations and the International Union for the Conservation of Nature defined 'conservation' as 'the rational use of the environment to achieve the highest quality of living for mankind.'[5]

But even in the early modern period there were some perhaps hypersensitive persons who were prepared to go further than this. For them it was increasingly difficult to accept the primacy of human needs when to do so involved inflicting pain on domestic animals or eliminating whole species of wild ones. In more recent times these difficulties have been widely perceived. Today there are writers of books who refer to the extermination of the wolf as a 'pogrom' or 'holocaust';[6] and the law journals carry articles on whether trees have rights.[7]

The early modern period had thus generated feelings which would make it increasingly hard for men to come to terms with the uncompromising methods by which the dominance of their species had been secured. On the one hand they saw an incalculable increase in the comfort and physical well-being or welfare of human beings; on the other they perceived a ruthless exploitation of other forms of animate life.

* Though Coleridge later described 'the vague misty, rather than mystic, confusion of God with the World & the accompanying nature-worship' as 'the trait in Wordsworth's poetic works that I most dislike as unhealthful, & denounce as contagious'; *Collected Letters of Samuel Taylor Coleridge*, ed. Earl Leslie Griggs (Oxford, 1956–71), v. 95.

There was thus a growing conflict between the new sensibilities and the material foundations of human society. A mixture of compromise and concealment has so far prevented this conflict from having to be fully resolved. But the issue cannot be completely evaded and it can be relied upon to recur. It is one of the contradictions upon which modern civilization may be said to rest. About its ultimate consequences we can only speculate.

ABBREVIATIONS

Unless otherwise stated, all books cited were published (or part-published) in London. Their titles have sometimes been shortened, and in addition the following abbreviations have been used:

AgHR	*Agricultural History Review*
Allen, *Naturalist*	David Elliston Allen, *The Naturalist in Britain* (1976)
Ann. Sci.	*Annals of Science*
Archaeol.	*Archaeological*
Bacon	*The Works of Francis Bacon*, ed. J. Spedding, R. L. Ellis and D. D. Heath (1857–9)
Bewick, *Quadrupeds*	*A General History of Quadrupeds. The figures engraved on wood by Thomas Bewick* (Newcastle upon Tyne, 1807; reprint 1970)
Bingley, *An. Biog.*	William Bingley, *Animal Biography, or Popular Zoology illustrated by authentic anecdotes* (5th edn, 1820)
Bingley, *Quadrupeds*	*Memoirs of British Quadrupeds, illustrative principally of their habits of life, instincts, sagacity, and uses to mankind* (1809)
BL	British Library
Bodl.	Bodleian Library
Byng	*The Torrington Diaries containing the Tours through England and Wales of the Hon. John Byng (later fifth Viscount Torrington) between the years 1781 and 1794*, ed. C. Bruyn Andrews (1934; reprint, 1970)
Coles, *Eden*	William Coles, *Adam in Eden* (1657)
Coles, *Simpling*	William Coles, *The Art of Simpling* (1656)
CSPD	*Calendar of State Papers Domestic*
CSPV	*Calendar of State Papers Venetian*
Dialect Dict.	*The English Dialect Dictionary*, ed. Joseph Wright (1898; reprint, 1970)
DNB	*Dictionary of National Biography*
EDS	English Dialect Society
EETS	Early English Text Society
EL	Everyman's Library

Ellis	William Ellis, *The Modern Husbandman* (1744–50)
Evelyn, *Diary*	*The Diary of John Evelyn*, ed. E. S. de Beer (Oxford, 1955)
Evelyn, *Sylva*	J[ohn] E[velyn], *Sylva, or a Discourse of Forest-Trees and the Propagation of Timber* (1664)
Evelyn/Hunter	John Evelyn, *Silva*, with notes by A Hunter (York, 1776)
Gerard, *Herball*, ed. Johnson	John Gerard, *The Herball or Generall Historie of Plantes*, enlarged and amended by Thomas Johnson (1633; reprint, 1975)
Goldsmith	Oliver Goldsmith, *An History of the Earth and Animated Nature* (1774)
Henrey	Blanche Henrey, *British Botanical and Horticultural Literature before 1800* (1975)
HMC	*Historical Manuscripts Commission*
Hobbes, *EW*	*The English Works of Thomas Hobbes*, ed. Sir William Molesworth (1839–45)
Hobbes, *LW*	*Thomae Hobbes Malmsburiensis Opera Philosophica*, ed. Sir William Molesworth (1839–45)
JHI	*Journal of the History of Ideas*
Jnl	*Journal*
Kalm	*Kalm's Account of his Visit to England on his way to America in 1748*, trans. Joseph Lucas (1892)
Lawrence, *Horses*	John Lawrence, *A Philosophical and Practical Treatise on Horses, and on the Moral Duties of Man towards the Brute Creation* (1798)
Locke Corr.	*The Correspondence of John Locke*, ed. E. S. de Beer (Oxford, 1976–)
Morton, *Northants.*	John Morton, *The Natural History of Northampton-shire* (1712)
OED	*The Oxford English Dictionary*, ed. James A. H. Murray *et al.* (Oxford, 1933)
Oldenburg	*The Correspondence of Henry Oldenburg*, ed. and trans. A. Rupert Hall and Marie Boas Hall (Madison, Milwaukee and London, 1965–)
Parkinson, *Paradisi*	John Parkinson, *Paradisi in Sole Paradisus Terrestris* (1629; reprint, 1904)
Pennant, *Zoology*	Thomas Pennant, *British Zoology* (new edn, 1812)
Pepys, *Diary*	*The Diary of Samuel Pepys*, ed. Robert Latham and William Matthews (1970–)
Plot, *Oxon.*	Robert Plot, *The Natural History of Oxford-shire* (Oxford, 1677)
Plot, *Staffs.*	Robert Plot, *The Natural History of Stafford-shire* (Oxford, 1686)

ABBREVIATIONS

Raven, *Naturalists*	Charles E. Raven, *English Naturalists from Neckam to Ray* (Cambridge, 1947)
Raven, *Ray*	Charles E. Raven, *John Ray Naturalist. His Life and Works* (2nd edn, Cambridge, 1950)
Ray, *Willoughby*	John Ray, *The Ornithology of Francis Willughby* (1678)
Ray, *Wisdom*	John Ray, *The Wisdom of God manifested in the Works of Creation* (1691)
Rev.	*Review*
R.O.	Record Office
Soc.	*Society*
Sternberg	Thomas Sternberg, *The Dialect and Folk-Lore of Northamptonshire* (1851)
Swainson, *Birds*	Charles Swainson, *The Folklore and Provincial Names of British Birds* (Folk-Lore Soc., 1886)
Topsell	Edward Topsell, *The Historie of Foure-Footed Beastes* (1607)
VCH	*Victoria County History*

NOTES

INTRODUCTION

1. *Clio, a Muse* (1913), 161.
2. *The Journal of William Beckford in Portugal and Spain 1787–1788*, ed. Boyd Alexander (1954), 99–100.
3. Cf. Paul Fussell, *The Great War and Modern Memory* (1975), chap. 7, and Martin J. Wiener, *English Culture and the Decline of the Industrial Spirit* (Cambridge, 1981), chaps. 3 and 4.
4. G. M. Trevelyan, *The Call and Claims of Natural Beauty* (1931); *id., Must England's Beauty Perish?* (1929); *England under Queen Anne* (1930–34), i. 25; National Trust, *Report*, 1961–2 (1962), 6–7.
5. Samuel Purchas, *Hakluytus Posthumus or Purchas his Pilgrims*, xix (Glasgow, 1906), 218–24; Chester E. Eisinger, 'The Puritans' Justification for Taking the Land', *Essex Institute Hist. Collns.*, lxxxiv (1948); Roy Harvey Pearce, *Savagism and Civilisation* (Baltimore, 1965), 21; Alden T. Vaughan, *New England Frontier* (Boston, 1965), 110–12. Cf. John Locke, *Two Treatises of Government*, ed. Peter Laslett (Cambridge, 1960), 308–10, 315–16 (i, paras 32–5, 42).
6. Eric Ashby, *Reconciling Man with the Environment* (1978), 3; George Steiner in *The Listener*, 97 (28 April 1977), 537.

CHAPTER I HUMAN ASCENDANCY

i. Theological Foundations

1. Aristotle, *Politics*, 1256a–b; Cicero, *De Natura Deorum*, ii. 14, 61–5; John Passmore, *Man's Responsibility for Nature* (1974), 14–18.
2. Arnold Williams, *The Common Expositor* (Chapel Hill, 1948), 81, 133–4.
3. E.g. Andrew Willet, *Hexapla in Genesin* (Cambridge, 1605), 16; Henry Ainsworth, *Annotations upon the Five Bookes of Moses* (1639), 39; Matthew Henry, *A Commentary on the Holy Bible* (1710; new edn, n.d.), i. 7. On meat-eating, see below, p. 289.
4. Thomas Wilcox, *A Right Godly and Learned Exposition, upon the Whole Booke of Psalmes* (1586), 19; John Downame, *Lectures upon the Foure First Chapters of Hosea* (1608), 282–3; Samuel Purchas, *Hakluytus Posthumus or Purchas his Pilgrims*, xix (Glasgow, 1906), 219–20; George Walker, *God made Visible in his Workes* (1641), 161–2; Thomas Manton, *A Practical Commentary ... on the Epistle of James* (2nd edn, 1653), 375.

5. *The Workes of Mr Williã Cowper late Bishop of Galloway* (2nd edn, 1629), 115. Also Thomas Draxe, *The Earnest of our Inheritance* (1613), 15; Benjamin Parker, *Philosophical Meditations* (1734–5), ii. 38–40.
6. Jeremiah Burroughes, *Gospel Reconciliation* (1657), 6; Richard Bentley, *The Folly of Atheism* (1692), 27; Bacon, vi. 747.
7. *The Works of... Robert Boyle* (1744), iv. 517.
8. Sir Matthew Hale, *The Primitive Origination of Mankind* (1677), 68; Downame, *Lectures upon Hosea*, 282. Cf. *The Sermons of M. John Calvin upon ... Deuteronomie*, trans. Arthur Golding (1583), 776; Arthur O. Lovejoy, *The Great Chain of Being* (1936; New York, 1960), 186–7.
9. Willet, *Hexapla in Genesin*, 16.
10. Jeremiah Burroughes, *An Exposition of the Prophesie of Hosea* (1643), 576. The original source of this image seems to have been Philo, *De Opificio Mundi*, 84–5. For its long survival, see, e.g., *History of the Robins, designed for the Instruction of Children* (2nd edn, Dublin, 1821), 147.
11. Philip Doddridge, *A Course of Lectures on the Principal Subjects in Pneumatology, Ethics, and Divinity* (1763), 133; Wilcox, *Exposition upon the* Psalmes, 546.
12. J. C[ockburn?], *An Enquiry into the Nature, Necessity, and Evidence of Christian Faith*, i (1696), 39; *The Works of the late Reverend and Pious Mr Tho. Gouge* (1706), 456.
13. Pennant, *Zoology*, ii. 259 (following William Borlase, *The Natural History of Cornwall* (Oxford, 1758), 289–90); George Cheyne, *Philosophical Principles of Religion* (2nd edn, 1715), i. 359.
14. *The Works of James Pilkington*, ed. James Scholefield (Parker Soc., Cambridge, 1842), 92; [John Day], *Day's Descant on Davids Psalmes* (Oxford, 1620), 213–14; Walker, *God made Visible*, 151–2; Goldsmith, iv. 356–7.
15. *The Prose Works of William Byrd of Westover*, ed. Louis B. Wright (Cambridge, Mass., 1966), 293; *Batman upon Bartholome His Booke De Proprietatibus* (1582), fol. 338; *The Natural History of English Song-Birds* (with figures by Eleazar Albin) (1737), 'To the Reader'.
16. George Owen of Henllys, *The Description of Penbrokshire*, ed. Henry Owen (Cymmrodorion Rec. Ser., 1892), 126; Henry More, *An Antidote against Atheism* (2nd edn, 1655), 116; William Swainson, *A Treatise on the Geography and Classification of Animals* (n.d.), 262; William Kirby, *On the Power, Wisdom and Goodness of God as manifested in the Creation of Animals and in their History, Habits, and Instincts* (1835), ii. 316.
17. More, *Antidote against Atheism*, 82–3; Coles, *Simpling*, 93 (plagiarizing More, *op. cit.*, 97).
18. Benjamin Rush, cit. Daniel J. Boorstin, *The Lost World of Thomas Jefferson* (New York, 1948), 50.
19. Thomas Robinson, *A Vindication of the Philosophical and Theological Exposition of the Mosaick System* (1709), 77; More, *Antidote against Atheism*, 117; William King, *An Essay on the Origin of Evil* (1731), 118–19; [William Wollaston], *The Religion of Nature Delineated* (1722; 5th edn, 1731), 34–5; *The*

Theological, Philosophical, and Miscellaneous Works of the Rev. William Jones (1801), xii. 461.

20. [Day], *Day's Descant*, 213.
21. Samuel Pufendorf, *Of the Law of Nature and Nations* (Eng. trans., 1710), 285–6.
22. Lancelot Andrewes, *A Pattern of Catechistical Doctrine* (1650; Oxford, 1846), 217; William Tyndale, *Doctrinal Treatises*, ed. Henry Walter (Parker Soc., Cambridge, 1848), 202.
23. William Ames, *Conscience with the Power and Cases thereof* (n. pl., 1639), iv. 222; [Samuel Gott], *The Divine History of the Genesis of the World* (1670), 424; Ezekiel Hopkins, *An Exposition on the Ten Commandments* (1692), ii. 3.
24. More, *Antidote against Atheism*, 125; *The Theological Works of Isaac Barrow*, ed. Alexander Napier (Cambridge, 1859), ix. 46; Hobbes, *EW*, iii. 125.
25. J. Ovington, *A Voyage to Surat in the Year 1689*, ed. H. G. Rawlinson (1929), 202; cf. Henry Lord, *A Display of Two Forraigne Sects in the East Indies* (1630), 41–56; John Hall of Richmond, *Of Government and Obedience* (1654), 140.
26. John Rawlinson, *Mercy to a Beast* (Oxford, 1612), 33; Andrewes, *Pattern of Catechistical Doctrine*, 217. Cf. Augustine, *De Civitate Dei*, i. 19; Aquinas, *Summa Theologica*, ii. 2.64.1.
27. John Levett, *The Ordering of Bees* (1634), 41; Doddridge, *Course of Lectures*, 130.
28. Thomas Fuller, *The Holy State* (2nd edn, Cambridge, 1648), 171; William Somervile, *The Chase* (1735), in *The Works of the English Poets*, ed. Alexander Chalmers, xi (1810), 166.
29. *Works of Robert Boyle*, iv. 363; Robert Charles Hope, *The Legendary Lore of the Holy Wells of England* (1893), xix-xx. Cf. Peter Brown, *The Cult of the Saints* (1981), 125–6; below, pp. 214–15.
30. Lynn White, Jr, 'The Historical Roots of our Ecologic Crisis', *Science*, 155 (10 March 1967), reprinted as chapter 5 of his *Machina ex Deo: Essays in the Dynamism of Western Culture* (1968).
31. 'Philotheos Physiologus' (Thomas Tryon), *The Country-Man's Companion* (n.d. [1683]), 151–3; David McClellan, *Marx's Grundrisse* (1971), 94–5; Marx, *Early Writings*, trans. Rodney Livingstone and Gregor Benton (Harmondsworth, 1975), 239. Cf. Vernon Lee, *Euphorion* (2nd edn, 1885), 128ff.
32. Thomas Morton, *New English Canaan* (1632), 34, in *Tracts and other Papers*, ed. Peter Force, ii (Washington, 1838). For discussion of White's thesis, see Jean Dorst, *Before Nature Dies*, trans. Constance D. Sherman (1970), chap. 1; *Western Man and Environmental Ethics*, ed. Ian G. Barbour (Reading, Mass., 1973); *Ecology and Religion in History*, ed. David and Eileen Spring (1974); Passmore, *Man's Responsibility for Nature*; *Man and Nature*, ed. Hugh Montefiore (1975), 47–9, 155–8.
33. Edward Pococke, *A Commentary on the Prophecy of Hosea* (Oxford, 1685), 95, 97.
34. Michael Cope, *A Godly and Learned Exposition upon the Proverbes of Solomon*, trans. M.O. (1580), fol. 207; Henry Ainsworth, *Annotations upon the Five Bookes of Moses* (1627), v. 92.

35. Passmore, *Man's Responsibility for Nature*, 12–13; though cf. C. F. D. Moule, *Man and Nature in the New Testament* (1964).

ii. The Subjugation of the Natural World

1. Pierre Chaunu, *L'Expansion européenne du XIIIe au XVe siècle* (Paris, 1969), 336; *id.*, *Histoire science sociale* (Paris, 1974), 260.
2. Fernand Braudel, *Capitalism and Material Life, 1400–1800*, trans. Miriam Kochan (1973), 67, 248–9, 133.
3. P. K. O'Brien, 'Agriculture and the Industrial Revolution', *Econ. Hist. Rev.*, 2nd ser., xxx (1977), 169.
4. P. R. Edwards, 'The Horse Trade of the Midlands in the Seventeenth Century', *AgHR*, 27 (1979), 91; Joan Thirsk, *Horses in Early Modern England* (Reading, 1978); J. A. Perkins, *The Ox, the Horse, and English Farming, 1750–1850* (Working Paper in Economic History, Univ. of New South Wales, 1975); *Probate Inventories and Manorial Excepts of Chetnole, Leigh and Yetminster*, ed. R. Machin (Bristol, 1976), 16.
5. *CSPV, 1557–8*, 1671–2; *CSPV, 1617–19*, 102; Thomas Muffett, *Healths Improvement*, enlarged by Christopher Bennet (1655), 50; Henry Peacham, *The Worth of a Peny* (1667), 23. Cf. [Richard Morison], *A Remedy for Sedition* (1536), sigs. Eiv^{v}-Fi; John Aylmer, *An Harborowe* (1559), sigs. P3^{v}–4; [Edward Chamberlayne], *Angliae Notitia* (1669), 15; John Dunton, *The Athenian Oracle* (1703–4), iii. 185; John Ashton, *Social Life in the Reign of Queen Anne* (1897), 141.
6. Downame, *Lectures upon Hosea*, 2nd pagination, 214 (cf. Goldsmith, iii. 8–9); Gregory King in *Seventeenth-Century Economic Documents*, ed. Joan Thirsk and J. P. Cooper (Oxford, 1972), 784–5; Daniel A. Baugh, *British Naval Administration in the Age of Walpole* (Princeton, 1965), 407.
7. John Laurence, *A New System of Agriculture* (1726), 131; Kalm, 14, 15.
8. James Hart, *KΛINIKH, or the Diet of the Diseased* (1633), 71; Dunton, *Athenian Oracle*, ii. 413–14. Cf. Muffett, *Healths Improvement*, 59; William Cobbett, *Rural Rides*, ed. George Woodcock (Harmondsworth, 1967), 409.
9. John Weemse, *An Explication of the Iudiciall Laws of Moses* (1632), 198.
10. Carl Bridenbaugh, *Vexed and Troubled Englishmen* (Oxford, 1968), 95n; *Locke Corr.*, ii. 8; John Fuller, 'Carving Trifles', *Procs. Brit. Acad.*, lxii (1976), 279–80; M. Dorothy George, *English Political Caricature to 1792* (Oxford, 1959), 107–8, 114, 177.
11. [Will. Rabisha], *The Whole Body of Cookery Dissected* (1661), 241. On the importance of carving, Hannah Woolley, *The Queen-Like Closet* (5th edn, 1684), 258; *The Letters of the Earl of Chesterfield to his Son*, ed. Charles Strachey (1901), i. 94, 269; ii. 89, 118; John Trusler, *The Honours of the Table* (2nd edn, 1791), 24ff.; John Hodgkin, 'Proper Terms II', *Trans. Philological Soc.* (1907–10).
12. Charles Webster, *The Great Instauration* (1975), chap. v; William Leiss, *The Domination of Nature* (New York, 1972), chap. 3; *Works of Robert Boyle*, v. 469; Joseph Glanvill, *Plus Ultra* (1668), 87.

13. William Forsyth, *A Treatise on the Culture and Management of Fruit-Trees* (2nd edn, 1803), 385.
14. Sir John Colbatch, *A Dissertation concerning Mistletoe* (1719), 2–4. Cf. below, p. 271.
15. John Stow, *A Survey of the Cities of London and Westminster*, ed. John Strype (1720), i. 161; *Oldenburg*, v. 423.
16. *The Advice of W. P. to Mr Samuel Hartlib* (1648), in *The Harleian Miscellany*, ed. J. Malham (1808–11), vi. 156; Samuel Hartlib, *The Compleat Husband-Man* (1659), 72; *The Collected Poems of Christopher Smart*, ed. Norman Callan (1949), i. 274; William Swainson, *A Preliminary Discourse on the Study of Natural History* (1834), 136. Cf. E. A. J. Johnson, *Predecessors of Adam Smith* (1937), 123; James Anderson, *Essays relating to Agriculture and Rural Affairs* (3rd edn, Edinburgh, 1784), ii. 277–92; Charles Darwin, *The Origin of Species* (World's Classics, 1902), 128; Eric Kerridge, *The Agricultural Revolution* (1967), chap. 9.
17. Kirby, *On the Power, Wisdom and Goodness of God*, i. 1.
18. J. W[orlidge], *Systema Agriculturae* (1669), 24. Cf. Robert Sharrock, *The History of the Propagation and Improvement of Vegetables* (2nd edn, 1677).
19. *The Flower-Garden Display'd* (2nd edn 1734), 139. Cf. Hartlib, *Compleat Husband-Man*, 63; William T. Stearn, 'The Origin and later Development of Cultivated Plants', *Jnl of the Royal Horticultural Soc.*, xc (1965).
20. Arthur B. Ferguson, *Clio Unbound* (Durham, N.C., 1979), 366; Ronald L. Meek, *Social Science and the Ignoble Savage* (Cambridge, 1976), 133; *Works of the Rev. William Jones*, vi. 43; Kirby, *On the Power, Wisdom and Goodness of God*, 521. For the origin of this idea, Plato, *Protagoras*, 322.
21. Clarence J. Glacken, *Traces on the Rhodian Shore* (Berkeley and Los Angeles, 1967), 676; Bewick, *Quadrupeds*, 324–5 (following Goldsmith, iii. 374–5).
22. Johann Gottfried von Herder, *Reflections on the Philosophy of the History of Mankind*, ed. Frank E. Manuel (1968), 53–5; Henry Home, Lord Kames, *Sketches of the History of Man* (Glasgow, 1817), i. 55; Adam Smith, *Lectures on Jurisprudence*, ed. R. L. Meek *et al.*, 15; Edward Gibbon, *The History of the Decline and Fall of the Roman Empire*, ed. J. B. Bury (1906 edn), v. 315.
23. Benjamin Farrington, *The Philosophy of Francis Bacon* (Liverpool, 1964), 92; Abraham Cowley, *Several Discourses*, ed. Harry Christopher Minchin (1904), 52; Somervile, *The Chase*, 155.
24. Thomas Carew, 'On his Entertainment at Saxham, 1634'. Cf. Ben Jonson, 'To Penshurst'; Rowland Watkyns, *Flamma sine Fumo* (1662), ed. Paul C. Davies (Cardiff, 1968), 18.
25. *CSPV, 1617–19*, 260; Margaret Cavendish, Marchioness of Newcastle, *The Philosophical and Physical Opinions* (1655), 100–101. On hunting rituals see D. H. Madden, *The Diary of Master William Silence* (new edn, 1907), 61–4; James Obelkevich, *Religion and Rural Society* (Oxford, 1976), 42–3.
26. E.g. Stow, *Survey of London*, ed. Strype, i(iii), 285.
27. Sir Thomas Elyot, *The Boke named the Governour*, ed. H. H. S. Croft (1883), i. 181–2. On horsemanship see Gervase Markham, *Countrey Contentments* (1615), i. 35ff.; Ruth Kelso, *The Doctrine of the English Gentleman in the Sixteenth*

Century, Univ. of Illinois Studs. in Lang. and Lit., xiv (1929), 154–6; Stephen Orgel, *The Illusion of Power* (Berkeley and Los Angeles, 1975), 75–7.

28. Thirsk, *Horses in Early Modern England*; *CSPV, 1557–8*, 1672; *The Economic Writings of Sir William Petty*, ed. C. H. Hull (Cambridge, 1899), i. 203; *Tracts and Other Papers*, ed. Force, ii (11). 9. Cf. *Seventeenth-Century Economic Documents*, ed. Thirsk and Cooper, 428.
29. Edmund S. Morgan, *American Slavery, American Freedom* (New York, 1975), 231–2.
30. E. Lankester, *The Uses of Animals in Relation to the Industry of Man* (n.d.), 272.

iii. Human Uniqueness

1. Aristotle, *De Anima*; C. S. Lewis, *The Discarded Image* (Cambridge, 1967), 152–65; Robert Burton, *The Anatomy of Melancholy* (EL, 1932), i. 154–5.
2. Aristotle, *Hist. An.*, 608^{b}; Robinson, *Vindication of the Mosaick System*, 81; Hale, *Primitive Origination of Mankind*, 64.
3. Burroughes, *Gospel Reconciliation*, 6.
4. Aristotle, *Politics*, 1253^{a}; *The Remaining Medical Works of Thomas Willis*, trans. S. P[ordage] (1681), 117; *Boswell's Life of Johnson*, ed. George Birkbeck Hill, revised by L. F. Powell (Oxford, 1934–64), iii. 245; v. 33n (cf. Claude Lévi-Strauss, *Le Cru et le Cuit* (Paris, 1964); *The Parliamentary History of England*, xvii (1813), 782; Thomas Love Peacock, *Headlong Hall* (1816), chap. 12.
5. Roy Pascal, *The Social Basis of the German Reformation* (1933), 161; *Church and State through the Centuries*, ed. Sidney Z. Ehler and John B. Morrall (1954), 326.
6. Plato, *Timaeus*, 90; Ovid, *Metamorphoses*, i. 84–6; Willet, *Hexapla in Genesin*, 107; Helkiah Crooke, *ΜΙΚΡΟΚΟΣΜΟΓΡΑΦΙΑ. A Description of the Body of Man* (2nd edn, 1631), 646; James Tyrrell, *A Brief Disquisition of the Law of Nature* (1692), 79; Herschel Baker, *The Image of Man* (New York, 1961 edn), 298 n18.
7. Aristotle, *De Part. An.*, 673^{a}; *Hist. An.*, 518^{a}, 492^{a}; H. C. Baldry, *The Unity of Mankind in Greek Thought* (Cambridge, 1965), 89–90.
8. Hart, *Diet of the Diseased*, 36 (following Pliny, *Nat. Hist.*, xi. 37).
9. Uvedale Price, *Essays on the Picturesque* (1810), iii. 223.
10. Ray, *Wisdom*, 191; Ben Jonson, *Timber* (Temple Classics, n.d.), 93; Pepys, *Diary*, viii. 554; James Anderson, *Recreations in Agriculture, Natural-History, Arts, and Miscellaneous Literature* (1799–1802), i. 9 (2nd pagination). Cf. Cicero, *De Inventione*, i. 4.
11. Richard Cumberland, *A Treatise of the Laws of Nature*, trans. John Maxwell (1727), 93; Robert Lovell, *ΠΑΝΖΩΟΡΥΚΤΟΛΟΓΙΑ Sive Panzoologico-mineralogia. Or a Compleat History of Animals and Minerals* (Oxford, 1661), intro.; Gervase Markham, *Markhams Maister-Peece* (1610), 57. Also Andrew Snape, *The Anatomy of an Horse* (1683), 105
12. For representative opinions on this large subject see Aristotle, *Hist. An.*, 488^{b}; Aquinas, *Summa Theologica*, i. 78. 4; Hobbes, *EW*, iii. 44, 48, 664; vii. 467; *id.*, *LW*, ii. 88–9; iii. 527; John Locke, *An Essay concerning Human Understanding*,

ed. Peter H. Nidditch (Oxford, 1975), 159–60 (ii, chap. xi); Dunton, *Athenian Oracle*, i. 140; iii. 75; William Smellie, *The Philosophy of Natural History* (Edinburgh, 1790–99), ii. 457; [James Burnet, Lord Monboddo], *Antient Metaphysics* (1779–99), iii, appendix, chap. iii.

13. Aristotle, *Politics*, 1254[b]; Aquinas, *Summa Theologica*, ii(1). 6; ii(2). 95. 7; King, *Essay on the Origin of Evil*, 161–2.
14. *Proceedings in the Parliaments of Elizabeth I*, i, ed. T. E. Hartley (Leicester, 1981), 240; Richard Baxter, *Compassionate Counsel to all Young-Men* (1681), 69; John Howe, *The Living Temple* (1675), 22–3; George Berkeley, *Alciphron* (1732), 5th dialogue, 28; Vicesimus Knox, *Lucubrations* (1788), no. 135.
15. John Chishull, *Two Treatises* (1654), sig. A5; M[atthew] S[mith], *A Philosophical Discourse of the Nature of Rational and Irrational Souls* (1695), 21.
16. *Antoniana Margarita . . . per Gometium Pereiram* (Medina del Campo, 1554) (discussed by Pierre Bayle, *Dictionaire historique et critique* (2nd edn, Rotterdam, 1702), 'Pereira'); René Descartes, *Discours de la Méthode* (1637), v; *id.*, *Méditations métaphysiques* (1641), vi; *Œuvres de Descartes*, ed. Charles Adam and Paul Tannery (Paris, 1897–1913), iii. 85; iv. 574–5; v. 275–9. See in general Leonora D. Cohen, 'Descartes and Henry More on the Beast-Machine', *Ann. Sci.*, i (1936); Leonora Cohen Rosenfield, 'Un Chapitre de l'histoire de l'animal-machine (1645–1749)', *Revue de littérature comparée*, 17 (1937); *id.*, *From Beast-Machine to Man-Machine. Animal Soul in French Letters from Descartes to La Mettrie* (New York, 1941); Hester Hastings, *Man and Beast in French Thought of the Eighteenth Century* (1936), 19–63; Albert G. A. Balz, *Cartesian Studies* (New York, 1951), 106–57; Robert M. Young, 'Animal Soul', in *The Encyclopaedia of Philosophy*, ed. Paul Edwards (New York, 1967), i; J. S. Spink, *French Free-Thought from Gassendi to Voltaire* (1960), chap. xi; Thomas H. Huxley, 'On the Hypothesis that Animals are Automata, and its History', in *Method and Results* (1893).
17. See Aram Vartanian, *Diderot and Descartes* (Princeton, 1953); *id.*, *La Mettrie's L'Homme Machine* (Princeton, 1960); *id.*, 'Man-Machine from the Greeks to the Computer', *Dictionary of the History of Ideas*, ed. Philip P. Wiener (New York, 1973–4), iii.
18. *Œuvres de Descartes*, iv. 574–6; v. 276–8.
19. Anthony Le Grand, *An Entire Body of Philosophy, according to the Principles of the famous Renate Des Cartes*, trans. Richard Blome (1694), ii. 252.
20. As was noted by Descartes, *Discours*, v; *Œuvres*, iii. 121. Cf. Le Grand, *op. cit.*, ii. 236; John Norris, *An Essay towards the Theory of the Ideal or Intelligible World* (1701–4), ii. 83–6.
21. Aquinas, *Summa Theologica*, i. 78. 4; ii(1). 17 (as pointed out by John Rodman, 'The Dolphin Papers', *The North American Rev.*, 259 (Spring 1974), 21).
22. Le Grand, *Entire Body of Philosophy*, i. 255–6; ii. 234–5.
23. Gottfried Wilhelm Leibniz, *Philosophical Papers and Letters*, trans. and ed. Leroy E. Loemker (2nd edn, Dordrecht, 1969), 588.
24. *Œuvres*, v. 279 (trans. in *Ann. Sci.*, i (1936), 53).
25. Norris, *An Essay*, ii. 74.

26. Bodl., MS. Locke f. 6, p. 26.
27. *Discours,* vi ('maîtres et possesseurs de la nature').
28. Cf. the bibliography of Ezra Abbott in William Rounseville Alger, *A Critical History of the Doctrine of a Future Life* (4th edn, New York, 1867), appendix.
29. Sir Kenelm Digby, *Two Treatises* (1645), i. 399–400 (though Bayle, *Dictionaire*, 'Rorarius', 2609, denied that Digby was of Descartes's opinion). For others see Henry Power, *Experimental Philosophy* (1664), sig. b2[v]; Tim. Nourse, *A Discourse upon the Nature and Faculties of Man* (1686), 77; Le Grand, *Entire Body of Philosophy*, part iii; Norris, *An Essay*, ii, chap. 2; [F.B.], *A Letter concerning the Soul and Knowledge of Brutes* (1721); Bernard Mandeville, *The Fable of the Bees*, ed. F. B. Kaye (Oxford, 1924), i. 181n. Cf. Wallace Shugg, 'The Cartesian Beast-Machine in English Literature (1663–1750)', *JHI*, 29 (1968).
30. *Œuvres de Descartes*, v. 243 (*'ab internicina illa et jugulatrice sententia'*).
31. *The Works of John Locke* (12th edn, 1824), ix. 283; Ray, *Wisdom*, 38–9; Raven, *Ray*, 374–5; *The Works of . . . Henry St John, Lord Viscount Bolingbroke* (new edn, 1809), v. 344; Howe, *Living Temple*, 89.
32. Evelyn, *Diary*, iii. 234; Richard Meggott, *A Sermon preached at White-Hall* (1683), 11.
33. Goldsmith, iv. 203–4; Bingley, *Quadrupeds*, 2; Harold E. Gruber, *Darwin on Man* (1974), 447.

iv. Maintaining the Boundaries

1. Plato, *Republic*, 571.
2. *Works of Robert Boyle*, v. 553; Thomas Carlyle, *The Letters and Speeches of Oliver Cromwell*, ed. S. C. Lomas (1904), ii. 541; L. Tyerman, *Wesley's Designated Successor* (1882), 412.
3. Michael MacDonald, *Mystical Bedlam* (Cambridge, 1981), 203; *The Diary of Ralph Josselin*, ed. Alan Macfarlane (Brit. Acad., 1976), 349. Cf. George Lyman Kittredge, *Witchcraft in Old and New England* (1929; reprint, New York, 1956), chap. x.
4. Clifford Geertz, 'Deep Play: Notes on the Balinese Cockfight', *Daedalus* (Winter 1972), 7. Cf. Isaac D'Israeli, *Curiosities of Literature* (n.d.), 61; *Social Aspects of the Human Body*, ed. Ted Polhemus (Harmondsworth, 1978), 191.
5. John Rutty, *A Spiritual Diary and Soliloquies* (1776), i. 134, 137 (also 103, 109). Cf. Thomas Granger, *A Familiar Exposition or Commentarie on Ecclesiastes* (1621), 275.
6. Goldsmith, iii. 6–7.
7. Desiderius Erasmus, *De Civilitate Morum Puerilium*, trans. Robert Whittinton (1540), sigs. B7[v], C1[v], C2[v]; *CSPV, 1617–19*, 319.
8. Erasmus, *op. cit.*, sigs. A8, A6, C7[v]–8, A4[v]; Henry Fielding, *The History of Tom Jones* (1749), ix. 5. Cf. C. J. Rawson, *Henry Fielding and the Augustan Ideal under Stress* (1972), 29.
9. E.g. John Hartcliffe, *A Treatise of Moral and Intellectual Virtues* (1691), 197; *The Works of . . . Edmund Burke* (Bohn edn, 1854–70), vii. 190.

10. *Diary of Cotton Mather 1681–1708* (Massachusetts Hist. Soc. Collns., 7th ser., vii, Boston, 1911), 357; *ibid., 1709–1724* (*ibid.*, viii, Boston, 1912), 69.
11. E.g. Richard Baxter, *A Holy Commonwealth* (1659), 212.
12. Winthrop D. Jordan, *White over Black* (Baltimore, 1969), 33; Charles Barber, *Early Modern English* (1976), 157–8. Cf. George Coffin Taylor, 'Shakespeare's Use of the Idea of the Beast in Man', *Studs. in Philology*, 42 (1945).
13. Bartholomew Batty, *The Christian Mans Closet*, trans. William Lowth (1581), fol. 95v.
14. Beryl Rowland, *Animals with Human Faces* (1974), *passim*; *The Works of Gerrard Winstanley*, ed. George H. Sabine (Ithaca, N.Y., 1941), 519; Jeremy Collier, *A Short View of the Immorality, and Profaneness of the English Stage* (1698), 6; Jordan, *White over Black*, 238–9.
15. John Stuart Mill, *Essays on Ethics, Religion and Society*, ed. J. M. Robson (1969), 394.
16. MacDonald, *Mystical Bedlam*, 130; Adam Hill, *The Crie of England* (1595), 38; John Block Friedman, *The Monstrous Races in Medieval Art and Thought* (1981), 31.
17. Bacon, ii. 550–1. Cf. J[ohn] B[ulwer], *Anthropometamorphosis* (1653), 474, and C. R. Hallpike, *The Foundations of Primitive Thought* (Oxford, 1979), 153.
18. Weemse, *Explication of the Iudiciall Laws of Moses*, 98; Sir Edward Coke, *Institutes of the Laws of England* (1794–1817 edn), iii. 63.
19. George Abbot, *An Exposition upon the Prophet Ionah* (1600), 549. Cf. Henry Thomas Buckle, *History of Civilization in England* (World's Classics, 1903–4), iii. 265n; John E. Mason, *Gentlefolk in the Making* (1935; reprint, New York, 1971), 81–2; *Englishmen at Rest and Play*, ed. Reginald Lennard (Oxford, 1931), 68–9. For a contrary view see Christofer Middleton, *A Short Introduction for to Learne to Swimme* (1595), sig. A4.
20. G. Gregory, *The Economy of Nature* (1796), iii. 556.
21. William Prynne, *Histrio-Mastix* (1633), 892–3; E. C. Cawte, *Ritual Animal Disguise* (Cambridge and Ipswich, 1978), esp. 21, 79, 181, 209; Topsell, 463.
22. *Philosophical Trans.*, i (1665 and 1666), 519; Richard D. French, *Antivivisection and Medical Science in Victorian Society* (1975), 387–8.
23. Richard Capel, *Tentations* (1633), 356.
24. Gervase Babington, *Comfortable Notes upon the Bookes of Exodus and Leviticus* (1604), 342. Cf. Peter Barker, *A Iudicious and Painefull Exposition upon the Ten Commandements* (1624), 270; Andrew Willet, *Hexapla in Leviticum* (1631), 434; James Usher, *A Body of Divinity* (1645), 280.
25. 25 Hen. VIII, c. 6 (1533–4), renewed by 28 Hen. VIII, c. 6 (1536); 32 Hen. VIII, c. 3 (1540); 2 & 3 Edw. VI, c. 29 (1548), lapsing in 1553 and renewed by 5 Eliz., c. 17 (1562–3) until 24 & 25 Vic., c. 100 (1861). For incest see Victor Bailey and Sheila Blackburn, 'The Punishment of Incest Act 1908', *Criminal Law Rev.* (1979).
26. Erasmus, *De Civilitate Morum Puerilium*, sig. C2v; Capel, *Tentations*, 356.
27. Crooke, *Description of the Body of Man*, 209; Philip Stubbes, *A Christall Glasse for Christian Women* (1618), sigs. A4v–B1. Cf. Topsell, 105–6; Kittredge, *Witchcraft*, chap. x.

28. Raymond Firth, *Elements of Social Organization* (1951), 199.
29. Stuart Clark, 'King James's *Daemonologie*', in *The Damned Art*, ed. Sydney Anglo (1977), 177.
30. *All the Workes of Iohn Taylor the Water-Poet* (1630), 232; Pepys, *Diary*, vi. 290. Cf. [Sir James Stewart], *Jus Populi Vindicatum* (1669), 239 ('drunk as a beast').
31. S. T. Coleridge, *Lay Sermons*, ed. R. J. White (1972), 183n. Cf. 'Phylotheus Physiologus' [Thomas Tryon], *Monthly Observations for the Preserving of Health* (1688), 7. For reflections on this theme see John Berger, 'Animal World', *New Soc.*, 25 Nov. 1971; Mary Midgley, 'The Concept of Beastliness', *Philosophy*, 48 (1973), and *id.*, *Beast and Man* (Hassocks, 1979) (and review by John Benson in *The Listener*, 102 (2 Aug. 1979)).

v. Inferior Humans

1. Mary Douglas, *Implicit Meanings* (1975), 289; Rodney Needham, *Primordial Characters* (Charlottesville, 1978), 5.
2. J[ohn] [Rogers], *The Displaying of an Horrible Secte of Grosse and Wicked Heretiques* (1578), sig. Ivii[v].
3. Gibbon, *Decline and Fall*, v. 314.
4. Robert Gray, *A Good Speed to Virginia* (1609), cit. Alden T. Vaughan, ' "Expulsion of the Savages" . . .' *William and Mary Quarterly*, xxxv (1978), 61; Edward, Earl of Clarendon, *Miscellaneous Works* (2nd edn, 1751), 195; T[homas] H[erbert], *A Relation of Some Yeares Travaile* (1634), 16–17; *The Wild Man Within*, ed. Edward Dudley and Maximillian E. Novak (Pittsburgh, 1972), 188; R. W. Frantz, *The English Traveller and the Movement of Ideas, 1660–1732* (1934; reprint, New York, 1968), 37. See also Margaret T. Hodgen, *Early Anthropology in the Sixteenth and Seventeenth Centuries* (Philadelphia, 1964), 410–13; Jordan, *White over Black*, *passim*; James Walvin, *Black and White* (1973), 163, 168.
5. *Wild Man Within*, 89; [Robert Johnson], *Nova Britannia* (1609), sig. B4; *The Jamestown Voyages*, ed. Philip L. Barbour (Hakluyt Soc., 1969), i. 134; Karen Ordahl Kupperman, 'English Perceptions of Treachery, 1583–1640', *Hist. Jnl*, 20 (1977), 265; *The Works of the Reverend Mr Edm. Hickeringill* (1709), ii. 446. In *Settling with the Indians* (1980), 176, Karen Ordahl Kupperman stresses the relative untypicality of these views.
6. Nicholas P. Canny, 'The Ideology of English Colonization', *William and Mary Quarterly*, xxx (1973), 588; Sir William Petty, *The Political Anatomy of Ireland* (1691; reprint, Shannon, 1970), 27; David Beers Quinn, *The Elizabethans and the Irish* (Ithaca, N.Y., 1966), 169.
7. Bulwer, *Anthropometamorphosis*, 411. Cf. L. Perry Curtis, Jr, *Apes and Angels* (Newton Abbot, 1971), esp. chap. 4.
8. John Moore, *A Mappe of Mans Mortalitie* (1617), 43. Cf. Aristotle, *Ethics*, 111[b]; *id.*, *Hist. An.*, 588[a–b]; Joseph Butler, *The Analogy of Religion* (1736; World's Classics, 1907), i. 1. 21; MacDonald, *Mystical Bedlam*, 43.
9. George Fox, *A Collection of Many Select and Christian Epistles*, ii (1698), 310; *Works of Gerrard Winstanley*, 576.

10. *HMC*, vii. 623; Essex R.O., D/ACA 16, fol. 122; Northants. R. O., Peterborough Diocesan Records, 43 (correction book, 1611–15), fol. 214v; William C. Braithwaite, *The Second Period of Quakerism* (1919), 270. On the (false) tradition that the doctrine had been upheld by the early Church, see Herbert Thurston, 'Has a Council denied that Women have Souls?', *The Month*, 559 (Jan. 1911).
11. John White, *The First Century of Scandalous, Malignant Priests* (1643), 50; F. G. Emmison, *Elizabethan Life: Morals and the Church Courts* (Chelmsford, 1973), 160.
12. Dorothy McLaren, 'Fertility, Infant Mortality and Breast Feeding in the Seventeenth Century', *Medical History*, 22 (1978); *Jane Austen's Letters*, ed. R. W. Chapman (2nd edn, Oxford, 1952), 488.
13. Alexander Murray, *Reason and Society in the Middle Ages* (Oxford, 1978), 241–2; Baker, *Image of Man*, 273.
14. Sir Thomas Pope Blount, *A Natural History* (1693), sig. A6.
15. *Economic Writings of Petty*, i. 275; Bodl., Ashmole MS. 240, fol. 282 (William Lilly); *The Journeys of Celia Fiennes*, ed. Christopher Morris (1947), 265.
16. *Letters and Papers, Foreign and Domestic of the Reign of Henry VIII*, ed. J. S. Brewer *et al.* (1862–1932), xi. 780(2); Owen, *Description of Penbrokshire*, 12; Margaret Gray, *The History of Bury* (Bury, 1970), 7; James Browne, *Travels over England, Scotland, and Wales* (1700), 115; Tyerman, *Wesley's Designated Successor*, 259. For other views of the common people as beastlike see Christopher Hill, *Change and Continuity in Seventeenth-Century England* (1974), chap. 8.
17. Christopher Hill, *Puritanism and Revolution* (1958), 227.
18. Michel Foucault, *Madness and Civilization*, trans. Richard Howard (1967), 72. Cf. Charlotte Bronte, *Jane Eyre* (1847), chap. 26; MacDonald, *Mystical Bedlam*, 179; Robert Chambers, *Vestiges of the Natural History of Creation* (1887 edn), 254.
19. Paul Slack, 'Poverty and Politics in Salisbury, 1597–1666', in *Crisis and Order in English Towns, 1500–1700*, ed. Peter Clark and Paul Slack (1972), 167; *The Morning Exercises at Cripplegate, St Giles in the Fields, and in Southwark* (5th edn, by James Nichols, 1844–5), i. 232. Cf. [Robert Shelford], *Lectures or Readings* (1602), 26; *Considerations concerning Common Fields* (1654), 24; M. G. Jones, *The Charity School Movement* (Cambridge, 1938), 145; Kupperman, *Settling with the Indians*, 135.
20. Jordan, *White over Black*, 60; Fynes Moryson, *An Itinerary* (Glasgow, 1907–8), ii. 95–6. Cf. Richard S. Dunn, *Sugar and Slaves* (1973), 77; Walvin, *Black and White*, 42–3.
21. Dunn, *op. cit.*, 252; Arthur Zilversmit, *The First Emancipation* (1967), 7.
22. F. O. Shyllon, *Black Slaves in Britain* (1974), 9.
23. See George M. Fredrickson, *White Supremacy* (New York, 1981), 73–5.
24. Samuel Pyeatt Menefee, *Wives for Sale* (Oxford, 1981), 2, 70–71.
25. Timothy Nourse, *Campania Foelix* (1700), 15–16.
26. To George Dyer, undated; *Christie's Sale Catalogue*, 2 April 1975, no. 101.
27. André G. Haudricourt, 'Domestication des animaux, culture des plantes et traitement d'autrui', *L'Homme*, ii (1962).

28. Ainsworth, *Annotations*, i. 7.
29. *Memoirs of Denzil Lord Holles* (1699), 1.
30. Borlase, *Natural History of Cornwall*, 291; Goldsmith, iv. 94.
31. Sir Frederick Pollock and Frederic William Maitland, *The History of English Law* (2nd edn, Cambridge, 1911), ii. 449; James L. Axtell, 'The Scholastic Philosophy of the Wilderness', *William and Mary Qtly.*, xxix (1972), 344.
32. T. C. Smout, *A History of the Scottish People, 1560–1830* (1969), 113; *Diary of Thomas Burton*, ed. John Towill Rutt (1828), ii. 210–11.
33. John Locke, *Two Treatises of Government*, ed. Peter Laslett (Cambridge, 1960), 401 (ii, para 172); Essex R.O., Q/SBa 2/91, cit. J. A. Sharpe, 'Crime in the County of Essex, 1620–1680' (Oxford D.Phil. thesis, 1978), 175.
34. Friends' House, MS. Portfolio i. 20, cit. Barry Reay, 'Early Quaker Activity and Reactions to it, 1652–1664' (Oxford D.Phil. thesis, 1980), 151.
35. *Politics, Religion and Literature in the Seventeenth Century*, ed. William Lamont and Sybil Oldfield (1975), 61–2; Nehemiah Wallington, *Historical Notices of Events*, ed. R. Webb (1869), ii. 243.
36. Murray, *Reason and Society*, 236; Karl Marx, *Grundrisse*, trans. Martin Nicolaus (Harmondsworth, 1973), 606.
37. *The Complete Works of St Thomas More*, viii(1), ed. Louis A. Schuster *et al.* (1973), 307–8; *M. Derings Workes* (1597), sig. B6v; Peter Heylyn, *Aërius Redivivus* (2nd edn, 1672), 285; Richard L. Greaves, *Society and Religion in Elizabethan England* (Minneapolis, 1981), 78–80.
38. E.g. Emmison, *Elizabethan Life: Morals and the Church Courts*, 142; *Middlesex County Records*, ed. John Cordy Jeaffreson (1886–92), iii. 179; *CSPD, 1611–18*, 361, 262, 540; *ibid., 1631–3*, 256; Thomas Edwards, *Gangraena* (1646), iii. 18.
39. John Vicars, *Jehovah-Jireh. God in the Mount* (1644–6), i. 430–31.
40. Natalie Zemon Davis, *Society and Culture in Early Modern France* (Stanford, 1975), chap. 6. Cf. R. W. Scribner, *For the Sake of Simple Folk* (1981), 74–7.
41. Nourse, *Campania Foelix*, 197–8; Walvin, *Black and White*, 177, 182; *Regall Tyrannie Discovered* (1647), 11.
42. *CSPD, 1595–7*, 344; *The Political Works of James Harrington*, ed. J. G. A. Pocock (Cambridge, 1977), 240; *The Memoirs of Edmund Ludlow*, ed. C. H. Firth (Oxford, 1894), i. 206–7.
43. Burton, *Diary*, i. lxx; Locke, *Two Treatises*, 407 (ii. para. 181).
44. Benjamin Vincent, *Haydn's Dictionary of Dates* (25th edn, 1910), 1239 (no source given); [J. Percival], *The Morality of Cumberland and Westmoreland* (1865), 23; *Local Population Studies*, 12 (1974), 26; *The Lisle Letters*, ed. Muriel St Clare Byrne (1981), v. 553.
45. *Works of Gerrard Winstanley*, 612.
46. 13 Ric. II, st. 1, c. 13 (1389–90); *Proceedings in Parliament, 1610*, ed. Elizabeth Read Foster (1966), i. 51. On the operation of the game laws between 1671 and 1831 see P. B. Munsche, *Gentlemen and Poachers* (Cambridge, 1981).
47. Hugo Grotius, *De Jure Belli et Pacis*, ed. William Whewell (Cambridge, n.d.), ii. 3. 5; Pufendorf, *Law of Nature and Nations*, 314. On the complex legal history

of this issue see W. S. Holdsworth, *A History of English Law*, vii (n.d.), 490–95.

48. Edwards, *Gangraena*, iii. 20; William Blackstone, *Commentaries on the Laws of England* (new edn, 1813), iv. 378 (iv. chap. 23).
49. E. P. Thompson, *Whigs and Hunters* (1975), 162; D. J. V. Jones, 'The Poacher', *Historical Jnl*, 22 (1979), 839. Cf. *Albion's Fatal Tree*, ed. Douglas Hay *et al.* (1975), 207, and Munsche, *Gentlemen and Poachers*, 6–7, 63–4.
50. Mary Wollstonecraft, *The Rights of Woman* (EL, 1929), 190. Cf. G. G. Coulton, *The Medieval Village* (Cambridge, 1925), 246.

CHAPTER II NATURAL HISTORY AND VULGAR ERRORS

i. Classification

1. Urban T. Holmes, 'Gerald the Naturalist', *Speculum*, xi (1936); William Worcestre, *Itineraries*, ed. John H. Harvey (Oxford, 1969), 75–7, 135, 137; A. C. Seward, 'The Foliage, Flowers and Fruit of Southwell Chapter House', *Procs. Cambridge Antiqn. Soc.*, 35 (1935); Joan Evans, *Nature in Design* (1933), 64–5, 97; G. Evelyn Hutchinson, 'Attitudes towards Nature in Medieval England', *Isis*, 65 (1974); M. R. James, 'An English Medieval Sketch-Book', *Walpole Soc.*, 13 (1924–5); Brunsdon Yapp, *Birds in Medieval Manuscripts* (1981).
2. Joan Evans, *A History of the Society of Antiquaries* (Oxford, 1956), 122.
3. As, for example, in the fourteenth-century bird illustrations in Cambridge; Francis Klingender, *Animals in Art and Thought to the End of the Middle Ages*, ed. Evelyn Antal and John Harthan (1971), 422–3. Cf. the comments of E. H. Gombrich on the Southwell foliage in *The Sense of Order* (1979), 189.
4. It is admirably told in Raven, *Naturalists*, and Raven, *Ray*. Canon Raven's perspective, however, was distinctly 'whiggish'; his concern was the emergence of 'modern' natural science and he was intolerant of older and outmoded ways of thinking. Allen, *Naturalist*, is a learned and elegant book, furnished with an excellent bibliography, but rather brief on the period before 1700.
5. Coles, *Simpling*, 15.
6. John Parkinson, *Theatrum Botanicum* (1640), sig. (a) 4; *The Grete Herball* (1529 edn), sig. Liii^{v}. On the herbals see Eleanour Sinclair Rohde, *The Old English Herbals* (1922) and Agnes Arber, *Herbals* (2nd edn, Darien, Conn., 1970).
7. John Lyon, 'The "Initial Discourse" to Buffon's *Histoire naturelle*', *Jnl Hist. of Biology*, 9 (1976), 161–2; Topsell, sig. A4; Georges Gusdorf, *Les Sciences humaines et la pensée occidentale*, ii (Paris, 1967), 461.
8. *Medieval Handbooks of Penance*, ed. John T. McNeill and Helena M. Gamer (New York, 1938), 40, 120, 157; Harold Barclay, *The Role of the Horse in Man's Culture* (1980), 74–5, 133; *The Whole Works of Jeremy Taylor*, ed. Reginald Heber, rev. Charles Page Eden (1847–54), ix. 362; A. S. Bicknell, 'Hippophagy', *Jnl of Soc. of Arts*, xvi (1867–8); *Animals in Folklore*, ed. J. R. Porter and W. M. S. Russell (Folklore Soc., 1978), 88.
9. 2 & 3 Edw. VI, c. 19 (1548); Thomas Wilson, *A Commentarie upon the Most*

Divine Epistle of S. Paul to the Romanes (1614), 1185; A Prebendary of York, *An Enquiry about the Lawfulness of Eating Blood* (1733), 69.

10. John Fines, '"Judaising" in the Period of the English Reformation', *Trans. Jewish Hist. Soc.*, xxi (1962–7), 323; *The Reports of . . . Sr Henry Hobart* (4th edn, 1678), 236; Thomas Edwards, *Gangraena* (1646), ii. 2; Nathanael Homes, *Daemonologie* (1650), 193.
11. Ray, *Willoughby*, sig. a 2. Cf. Joannes Amos Comenius, *Orbis Sensualium Pictus* (3rd edn, 1672), 47; Louis Lemery, *A Treatise of Foods* (Eng. trans., 1704), 155–6.
12. Thomas Muffett, *Healths Improvement*, enlarged by Christopher Bennet (1655), 102; Richard S. Dunn, *Sugar and Slaves* (1973), 275; William Smellie, *The Philosophy of Natural History* (Edinburgh, 1790–99), ii. 251.
13. W. Robertson Smith, *Lectures on the Religion of the Semites* (Edinburgh, 1889), 265–8, 277–8, 286; Pliny, *Nat. Hist.*, viii. 45; Fynes Moryson, *An Itinerary* (Glasgow, 1907–8), ii. 82; iv. 200–201; *Advertisements for Ireland*, ed. George O'Brien (Dublin, 1923), 8–9; Robert Trow-Smith, *A History of British Livestock Husbandry* (1957), 188–9.
14. Julius Caesar, *De Bello Gallico*, v. 12; Andrew Boorde, *The Fyrst Boke of the Introduction of Knowledge*, ed. F. J. Furnivall (EETS, 1870), 275; *Year Books of Edward II, viii (Eyre of Kent 6 & 7 Edward II, iii)*, ed. William Craddock Bolland (Selden Soc., 1913), xlix–li; *The Gospelles of Dystaves* (1507), sig. a vii; John Bulwer, *Anthropometamorphosis* (1653), 175.
15. Bingley, *An. Biog.*, i. 86.
16. James Hart, *ΚΛΙΝΙΚΗ, or the Diet of the Diseased* (1635), 85; [Richard Morison], *A Remedy for Sedition* (1536), sig. Eiv^{v}; Izaak Walton, *The Compleat Angler* (1653), pt 1, chap. 20; John Ray, *Observations Topographical, Moral, and Physiological; made on a Journey* (1673), 404; Thomas Cook, *Anecdotes of Mr Hogarth* (1803), 301. On generation from putrefaction, below, pp. 87–8.
17. Topsell, 144; Bacon, ii. 625; Bingley, *Quadrupeds*, 4.
18. Pennant, *Zoology*, iii. 95; *The Regulations and Establishment of the Household of Henry Algernon Percy, the Fifth Earl of Northumberland* (1905), 409.
19. Charles Eliot Norton, *The Poet Gray as a Naturalist* (Boston, 1903), 26–37; A. W. Oxford, *English Cookery Books* (1913), 59; Charles Stevens and John Liebault, *Maison Rustique*, trans. Richard Surflet, ed. Gervase Markham (1616), 506; Bacon, ii. 625; Muffett, *Healths Improvement*, 51, 77–8, 102; *The Miscellaneous Writings of Sir Thomas Browne*, ed. Geoffrey Keynes (new edn, 1946), 406; J[oshua] Childrey, *Britannia Baconica* (1661), 19.
20. Bacon, ii. 625; William Borlase, *The Natural History of Cornwall* (Oxford, 1768), 284; Moryson, *An Itinerary*, iv. 98; Ray, *Willoughby*, 196.
21. *The Animal Kingdom, or Zoological System of the celebrated Sir Charles Linnaeus*, trans. Robert Kerr (1792), 129–35.
22. Aristotle, *De Part. An.*, 643^{b} (though cf. *Hist. An.*, 488^{a}).
23. Robert Lovell, *ΠΑΝΖΩΟΡΥΚΤΟΛΟΓΙΑ. Sive Panzoologicomineralogia* (Oxford, 1661), intro.
24. Thomas Mouffet, *The Theater of Insects*, appended to Edward Topsell, *The*

History of Four-Footed Beasts, rev. by J. R. (1658), 893. Cf. Pliny, *Nat. Hist.*, viii. 56.

25. W. S. Holdsworth, *A History of English Law*, vii (1925), 489–90; Sir Edward Coke, *Institutes of the Laws of England* (1794–1817 edn), iii. chap. 47; *The First Part of the Reports of S[r] George Croke*, rev. Sir Harbottle Grimston (1669), 125–6.
26. P. M. North, *The Modern Law of Animals* (1972), 9.
27. Topsell, 4; *The Workes of . . . Gervase Babington* (1615), i. 246; Pennant, *Zoology*, ii. 283.
28. Henry More, *An Antidote against Atheism* (2nd edn, 1655), 120; *Oldenburg*, vii. 342; Jonathan Swift, *Gulliver's Travels* (1726), pt ii, chap. i; [John Newton and William Cowper], *Olney Hymns* (Glasgow, 1829), 276.
29. [Nicholas Cox], *The Gentleman's Recreation* (1677; reprint, East Ardsley, 1973), iv. 32.
30. Pennant, *Zoology*, iii. 3, 504; Goldsmith, vii. 91; *Batman upon Bartholome, His Booke De Proprietatibus Rerum* (1582), fol. 351; *Oldenburg*, vii. 26; *The Works of Michael Drayton*, ed. J. William Hebel (Oxford, 1961), iv. 33; Tobias Crisp, *Christ Alone Exalted* (5th edn, 1816), ii. 62; Edward Bury, *The Husbandmans Companion* (1677), 216.
31. Edmund Leach, 'Anthropological Aspects of Language', in *New Directions in the Study of Language*, ed. Eric H. Lenneberg (Cambridge, Mass., 1964), 40–42.
32. *Philosophical Letters between the late learned Mr Ray and Several of his Ingenious Correspondents*, ed. W. Derham (1718), 242; Richard Bradley, *A Philosophical Account of the Works of Nature* (1721), 121; Goldsmith, viii. 3; Pennant, *Zoology*, iii. 17; Bingley, *An. Biog.*, iv. 228; i. 80.
33. Thomas Brooks, *The Crown & Glory of Christianity* (1662), 93.
34. John Hill, *An History of Animals* (1752), 323, 221, 552, 541.
35. Lovell, *Panzoologicomineralogia*, intro.; Bingley, *An. Biog.*, i. 25 (cf. Pennant, *Zoology*, ii. 367).
36. Aristotle, *Hist. An.*, 488[b], 588[a], 610[b]; Ray, *Willoughby*, opposite 55; Goldsmith, v. 162–3.
37. Cox, *Gentleman's Recreation*, iii. 61.
38. [Oliver] St John, *An Argument of Law concerning the Bill of Attainder of High-Treason of Thomas Earle of Strafford* (1641), 72.
39. Beryl Rowland, *Blind Beasts* ([Kent, Ohio], 1971), 128.
40. Joan Thirsk, *Horses in Early Modern England* (Reading, 1978), 24 (an indispensable account). See also P. R. Edwards, 'The Horse Trade of the Midlands in the Seventeenth Century', *AgHR*, 27 (1979). For contemporary accounts of different types of horse, see, e.g., Topsell, 285–93; Gervase Markham, *Countrey Contentments* (1615), i. 67; Bradley, *A Philosophical Account*, 93–4; Richard Berenger, *The History and Art of Horsemanship* (1771), i. 178–81.
41. See C. M. Prior, *Early Records of the Thoroughbred Horse* (1924); Peter Willett, *An Introduction to the Thoroughbred* (rev. edn, 1975); and the comments of Ronald Paulson, *Emblem and Expression* (1975), 165.
42. Prior, *ibid.*, 141.
43. Eric Kerridge, *The Agricultural Revolution* (1967), chap. ix (cattle, sheep);

The Berkeley Manuscripts, ed. Sir John Maclean (Gloucester, 1883–5), ii. 363 (foxhounds); William Ramesey, *Mans Dignity* (1661), 54 (pigeons).

44. Cox, *Gentleman's Recreation*, i. 27; *Sotheby's Sale Catalogue*, 22 June 1976, item no. 161.
45. *Seventeenth-Century Economic Documents*, ed. Joan Thirsk and J. P. Cooper (Oxford, 1972), 331; Anthony Ashley Cooper, 3rd Earl of Shaftesbury, *Characteristicks* (1737), iii. 218.
46. William Cavendish, *A New Method and Extraordinary Invention to Dress Horses* (1667), sig. (b)2; *Tracts on Liberty in the Puritan Revolution*, ed. William Haller (New York, 1934), ii. [48–9].
47. Cf. Arthur O. Lovejoy, *The Great Chain of Being* (1936; New York, 1960); William F. Bynum, 'The Great Chain of Being after Forty Years', *Hist. of Sci.*, xiii (1975).
48. E.g. *The Petty Papers*, ed. Marquis of Lansdowne (1927), ii. 21–30; *The Petty-Southwell Correspondence*, ed. Marquis of Lansdowne (1928; reprint, New York, 1967), 43–6; M. F. Ashley Montagu, *Edward Tyson* (Philadelphia, 1943), 242; Topsell, 6.
49. William Swainson, *A Preliminary Discourse on the Study of Natural History* (1834), 76.
50. John Caius, *Of English Dogges*, trans. Abraham Fleming (1576), 24 (and similar stories in Raven, *Naturalists*, 6–7; Thomas Fuller, *The Worthies of England*, ed. John Freeman (1952), 488).
51. Cf. Claude Lévi-Strauss, *Totemism*, trans. Rodney Needham (Harmondsworth, 1969); Rodney Needham, *Primordial Characters* (Charlottesville, Va., 1978), 4–5, 39; Barry Barnes and Steven Shapin, 'Where is the Edge of Objectivity?', *Brit. Jnl for the Hist. of Sci.*, x (1977); Yi-Fu Tuan, *Topophilia* (Englewood Cliffs, N.J., 1974), 18; Mary Douglas, *Implicit Meanings* (1975), 285; Marshall Sahlins, *The Use and Abuse of Biology* (1977), 101.
52. Karl Figlio, 'Chlorosis and Chronic Disease in Nineteenth-Century Britain', *Social History*, 3 (1978), 169 (my italics).
53. Bury, *Husbandmans Companion*, 81; R. T. Gunther, *Early Science in Oxford*, iii (Oxford, 1925), 476; William Turner, *A Compleat History of the Most Remarkable Providences* (1697), iv. 51.
54. *The Theological Works of ... Henry More* (1708), 33 (following Aristotle, *Hist. An.*, 614^{b} and Pliny, *Nat. Hist.*, x. 23).
55. Charlotte M. Yonge, *An Old Woman's Outlook in a Hampshire Village* (1896), 257–9; Goldsmith, v. 221; Sir Thomas Browne, *Notes and Letters on the Natural History of Norfolk* (1902), 96.
56. Smellie, *Philosophy of Natural History*, i. 420.
57. Richard Jobson, *The Golden Trade* (1623), 152.
58. Plot, *Staffs.*, 233.
59. George R. Jesse, *Researches into the History of the British Dog* (1866), i. 319; William Camden, *Britannia*, ed. Richard Gough (2nd edn, 1806), i. 4.
60. Joseph Warder, *The True Amazons* (3rd edn, 1716), v–vi (plagiarizing Moses Rusden, *A Further Discovery of Bees* (1679), sig. A2). Cf. Virgil, *Georgics*, iv;

Varro, *Rerum Rusticarum*, iii. 16. Hilda M. Ransome, *The Sacred Bee in Ancient Times and Folklore* (1937) is a valuable compilation.

61. John Levett, *The Ordering of Bees* (1634), 68; Rusden, *Further Discovery*, 3.
62. E.g. Charles Butler, *The Feminine Monarchie* (Oxford, 1609), sig. A3; *The Works of John Selden*, ed. David Wilkins (1726), iii. col. 928; John Thorley, *ΜΕΛΙΣΣΗΛΟΓΙΑ or, the Female Monarchy* (1744), 7, 11–12, 47, 48.
63. *A Supplement to Mr Chambers's Cyclopaedia* (1753), ii, 'Queen-Bee'. Cf. Ransome, *Sacred Bee*, 208.
64. Samuel Purchas, *A Theatre of Politicall Flying-Insects* (1657).
65. Rusden, *Further Discovery*, sigs. A2^v–3. Cf. 17, 21, 28.
66. Cited by Ellis, iii(3). 150. Cf. Morris Berman, *Social Change and Scientific Organization* (1978), 7.
67. *Patriarcha and other Political Works of Sir Robert Filmer*, ed. Peter Laslett (Oxford, 1949), 84. Cf. W. H. Greenleaf, *Order, Empiricism and Politics* (1964), 45; Michael Hudson, *The Divine Right of Government* (1647), 62; Ro. Grose, *Royalty and Loyalty* (1647), 4.
68. Joseph Caryl, *An Exposition ... upon ... Job* (1643–66), xi. 625–7; vi. 117. Cf. Peter Barker, *A Iudicious and Painefull Exposition upon the Ten Commandements* (1624), 206; Alexander Rosse, *Leviathan drawn out with a Hook* (1653), 19.
69. E.g. Muffett, *Theater of Insects*, 1025; Topsell, sig. A4; [Thomas Scott], *The Belgick Pismire* (1622), 17–18; Brooks, *Crown and Glory of Christianity*, 32; *Oldenburg*, vii. 51; Butler, *Feminine Monarchie*, sigs. b1, b5; Henry Power, *Experimental Philosophy* (1664), 25; *The Theological, Philosophical and Miscellaneous Works of the Rev. William Jones*, vi. 47.
70. *The Sermons of M. John Calvin upon ... Deuteronomie*, trans. Arthur Golding (1583), 560, 776; John Maplet, *A Greene Forest* (1567), fol. 104; Muffett, *Theater of Insects*, sig. Ffff4; Bury, *Husbandmans Companion*, 300; Topsell, 245; Henry Church, *Miscellanea Philo-Theologica* (1637), 84–5.
71. Pennant. *Zoology*, i. 101. Cf. Mrs Eliza Brightwen, *More about Wild Nature* (n.d.), 91 n3.
72. Joseph Taylor, *Ornithologia Curioso* (1807), 62–3.
73. [J. L. Knapp], *The Journal of a Naturalist* (3rd edn, 1830), 153, 170; Brightwen, *op. cit.*, 106.
74. Cf. John Berger, 'Animals as Metaphor', *New Soc.*, 39 (10 March 1977); Needham, *Primordial Characters*, 53.
75. E.g. Goldsmith, iii. 347; v. 220–21; Bewick, *Quadrupeds*, 162, 206; Bingley, *Quadrupeds*, 450. Beryl Rowland, *Animals with Human Faces* (1974) and *id.*, *Birds with Human Souls* (Knoxville, Tennessee, 1978) are useful alphabetical guides to the symbolic associations of animals and birds. See also Joshua Poole, *The English Parnassus* (1657), compiled out of the English poetry of the preceding century; John Swan, *Speculum Mundi* (2nd edn, Cambridge, 1644), 358–488; G. L. Remnant, *A Catalogue of Misericords in Great Britain* (Oxford, 1969), xxxv–xxxvii, 211–14; Arthur H. Collins, *Symbolism of Animals and Birds represented in English Church Architecture* (1913).

76. [Thomas Wilcox], *A Short, yet sound Commentarie ... on ... the Proverbes of Salomon* (1589), fol. 18ᵛ; George Cheyne, *An Essay on Regimen* (1740), 71–2.
77. See Rosemary Freeman, *English Emblem Books* (1967); Mario Praz, *Studies in Seventeenth-Century Imagery* (2nd edn, Rome, 1964).
78. Muffett, *Theater of Insects*, 951, 980; Thomas Scot, *Philomythie or Philomythologie* (2nd edn, 1616), sig. E5ᵛ; Richard Bradley, *New Improvements of Planting and Gardening* (2nd edn, 1718), iii. 59.
79. Roger Williams, 'A Key into the Language of America', ed. James Hammond Trumbull (Pubs. of the Narragansett Club, Providence, R.I., 1866), i. 130, 190–91; *The Diary of Ralph Josselin*, ed. Alan Macfarlane (Brit. Acad., 1976), 134.
80. There is a vast literature on the history of botanical classification. I have found the following helpful: M. Adanson, *Familles des Plantes* (Paris, 1763), i; Richard Pulteney, *Historical and Biographical Sketches of the Progress of Botany in England* (1790), ii; James Edward Smith, 'Introductory Discourse on the Rise and Progress of Natural History', *Trans. of the Linnaean Soc.*, i (1791); Julius von Sachs, *History of Botany*, trans. Henry E. F. Garnsey, rev. I. B. Balfour (Oxford, 1906); Sydney Howard Vines, 'Robert Morison and John Ray', in *Makers of British Botany*, ed. F. W. Oliver (Cambridge, 1913); Emile Callot, *La Renaissance des sciences de la vie au XVIᵉ siècle* (Paris, 1951); Michel Foucault, *The Order of Things* (Eng. trans., 1970), chap. 5; David Knight, *Ordering the World* (1981); A. G. Morton, *History of Botanical Science* (1981), chaps. 5–8.
81. John Woodward, *Fossils of All Kinds* (1728), ii. 2.
82. See Raven, *Ray*, esp. chap. 8; Phillip R. Sloan, 'John Locke, John Ray, and the Problem of the Natural System', *Jnl of the Hist. of Biology*, 5 (1972); Morton, *History of Botanical Science*, 201–7.
83. On it see James L. Larson, *Reason and Experience* (1971). For its acceptance in England, Carl Linnaeus, *Species Plantarum* (facsimile of 1st edn, 1753), with intro. by W. T. Stearn (Ray Soc., 1975–9), i. 79–80; Henrey, ii. 89–90.
84. Stearn, intro. to Linnaeus, *Species Plantarum*, 25.
85. *A System of Vegetables ... translated from the 13th edn ... of the Systema Vegetabilium of the late Professor Linneus ... by a Botanical Society at Lichfield* (Lichfield, 1782), i. 3–5.
86. Raven, *Ray*, 193; Sloan, 'John Locke, John Ray and the Problem of the Natural System', 4–5, 28–9; Morton, *History of Botanical Science*, 135–6, 201.
87. E.g. *The Letters of Sir William Jones*, ed. Garland Cannon (Oxford, 1970), ii. 776.
88. André Haudricourt, 'Botanical Nomenclature and its Translation', in *Changing Perspectives in the History of Science*, ed. Mikulás Teich and Robert Young (1973), 267.
89. *Miscellaneous Tracts relating to Natural History*, trans. Benjamin Stillingfleet (2nd edn, 1762), xxii–xxiii.
90. Émile Guyénot, *Les Sciences de la vie aux xviiᵉ et xviiiᵉ siècles* (Paris, 1941), bk i, chap. iv.
91. E.g. Lyon, '"Initial Discourse" to Buffon's *Histoire naturelle*', 164; and other writers cited by Daniel J. Boorstin, *The Lost World of Thomas Jefferson* (New York, 1948), 137.

92. Cf. Arthur B. Ferguson, *Clio Unbound* (Durham, N. C., 1979), 60.
93. Kitty W. Scoular, *Natural Magic* (Oxford, 1965), 8–9; Ray, *Willoughby*, sig. A4.
94. Freeman, *English Emblem Books*, 227–8; Herbert M. Atherton, *Political Prints in the Age of Hogarth* (Oxford, 1974), 29–30.
95. *The Complete Works of Benjamin Franklin*, ed. John Bigelow, viii (1888), 444.
96. *Works of Jeremy Taylor*, ix. 283–4; *Works of Rev. William Jones*, iii. 23.
97. Hobbes, *EW*, ii. 66–7; iii. 156–7; iv. 120–21. Cf. John Hall of Richmond, *Of Government and Obedience* (1654), 30, 34, 99–100, 168.
98. John Swammerdam, *The Book of Nature*, trans. Thomas Floyd, rev. John Hill (1758), 170; Hester Hastings, *Man and Beast in French Thought of the Eighteenth Century* (1936), 88–93. Also John Keys, *The Practical Bee-Master* (n.d., [? 1780]), 2.
99. [Henry Home of Kames], *Elements of Criticism* (6th edn, Edinburgh, 1785), ii. 186; *Letters of Hartley Coleridge*, ed. Grace Evelyn Griggs and Earl Leslie Griggs (1941 edn), 173.
100. Aristotle, *De Part. An.*, 645[a].
101. Mouffet, *Theater of Insects*, 1012; Sir Thomas Browne, *Religio Medici* (1643), i. paras. 15–16; Plato, *Gorgias*, 474[a].
102. Shaftesbury, *Characteristicks*, ii. 388; *The Life, Unpublished Letters, and Philosophical Regimen of Anthony, Earl of Shaftesbury*, ed. Benjamin Rand (1900), 121–2.
103. Ray, *Willoughby*, 170; Sir Charles Linné, *A General System of Nature*, trans. and ed. William Turton (1806), *passim*; Pennant, *Zoology*, i. 65; iii. 18–19.
104. Basil Taylor, *Stubbs* (2nd edn, 1975), intro.; Paulson, *Emblem and Expression*, 163–5, 179; Edmund Blunden, *Nature in English Literature* (1929), 23; *John Constable's Discourses*, ed. R. B. Beckett (Suffolk Recs. Soc., 1970), 12n.

ii. Vulgar Errors

1. B. A. L. Cranstone, 'Animal Husbandry: the evidence from ethnography', in *The Domestication and Exploitation of Plants and Animals*, ed. Peter J. Ucko and G. W. Dimbleby (1969), 251.
2. Some early ones are listed in John Russell Smith, *A Bibliographical List of the Works that have been published towards illustrating the Provincial Dialects of England* (1839). For others see the 'Select Bibliographical List', prefaced to *Dialect Dict.*
3. Sternberg, 14; William Dickinson, *A Glossary of Words and Phrases pertaining to the Dialect of Cumberland* (EDS, 1878), 20.
4. Francois Rabelais, *Gargantua and Pantagruel*, trans. Sir Thomas Urquhart (1653–9; World's Classics, 1934), bk iii, chap. 13. For similar lists see *The Boke of Saint Albans*, ed. William Blades (1881), sig. fv–vi; John Manwood, *A Treatise and Discourse of the Lawes of the Forrest* (1598), fols. 26[v]–8; Randle Holme, *The Academy of Armory* (Chester, 1688), ii. 131–6; and others cited by John Hodgkin, 'Proper Terms', *Trans. Philological Soc.* (1907–10).
5. Holme, *op. cit.*, ii. 131–2; Hodgkin, *art. cit.*

6. Claude Lévi-Strauss, *The Savage Mind* (Eng. trans., 1966), 3, 8; Godfrey Lienhardt, *Divinity and Experience* (Oxford, 1961), 16. Cf. Clifford Geertz, 'Common Sense as a Cultural System', *The Antioch Rev.*, 33 (1975), 20–21; G. E. Hutchinson, *The Itinerant Ivory Tower* (1953), 30.
7. *The Gentleman's Recreation* (1674), 197–8. Cf. J. C. Atkinson, *Forty Years in a Moorland Parish* (1907 edn), 338–9; Gerald E. H. Barrett-Hamilton, *A History of British Mammals* (1910–13), ii. 282–3.
8. F. N. L. Poynter, 'Nicholas Culpeper and his Books', *Jnl Hist. Medicine*, 17 (1962), 161–2. For eighteenth-century herbals, see Henrey, iii.
9. *Grete Herball*, sigs. Ui, Giv[v], Fiii, vi.
10. Robert Turner, *Botanologia* (1664), 70; Gerard, *Herball*, ed. Johnson, 906; John R. Wise, *The New Forest* (1863), 176; Worcestre, *Itineraries*, 125n.
11. Coles, *Eden*, 317; Parkinson, *Paradisi*, 472.
12. *The Life of Thomas Cooper by himself* (1897 edn), 18–19.
13. Pulteney, *Sketches of the Progress of Botany*, ii. 309
14. Parkinson, *Paradisi*, 468; J. W[orlidge], *Systema Horti-Culturae* (1677), 230, 224. Cf. Coles, *Simpling*, 48–50.
15. G. W. Francis, *The Little English Flora* (1839), 49, 88; *Agnus Castus, a Middle English Herbal*, ed. Gösta Brodin (Upsala, 1950), 168; Turner, *Botanologia*, 290 (cf. *King Lear*, iv. 6, lines 14–15).
16. A practice forbidden by 5 & 6 Edw. VI, c. 23 (1551–2). Cf. William Turner, *A New Herball* (London and Cologne, 1551–62), ii, fol. 11[v]; Gerard, *Herball*, ed. Johnson, 46.
17. Turner, *New Herball*, i. sig. Cii.
18. Gerard, *Herball*, ed. Johnson, 1582, 1584.
19. *Ibid.*, 1116, 1478; Coles, *Eden*, 575.
20. Gerard, *Herball*, ed. Johnson, 963; Sir Hans Sloane, *A Voyage to the Islands* (1707–25), i. xxv; Francis, *Little English Flora*, 78.
21. Turner, *New Herball*, ii. fol. 18; Gerard, *Herball*, ed. Johnson, 112; James Britten and Robert Holland, *A Dictionary of English Plant-Names* (EDS, 1886), 18.
22. Turner, *New Herball*, i. sig. Ciii; Coles, *Simpling*, 68; *Grete Herball*, sig. Bvi[v].
23. Raven, *Naturalists*, 66–7; *Turner on Birds*, ed. A. H. Evans (Cambridge, 1903), 119; *Miscellaneous Writings of Sir Thomas Browne*, 381.
24. Raven, *Naturalists*, 305, 317.
25. Eleazar Albin, *A Natural History of Birds* (1731–8), sig. A3; *DNB*, 'Curtis, William'; 'Banks, Sir Joseph'.
26. Turner, *New Herball*, i. sig. Aiii[v]; Gerard, *Herball*, ed. Johnson, 672, 676, 1060; Coles, *Simpling*, 4.
27. Morton, *Northants.*, 98; BL, Sloane MS. 3340, fol. 45[v].
28. *Memoirs of the Wernerian Natural History Soc.*, iv (for 1821–2) (Edinburgh, 1822), 219.
29. Butler, *Feminine Monarchie*, sig. a3; Hill, *History of Animals*, 434–5.
30. Samuel Hartlib, *The Compleat Husband-Man* (1659), 63, 68–71; *Miscellaneous Tracts*, trans. Stillingfleet, 370.
31. *Suffolk in the XVIIth Century*, ed. Lord Francis Hervey (1902), 34; Edward Moor,

Suffolk Words and Phrases (1823), 365; Knapp, *Journal of a Naturalist*, 309; Maude Robinson, *A South Down Farm in the Sixties* (1947 edn), 45.

32. Stephen Glover, *The History of the County of Derby*, ed. Thomas Noble (1829), i. 137; T. D. Brushfield in *Trans. Devonshire Assoc.*, xxix (1897), 313–14; Barrett-Hamilton, *British Mammals*, ii. 68n.
33. Swainson, *Preliminary Discourse*, 144; Glover, *op. cit.*, i. 178; *The Garden Book of Sir Thomas Hanmer*, with intro. by Eleanour Sinclair Rohde (1933), xviii; Barrett-Hamilton, *op. cit.*, ii. 100–101.
34. Thomas Dawson, *The Good Huswifes Jewell* (1596), fol. 44; Sternberg, 157; *Gilbert White's Journals*, ed. Walter Johnson (1931; reprint, Newton Abbot, 1970), 357; Evelyn, *Sylva*, 46; Stephen Switzer, *Ichnographia Rustica* (1718), i. 220–21.
35. Rusden, *Further Discovery*, 56, 64; Warder, *True Amazons*, 5, 23.
36. Raven, *Naturalists*, 194–5; Ray, *Willoughby*, 60; *The Journeys of Celia Fiennes*, ed. Christopher Morris (1947), xxxviii.
37. [Sir Peter Pett], *The Happy Future State of England* (1688), 58; Bingley, *An. Biog.*, iii. 171; Rowland Hill, *Journal of a Tour through the North of England* (1799), 86n–87n; Robert Willan, *A Glossary of Words used in the West Riding*, ed. Walter W. Skeat (EDS, 1873), 96.
38. Barrett-Hamilton, *British Mammals*, ii. 37.
39. *Gilbert White's Journals*, 341; 'A Description of Cleveland', *The Topographer and Genealogist*, ii (1853), 419, 418. Cf. *The Diary of Ralph Thoresby*, ed. Joseph Hunter (1830), i. 144–5; Childrey, *Britannia Baconica*, 160.
40. Goldsmith, vi. 2–3; Swainson, *Birds*, 17–18; Barrett-Hamilton, *British Mammals*, ii. 240–42; Sir Thomas Browne, *Pseudodoxia Epidemica* (1646), iii. chap. 5; Charles Leigh, *The Natural History of Lancashire, Cheshire, and the Peak* (Oxford, 1700), i. 148; Joseph Hunter, *The Hallamshire Glossary* (1829), 41; Sternberg, 79.
41. *Agnus Castus*, 124; Turner, *New Herball*, i. sig. ci^{v}; ii. fols. 115, 162^{v}; Plot, *Staffs.*, 223; Coles, *Simpling*, 66–7; Coles, *Eden*, 20, 27, 45; Gerard, *Herball*, ed. Johnson, 1001.
42. Evelyn/Hunter, 396–7; Evans, *Pattern under the Plough*, 132; John Smith, *Englands Improvement Revived* (1670), 181; Mrs Stone, *God's Acre* (1858), 257.
43. John H. Harvey, 'Medieval Plantsmanship in England', *Garden History*, i (1972).
44. Evelyn/Hunter, 223; Cyril E. Hart, *The Free Miners of the Royal Forest of Dean* (Gloucester, 1953), 75–6.
45. Plot, *Staffs.*, 207; Robert Holland, *A Glossary of Words used in the County of Chester* (EDS, 1884–6), 390.
46. Britten and Holland, *Dictionary of English Plant-Names*, 342–3; Nich[olas] Culpeper, *The English Physitian Enlarged* (1653), 163 (denied by Turner, *Botanologia*, 16).
47. Gerard, *Herball*, ed. Johnson, 845–6.
48. Plot, *Staffs.*, 222–3; Gilbert White, *The Natural History and Antiquities of Selborne* (1789; ed. James Fisher, Harmondsworth, 1941), letters to Barrington, xxviii; Evelyn/Hunter, 157n.

49. Sternberg, 161; Geoffrey Grigson, *A Dictionary of English Plant Names* (1974), 68.
50. Turner, *Botanologia*, 114; Keith Thomas, *Religion and the Decline of Magic* (Harmondsworth, 1973), 47, 213–14, 422, 756.
51. Smith, *Englands Improvement*, 210; John Ashton, *Social Life in the Reign of Queen Anne* (new edn, 1897), 42; Charles Smyth, *Church and Parish* (1955), 64; Gordon Huelin in *Guildhall Studs. in London History*, iii (1978), 166–7; J. G. Frazer, *The Magic Art* (3rd edn, 1911), ii. 52, 59; Charles Phythian-Adams, *Local History and Folklore* (1975), and *id.*, 'Rural Culture', in *The Victorian Countryside*, ed. G. E. Mingay (1981); below, pp. 230–31.
52. Swainson, *Birds*, 7, 14, 33, 34; *id.*, *A Handbook of Weather Folk-Lore* (1873), 228–57; Gervase Markham, *The English Husbandman* (1635), 11; *Gospelles of Dystaves*, sigs. Ci, Cvii[v]; *The Works of Symon Patrick*, ed. A. Taylor (Oxford, 1858), viii. 636.
53. Borlase, *Natural History of Cornwall*, 324; White, *Natural History of Selborne*, letters to Barrington, xlvii.
54. Bacon, ii. 516; Gerard, *Herball*, ed. Johnson, 1341.
55. *Gospelles of Dystaves*, sigs. Biv, Eiii[v], Eiv[v]; Elizabeth Mary Wright, *Rustic Speech and Folk-Lore* (1913), 219.
56. Thomas, *Religion and the Decline of Magic*, 747; *Gospelles of Dystaves*, sig. Ciii[v]; *Poems of William Browne*, ed. Gordon Goodwin (n.d., *c.* 1893), ii. 280; Goldsmith, vi. 4–5; vii. 349; Knapp, *Journal of a Naturalist*, 173; Browne, *Pseudodoxia Epidemica*, ii. chap. 7; William Horman, *Vulgaria* (1519), fol. 19.
57. *The Letters of Mrs Gaskell*, ed. J. A. V. Chapple and Arthur Pollard (Manchester, 1966), 31–2.
58. J. W[orlidge], *Systema Agriculturae* (1669), 261; Power, *Experimental Philosophy*, 19. See in general Frank Gibson, *Superstitions about Animals* (1904); Thomas, *Religion and the Decline of Magic*, 745–51; Wright, *Rustic Speech and Folk-Lore*, chap. 13.
59. H. Kirke Swann, *A Dictionary of English and Folk-Names of British Birds* (1913), 198–9, 231; Swainson, *Birds*, 14–15, 52–3; Moor, *Suffolk Words and Phrases*, 435.
60. Margaret Cavendish, Marchioness of Newcastle, *Poems and Phancies* (2nd imp., 1664), 87; Thomas Jackson, *A Treatise containing the Originall of Unbeliefe* (1625), 177; George Smith, *Six Pastorals* (1770), 30.
61. Sternberg, 47; *The Guardian*, 61 (21 May 1713); Goldsmith, v. 380–81; vii. 349; Swann, *Dictionary*, 67–8; Wright, *Rustic Speech*, 219; Holland, *Glossary of Words in Chester*, 165, 249. See in general James Ritchie, *The Influence of Man on Animal Life in Scotland* (Cambridge, 1920), 237–40.
62. Swainson, *Birds*, 36–43; Swann, *Dictionary*, 262; Wise, *New Forest*, 282; Phythian-Adams, *Local History and Folklore*, 26.
63. Sternberg, 160.
64. Below, p. 147.
65. 'The Excursion', iv. lines 614–15.
66. Cf. the exchange between Hildred Geertz and the present author in *Jnl of Interdisciplinary History*, vi (1975), 71–109.
67. Browne, *Pseudodoxia Epidemica*, esp. i. chaps. 6–9; William Cobbett, *A Year's*

Residence in America (1818; Abbey Classics, n.d.), para. 201. Cf. Pliny, *Nat. Hist.*, viii. 22, 57; ix. 16; x. 12; xi. 24; xviii. 35; Evelyn/Hunter, 408.

68. As is argued by Edward A. Armstrong, *The Folklore of Birds* (1958), chaps. 2, 9.
69. J. Harvey Bloom, *Shakespeare's Garden* (1903), 26; William Langland, *Piers Plowman*, A-text, passus 1, lines 65–6; *Works of Drayton*, iv. 507.
70. Above, p. 74; Bradley, *A Philosophical Account*, 119; [W. Carr], *The Dialect of Craven* (2nd edn, 1828), i. 250; Swainson, *Birds*, 113; Pliny, *Nat. Hist.*, x. 9.
71. As is suggested by James Edmund Harting, *The Ornithology of Shakespeare* (1871), 104, and Armstrong, *Folklore of Birds*, 72–3.
72. George Perkins Marsh, *Man and Nature* (1864), ed. David Lowenthal (Cambridge, Mass., 1965), 354n.
73. E. Estyn Evans, *Irish Folk Ways* (1957), 297.
74. Worlidge, *Systema Agriculturae*, 262; Sternberg, 159; *Thoresby Diary*, i. 149; *The Hawkins' Voyages*, ed. Clements R. Markham (Hakluyt Soc., 1878), 150; George Gifford, *A Discourse of the Subtill Practises of Devilles* (1587), sigs. C1v–2; Swainson, *Birds*, 81; Wise, *New Forest*, 284; Yonge, *An Old Woman's Outlook*, 38–9, 76.
75. *The Works of Thomas Nashe*, ed. R. B. McKerrow (Oxford, 1966), ii. 172; *Journals of the House of Commons*, i. 983.
76. Wright, *Rustic Speech*, 217. Cf. Michael D. Jackson, 'Structure and Event: Witchcraft Confession among the Kuranko', *Man*, new ser., 10 (1975), 395.
77. Sternberg, 161–2; John Josselyn, *An Account of Two Voyages to New-England* (1674), 193 (cf. Edward Peacock, *A Glossary of Words used in the Wapentakes of Manley and Corringham, Lincolnshire* (EDS, 1887), 133).
78. E.g. *The Workes of . . . Gervase Babington* (1615), i. 495; Edward Elton, *An Exposition of the Epistle of Saint Paul to the Colossians* (3rd edn, 1637), 7, 152. See Thomas, *Religion and the Decline of Magic*, 747–8.
79. Paul S. Seaver, *The Puritan Lectureships* (Stanford, Calif., 1970), 347.
80. William Prynne, *Histrio-Mastix* (1633), 21; Gordon Huelin in *Guildhall Studies in London History*, iii (1978), 168; John Stow, *A Survey of the Cities of London and Westminster*, enlarged by John Strype (1720), ii(2). 66; Tho. Hall, *Funebria Florae* (2nd edn, 1661).
81. Edward Fisher, *A Chrisian* (*sic*) *Caveat* (5th edn, 1653), 63.
82. Morton, *Northants.*, 340.
83. Turner, *New Herball*, ii. fol. 46. Cf. *Grete Herball*, sig. Pviv; Parkinson, *Paradisi*, 377; Gerard, *Herball*, ed. Johnson, 351–2.
84. Bernard Capp, *Astrology and the Popular Press* (1979), 196; *The Morning Exercises at Cripplegate, St Giles in the Fields, and in Southwark* (5th edn, by James Nichols, 1845), v. 65.
85. Plot, *Staffs.*, 22; Childrey, *Britannia Baconica*, 54–5; Borlase, *Natural History of Cornwall*, 219; H[ugh] P[latt], *Floraes Paradise* (1608), 4, 150–51; Ray, *Observations*, 410.
86. Topsell, 712.
87. *True and Wonderfull. A Discourse relating a Strange and Monstrous Serpent* (1614), in *The Harleian Miscellany* (1808–11), iii. 227–31; *Clenennau Letters and*

Papers, ed. T. Jones Pierce, *National Lib. of Wales Jnl*, supplement, ser. iv, pt i (1947), 86. For the classical and mythological sources of such beliefs see Margaret W. Robinson, *Fictitious Beasts, a Bibliography* (1961); Rudolf Wittkower, 'Marvels of the East', *Jnl of the Warburg and Courtauld Institutes*, v (1942), 13; Heinz Mode, *Fabulous Beasts and Demons* (Eng. trans., 1975); Katharine Park and Lorraine J. Daston, 'Unnatural Conceptions', *Past & Present*, 92 (1981).

88. Parkinson, *Theatrum Botanicum*, 1611; Lovell, *Panzoologicomineralogia*, intro.; Thomas Birch, *The History of the Royal Society*, i (1756), 26, 35, 83.
89. Ray, *Willoughby*, sig. a1; Bradley, *Philosophical Account*, 73. Cf. Caryl, *Exposition upon Job*, xi. 346–58; Raven, *Ray*, 380; Odell Shepard, *The Lore of the Unicorn* (1930).
90. Ray, *Observations*, 27; Raven, *Ray*, 132n; Edward Tyson, *Orang-Outang, sive Homo Sylvestris* (1699), sig. A2.
91. Hill, *History of Animals*, 317. For earlier scepticism, Raven, *Naturalists*, 199.
92. Browne, *Pseudodoxia Epidemica*; John Aubrey, *Remaines of Gentilisme and Judaisme*, ed. James Britten (Folk-Lore Soc., 1881); John Ray, *A Collection of English Proverbs* (Cambridge, 1670); Henry Bourne, *Antiquitates Vulgares* (Newcastle, 1725).
93. *A Selection of the Correspondence of Linnaeus*, ed. Sir James Edward Smith (1821), i. 49; *Miscellaneous Tracts*, trans. Stillingfleet, 359; Smellie, *Philosophy of Natural History*, i. 128.
94. *Thraliana*, ed. Katherine C. Balderston (Oxford, 1942), i. 224.

iii. Nomenclature

1. For the history of English plant names the starting-point is William Turner, *Libellus De Re Herbaria* (1538) and *The Names of Herbes* (1548), both reissued in facsimile with introductions by James Britten, B. Daydon Jackson and W. T. Stearn (Ray Soc., 1965). Other guides are R. C. A. Prior, *On the Popular Names of British Plants* (3rd edn, 1879); John Earle, *English Plant Names from the Tenth to the Fifteenth Century* (Oxford, 1880); Britten and Holland, *Dictionary of English Plant-Names* ;Wright, *Rustic Speech*, chap. 21; Grigson, *Dictionary of English Plant Names*.
2. Wright, *Rustic Speech*, 203–4.
3. Gerard, *Herball*, ed. Johnson, 803, 805.
4. Josselyn, *Account of Two Voyages*, 73.
5. Parkinson, *Paradisi*, 300, 283.
6. Gerard, *Herball*, ed. Johnson, 897, 780.
7. *Turner on Birds*, ed. Evans, 19; Moor, *Suffolk Words*, 347. For bird names, see Swainson, *Birds*; Charles Louis Hett, *A Glossary of Popular, Local and Old-Fashioned Names of British Birds* (1902); Wright, *Rustic Speech*, 339–340.
8. Sternberg, 29; James Orchard Halliwell, *A Dictionary of Archaic and Provincial Words* (5th edn, 1845), ii. 650; Pennant, *Zoology*, iii. 14; Bingley, *An. Biog.*, iii. 161; William Cobbett, *The English Gardener* (1829; Oxford, 1980), 222.

9. *A Memoir of Thomas Bewick written by himself* (1979 edn), 15.
10. Raven, *Naturalists*, 127.
11. Turner, *New Herball*, ii. fol. 94ᵛ.
12. Culpeper, *English Physitian Enlarged*, 8, 27.
13. Gerard, *Herball*, ed. Johnson, 774; Turner, *Botanologia*, 216.
14. Gerard, *Herball*, ed. Johnson, 464–5.
15. Britten and Holland, *Dictionary of English Plant-Names*, vii.
16. J. C. Atkinson, *Play Hours and Half-Holidays* (1892 edn), 137.
17. N. F., *The Fruiterers Secrets* (1604), sig. Aij.
18. Parkinson, *Paradisi*, 247.
19. *Dialect Dict.*
20. Browne, *Notes and Letters on Natural History of Norfolk*, 76; Peter J. Schmitt, *Back to Nature* (New York, 1969), 34.
21. Parkinson, *Paradisi*, 571; Henrey, ii. 212.
22. Ellis, i(3). 147.
23. Nicholas Culpeper, *Pharmacopeia Londinensis* (1654), sig. A8ᵛ; Geoffrey Grigson, *The Englishman's Flora* (1955), 72; *The 'Critica Botanica' of Linnaeus*, trans. Sir Arthur Hort, rev. M. L. Green (Ray Soc., 1938), 55–6.
24. E.g. Turner, *Botanologia*, sig. a2; Coles, *Simpling*, 88–92.
25. Raven, *Ray*, 98–9, 159, 224, 464; *The Correspondence of John Ray*, ed. Edwin Lankester (Ray Soc., 1848), 187–8.
26. Coles, *Simpling*, sig. A2. Cf. R. S. Roberts, 'The Early History of the Import of Drugs into Britain', in *The Evolution of Pharmacy in Britain*, ed. F. N. L. Poynter (1965).
27. Thomas Short, *Medicina Britannica* (1746), vii. Cf. William Chafin, *A Second Edition of the Anecdotes and History of Cranbourn Chase* (1818), 59–60.
28. Britten and Holland, *Dictionary of English Plant-Names*, xviii. See, e.g., William Fox, *The Working Man's Model Family Botanic Guide or, Every Man his own Doctor* (10th edn, Sheffield, 1884).
29. Nearly all these names can be found in *Dialect Dict.* or Halliwell, *Dictionary of Archaic or Provincial Words*. For the others see Turner, *Botanologia*, 296; Stephen Blake, *The Compleat Gardeners Practice* (1664), 89.
30. Robert Smith, *The Universal Directory for Taking Alive and Destroying Rats* (1768), 187 (in the prefatory list of contents the bird's name has been tactfully shortened).
31. *The Diaries of Thomas Wilson, D.D., 1731–37 and 1750*, ed. C. L. S. Linnell (1964), 50 (where the name is given as Dunchey).
32. *Petty–Southwell Correspondence*, 306.
33. *'Critica Botanica' of Linnaeus*, 87.
34. In addition to references in note 83 above (p. 326), see John L. Helmer, 'The Early History of Binomial Nomenclature', *Huntia*, i (1964); William T. Stearn, *Botanical Latin* (2nd edn, Newton Abbot, 1973); Pulteney, *Sketches of the Progress of Botany*, ii. 347–52.
35. Despite the urgings of Thomas Martyn, *The Language of Botany* (1793). Cf. André Haudricourt, 'Botanical Nomenclature and its Translation'.

36. Josiah Frampton [William Gilpin], *Three Dialogues on the Amusements of Clergymen* (2nd edn, 1797), 185.
37. Cit. Susie I. Tucker, *Protean Shape* (1967), 73.
38. Stearn, *Botanical Latin*, 6; Swainson, *Preliminary Discourse*, 144.
39. 'Proserpina', in *The Works of John Ruskin*, ed. E. T. Cook and Alexander Wedderburn (1903–12), xxv; Anne Pratt, *The Flowering Plants and Ferns of Great Britain* (n.d. [1857]); W. Robinson, *The English Flower Garden* (1883; 8th edn, 1900), ix.

iv. Changing Perspectives

1. John Aubrey, *Brief Lives*, ed. Anthony Powell (1949), 266; *Forerunners of Darwin*, ed. Bentley Glass *et al.* (Baltimore, 1959), 39–40.
2. *Topographer and Genealogist*, ii (1853), 415. See in general Aristotle, *Hist. An.*, 539[a], 551[a]; Pliny, *Nat. Hist.*, xi. 32, 33; Judges, xiv. 8–9; Raven, *Naturalists*, 42, 131, 190, 211–12; Bacon, ii. 517, 529, 554–5, 557–9; Aram Vartanian, 'Spontaneous Generation', *Dictionary of the History of Ideas*, ed. Philip P. Wiener (New York, 1973–4), iv; John Farley, *The Spontaneous Generation Controversy* (1977).
3. Sir Robert Sibbald, *A Collection of Several Treatises* (Edinburgh, 1739), vi. 18; William Wollaston, *The Religion of Nature* (5th edn, 1731), 91; Ray, *Wisdom*, 221–2; William Derham, *Physico-Theology* (1713), 244n; Charles Owen, *An Essay towards a Natural History of Serpents* (1742), 4; [John Dunton], *The Athenian Oracle* (1703–4), ii. 449.
4. *Oldenburg*, v. 228; Plot, *Staffs.*, 24–5; Morton, *Northants.*, 338–40; William Cobbett, *Rural Rides* (Harmondsworth, 1967), 236; *Life of Thomas Cooper*, 21.
5. *Turner on Birds*, 27; Charles Homer Haskins, *Studies in the History of Mediaeval Science* (Cambridge, Mass., 1924), 263; Edward Heron-Allen, *Barnacles in Nature and Myth* (1928).
6. Gerard, *Herball*, ed. Johnson, 1588–9; Ralph Thoresby, *Ducatus Leodiensis* (1715), 445.
7. Richard Garnett, 'Defoe and the Swallows', *Times Literary Supplement*, 13 Feb. 1969; Aristotle, *Hist. An.*, 600[a]; Walton, *Compleat Angler*, i. iv; Boswell, *Johnson*, ii. 55; White, *Natural History of Selborne*, letters to Barrington, xii. For acceptance of migration see *Selection of the Correspondence of Linnaeus*, i. 45–50, 54–6, 59–62, 73, 76; W. Derham in *Philosophical Transactions* (1710 for 1708–9), 123–4; Pennant, *Zoology*, i. 553–65.
8. Benjamin Martin, *The Natural History of England* (1759–63), i. 19; B. Dew Roberts, *Mr Bulkeley and the Pirate* (1936), 33–4; Pennant, *Zoology*, i. 559–60; Atkinson, *Forty Years in a Moorland Parish*, 313; James Jennings, *Ornithologia* (1828), 82; Gibson, *Superstitions about Animals*, 113–17.
9. Johannes Jonstonus, *An History of the Wonderful Things of Nature*, trans. by a person of quality [John Rowland] (1657) (originally published at Amsterdam in 1632 as *Thaumatographia Naturalis*). The Bodl. shelfmark is 4Δ217.
10. Gavin Maxwell, *The House of Elrig* (1968 edn), 42.

11. John Denne, *God's Regard to Man in his Works of Creation* (1746), 5. Cf. *The Poetical Works of Soame Jenyns*, ed. Thomas Park (1807), 33; Pennant, *Zoology*, i. 66, 131; Smellie, *Philosophy of Natural History*, i. 388–98; Swainson, *Preliminary Discourse*, 172–3; Roy Porter, *The Making of Geology* (Cambridge, 1977), 190.
12. Knapp, *Journal of a Naturalist*, 102.
13. Karl Marx and Frederick Engels, *Selected Correspondence* (1956), 156–7 (letter of 18 June 1862). Cf. Robert M. Young, 'Malthus and the Evolutionists', *Past & Present*, 43 (1969) and 'The Historiographical and Ideological Context of the Nineteenth-Century Debate on Man's Place in Nature', in *Changing Perspectives in the History of Science*, ed. Teich and Young.
14. Charles Darwin, *The Origin of Species* (1859; World's Classics, 1902), 441, 72.
15. Lord Kames, cit. Gladys Bryson, *Man and Society* (Princeton, N.J., 1945), 167.
16. Sahlins, *Use and Abuse of Biology*, 101.
17. See Phythian-Adams, 'Rural Culture'.
18. *Collected Letters of Samuel Taylor Coleridge*, ed. Earl Leslie Griggs (Oxford, 1956–71), ii. 864.
19. See Hans Kelsen, *Society and Nature* (1946), 245–8.

CHAPTER III MEN AND ANIMALS

i. Domestic Companions

1. John Flavell, *Husbandry Spiritualized* (1669), 206.
2. Thomas Hardy, *Jude the Obscure* (1895), i. chap. 9. Cf. Ellis, ii(2). 97; ii(3). 118; Martin Lister, *A Journey to Paris in the year 1698* (1699), 157; Sir Hans Sloane, *A Voyage to the Islands* (1707–25), i. xvi; Kalm, 373; William Marshall, *The Rural Economy of the West of England* (1796; reprint, Newton Abbot, 1970), i. 247.
3. *Rural Economy in Yorkshire in 1641*, ed. C. B. Robinson (Surtees Soc., 1857), 24. Cf. Louis Lemery, *A Treatise of Foods* (Eng. trans., 1704), 139; John Arbuthnot, *An Essay concerning the Nature of Aliments* (Dublin, 1731), 38; Thomas Short, *New Observations* (1750), 157–8.
4. Thomas Muffett, *Healths Improvement*, enlarged by Christopher Bennet (1655), 45, 61; Robert Lovell, *ΠΑΝΖΩΟΡΥΚΤΟΛΟΓΙΑ sive Panzoologicomineralogia* (Oxford, 1661), 23; Emma Phipson, *The Animal-Lore of Shakespeare's Time* (1883), 136; John Ray, *Observations Topographical, Moral, and Physiological; made in a Journey* (1673), 361; John Houghton, *A Collection for Improvement of Husbandry and Trade*, v. 108 (24 Aug. 1694).
5. E.g. *Notes & Queries*, 155 (1928), 9, 86, 268, 302; *ibid.*, 197 (1952), 23–4; *York Civic Records*, ed. Angelo Raine, iv (Yorks. Archaeol. Soc., 1945), 53; F. G. Emmison, *Elizabethan Life: Home, Work and Land* (Chelmsford, 1976), 239–40; *Poverty in Early-Stuart Salisbury*, ed. Paul Slack (Wilts. Rec. Soc., 1975), 97; *Louth. Old Corporation Records*, ed. R. W. Goulding (Louth, 1891), 91; Charles Phythian-Adams, *Desolation of a City* (Cambridge, 1979), 77n.
6. Muffett, *Healths Improvement*, 67.

7. *ibid.*, 67.
8. Joannes Amos Comenius, *Orbis Sensualium Pictus* (3rd edn, 1672), 43. Cf. *The Miscellaneous Writings of Sir Thomas Browne*, ed. Geoffrey Keynes (1946), 382; Goldsmith, vi. 35.
9. Wm. B. Daniel, *Rural Sports* (1801–2), ii. 466–7; R. T. Gunther, *Early Science in Oxford*, iv (Oxford, 1925), 66.
10. *The Petty–Southwell Correspondence*, ed. Marquis of Lansdowne (1928), 210, 212; Bewick, *Quadrupeds*, 54.
11. W. G. Hoskins, *The Making of the English Landscape* (1955), 109.
12. *Deserted Medieval Villages*, ed. Maurice Beresford and John G. Hurst (1971), 236; David Beers Quinn, *The Elizabethans and the Irish* (Ithaca, N.Y., 1966), 70–71; Fynes Moryson, *An Itinerary* (Glasgow, 1907–8), iv. 236; E. Estyn Evans, *The Personality of Ireland* (Cambridge, 1973), 53.
13. W. R[ichards], *Wallography* (1682), 110–11; Iorwerth C. Peate, *The Welsh House* (Liverpool, 1944), 59, 79; J. Gwynn Williams, 'Witchcraft in Seventeenth-Century Flintshire, pt ii', *Flints. Hist. Soc. Pub.*, 27 (1975–6), 11.
14. William Smith and William Webb, *The Vale-Royall of England*, ed. Daniel King (1656), i. 19.
15. Joseph Hall, *Satires*, ed. Samuel Weller Singer (1824), 128 (v. i).
16. On this complex (and controversial) subject see Peate, *Welsh House*, chap. 4; *Culture and Environment*, ed. I. Ll. Foster and L. Alcock (1963), chaps. xvi, xvii, xviii and xx; Peate, 'The Long-House again', *Folk Life*, ii (1964); *Deserted Medieval Villages*, 104–7, 112–13, 176–7; M. W. Barley, *The English Farmhouse and Cottage* (1961), ii. 76–7, 119–20; Eric Mercer, *English Vernacular Houses* (1975), 34, 37–8; J. T. Smith, 'The Evolution of the English Peasant House to the late Seventeenth Century', *Jnl Brit. Archaeol. Assoc.*, 3rd ser., 33 (1970).
17. William Harrison, *The Description of England*, ed. Georges Edelen (Ithaca, N.Y., 1968), 199.
18. Mercer, *English Vernacular Houses*, 39, 44–5; *The Agrarian History of England and Wales*, iv, ed. Joan Thirsk (Cambridge, 1967), 749; Peate, *Welsh House*, 80–81; S. R. Jones, 'Devonshire Farmhouses, iii', *Trans. Devonshire Assoc.*, 103 (1971).
19. Barley, *English Farmhouse and Cottage*, ii. 15, 51; Bingley, *An. Biog.*, iii. 93–4.
20. For town regulations see, e.g., John Tickell, *The History of . . . Kingston upon Hull* (Hull, 1796), 277; *Some Municipal Records of the City of Carlisle*, ed. R. S. Ferguson and W. Nanson (Cumbs. and Westld. Antiqn. and Archaeol. Soc., 1887), 65, 278, 281, 297; *HMC, Hatfield*, xv. 228; *Glamorgan County History*, iv., ed. Glanmor Williams (Cardiff, 1974), 43; T. S. Willan, *Elizabethan Manchester* (Chetham Soc., 1980), 41, 51.
21. P. E. Jones, *The Worshipful Company of Poulters* (2nd edn, 1965), 82, 84; Joan Thirsk, *Economic Policy and Projects* (Oxford, 1978), 91.
22. Edwin Chadwick, *Report on the Sanitary Condition of the Labouring Population* (1842), ed. M. W. Flinn (Edinburgh, 1965), 189.
23. *The Memoirs of Sir Hugh Cholmley* (1787), 35–6; Goldsmith, iii. 180; Chadwick, *Report*, *passim*; Barbara A. Hanawalt in *Jnl Interdisciplinary History*, viii (1977),

14; *Records of Mediaeval Oxford*, ed. H. E. Salter (1912), 46; Beryl Rowland, *Animals with Human Faces* (1974), 37.

24. Priscilla Wakefield, *Instinct Displayed* (1811), 50; James Anderson, *Recreations in Agriculture* (1799–1802), i (2nd pagination), 63; William Smellie, *The Philosophy of Natural History* (Edinburgh, 1790–99), i. 466; William Gilpin, *Observations ... on ... the High-Lands of Scotland* (1789), i. 207.
25. Emmison, *Elizabethan Life: Home, Work and Land*, 52; University of York: the Borthwick Institute of Historical Research, *Classified Subject Index*, comp. J. S. Purvis (York, 1963), 218–19; *Wills and Administrations from the Knaresborough Court Rolls*, ed. Francis Collins (Surtees Soc., 1902–5), i. 240; J[oshua] Childrey, *Britannia Baconica* (1661), 12; Maude Robinson, *A South Down Farm in the Sixties* (1938), 5. On animal names in general see Claude Lévi-Strauss, *The Savage Mind* (Eng. trans., 1966), chap. 7.
26. *The Journal of George Fox*, ed. Norman Penney (Cambridge, 1911), i. 177; Flavell, *Husbandry Spiritualized*, 31, 200; George Culley, *Observations on Live Stock* (1786), 12; William Marshall, *Minutes, Experiments, Observations, and General Remarks, on Agriculture, in the Southern Counties* (new edn, 1799), i. 92; ii. 29; Dunton, *Athenian Oracle*, iii. 106.
27. *Hamlet*, i. 5; Beaumont and Fletcher, *The Maid of the Mill*, v. i; *Dialect Dict.*, *passim*; Edward Moor, *Suffolk Words and Phrases* (1823), 18; Roger Wilbraham, *An Attempt at a Glossary of some Words used in Cheshire* (1820), 55–6.
28. Pliny, *Nat. Hist.*, xi. 20; William Horman, *Vulgaria* (1519), fol. 175; *Rural Economy in Yorkshire*, 63; Ellis, vi. 172.
29. John Keys, *The Practical Bee-Master* (n.d. [?1780]), 124–5. Also John Laurence, *A New system of Agriculture* (1726), 159; John Thorley, *ΜΕΛΙΣΣΗΛΟΓΙΑ or, the Female Monarchy* (1744), 143.
30. William Charles Cotton, *My Bee Book* (1842), 121–2; John Mills, *An Essay on the Management of Bees* (1766), 39; Bryan J'Anson Bromwich, *The Experienced Bee-Keeper* (2nd edn, 1783), 19; Robert Holland, *A Glossary of Words used in the County of Chester* (EDS, 1884–6), 192.
31. [Thomas Powell?], *Humane Industry* (1661), 176. Cf. John Earle, *Microcosmographie*, ed. Edward Arber (1895), 49.
32. Chaucer, *The Friar's Tale*, line 1543; Moor, *Suffolk Words*, 166–7.
33. Marshall, *Rural Economy of West of England*, i. 116. On songs for oxen at work see George Ewart Evans, *The Horse in the Furrow* (1960; 1967), 41.
34. *Dialect Dict.*; Elizabeth Mary Wright, *Rustic Speech and Folk-Lore* (1913), 326; Samuel Pegge, *Anecdotes of the English Language* (2nd edn, 1814), 11–16.
35. Gervase Markham, *Countrey Contentments* (1615), i. 41; Thomas Blundeville, *A Newe Book containing the Arte of Ryding* (1560), sig. Bvi.
36. Evans, *Horse in Furrow*, 239–40, 262–7; William Youatt, *The Horse* (new edn by E. N. Gabriel, 1859), 457.
37. John Hildrop, *Free Thoughts upon the Brute-Creation* (1742), i. 6–7. Cf. Jean Meslier, *Textes*, ed. Roland Desné (Paris, 1973), 156–7.
38. Karl Marx and Frederick Engels, *Selected Works* (Moscow, 1951), ii. 77.
39. Anthony Ashley Cooper, Earl of Shaftesbury, *Characteristicks* (1737 edn), iii. 217.

40. *The Great American Gentleman. William Byrd of Westover in Virginia. His Secret Diary for the Years 1709–1712*, ed. Louis B. Wright and Marion Tinling (New York, 1963), 63.
41. William Ames, *Conscience with the Power and Cases thereof* (Eng. trans., 1639), iv. 194; *The Whole Works of ... Jeremy Taylor*, ed. Reginald Heber, rev. Charles Page Eden (1847–54), ix. 284; Exodus, xxi. 28.
42. John T. McNeill and Helena M. Gamer, *Medieval Handbooks of Penance* (New York, 1938), 208.
43. *The Laws and Liberties of Massachusetts* (1648; reprint, Cambridge, Mass., 1929), 5 (cf. Cotton Mather, *Magnalia Christi Americana* (Hartford, Conn., 1853), ii. 401, 405–7); *The Life and Times of Anthony Wood*, ed. Andrew Clark (Oxford Hist. Soc., 1891–1900), ii. 379.
44. *The Hawkins' Voyages*, ed. Clements R. Markham (Hakluyt Soc., 1878), 151; D. Harris Willson, *James VI and I* (1956), 182; *Notes & Queries*, 4th ser., xii (1873), 273; *The Diary of Thomas Isham*, trans. Norman Marlow, ed. Sir Gyles Isham (Farnborough, 1971), 81; *Surrey Archaeol. Collns.*, ix (1885–8), 201; Douglas Hay *et al.*, *Albion's Fatal Tree* (1975), 196; Ellis, iv(3). 124.
45. Sir Kenelm Digby, *A Late Discourse ... touching the Cure of Wounds* (1658), 117.
46. 3 Edw. I, c. 4 (1275); Dorothy Burwash, *English Merchant Shipping, 1460–1540* (reprint, Newton Abbot, 1969), 40.
47. Ellis, vi (2). 117.
48. Thomas Mouffet, *The Theater of Insects*, appended to Edward Topsell, *The History of Four-Footed Beasts*, rev. by J.R. (1658), 907; Thorley, *ΜΕΛΙΣΣΗΛΟΓΙΑ*, 31; Hilda M. Ransome, *The Sacred Bee* (1937), 221; Sternberg, 159.
49. J. W[orlidge], *Systema Agriculturae* (1669), 174; Moses Rusden, *A Further Discovery of Bees* (1679), sig. A7v; Ransome, *Sacred Bee*, 226–7; Ellis, ii(3). 182.
50. Wright, *Rustic Speech*, 281–2; J. C. Atkinson, *Forty Years in a Moorland Parish* (1907 edn), 126–8; James Obelkevich, *Religion and Rural Society* (Oxford, 1976), 296; Ransome, *op. cit.*, 219–20.
51. *The Petty Papers*, ed. Marquis of Lansdowne (1927), ii. 29.
52. Timothy Nourse, *Campania Foelix* (1700), 147. Cf. *The Journeys of Celia Fiennes*, ed. Christopher Morris (1947), 265.
53. In addition to *Dialect Dict.*, see Sternberg, 69, 120; Holland, *Glossary of Words in Chester*, 238; *Before the Bawdy Court*, ed. Paul Hair (1972), 55; Joseph Hunter, *The Hallamshire Glossary* (1829), 92–3.
54. Anthony G. Petti, 'Beasts and Politics in Elizabethan Literature', *Essays & Studies* (1963); Agnes Strickland, *Lives of the Queens of England* (new edn, 1864–5), iii. 321–2.
55. *Spectator*, 28 (2 Apr. 1711); Bryant Lillywhite, *London Signs* (1972); M. D. Anderson, *Animal Carvings in British Churches* (Cambridge, 1938), 17.

ii. Privileged Species

1. Morris Palmer Tilley, *A Dictionary of the Proverbs in England* (Ann Arbor, 1966), 187–8.
2. Horman, *Vulgaria*, fol. 248^{v}.
3. Flavell, *Husbandry Spiritualized*, 206. Cf. John Evelyn, *Acetaria* (1699), 141.
4. John Gay, *Poetry and Prose*, ed. Vinton A. Dearing and Charles E. Beckwith (Oxford, 1974), 150.
5. Thomas de Grey, *The Compleat Horse-Man* (3rd edn, 1656), sig. c1^{v}.
6. Benjamin Needler, *Expository Notes with Practical Observations* (1655), 125; P. Brydone, *A Tour through Sicily and Malta* (Dublin, 1773), i. 189.
7. Jonathan Swift, *Gulliver's Travels* (1726), iv. 4; John Stow, *A Survey of... London*, ed. John Strype (1720), i. 49.
8. Joan Thirsk, *Horses in Early Modern England* (Reading, 1978), 7.
9. Topsell, 281; de Grey, *Compleat Horse-Man*, sig. b2; *Suffolk in the XVIIth Century*, ed. Lord Francis Hervey (1902), 43; Gervase Markham, *Cavelarice, or the English Horseman* (1607), v. 45.
10. William Cavendish, Duke of Newcastle, *A New Method, and Extraordinary Invention, to Dress Horses* (1667), sig. (b) 2.
11. Markham, *Cavelarice*, ii. 96–7; *id.*, *Markhams Maister-Peece* (1610), 116.
12. Markham, *Cavelarice*, v. 45; William Cowper, 'Retirement'. Cf. Cavendish, *New Method*, 18, 42; Harry Harewood, *A Dictionary of Sports* (1835), 172.
13. Symon Latham, *Lathams Falconry* (1614), 5; Edmund Bert, *An Approved Treatise of Hawkes and Hawking* (1619), 22, 35, 52–3.
14. Moryson, *Itinerary*, iv. 169.
15. *Select Pleas of the Forest*, ed. G. J. Turner (Selden Soc., 1901), 145; John Manwood, *A Treatise and Discourse of the Lawes of the Forrest* (1598), fol. 93^{v}; Sir Edward Coke, *Institutes of the Laws of England*, iv. chap. 73. Cf. Ellis, iv(2). 118–31.
16. E.g. W. J. Monk, *History of Witney* (Witney, 1894), 113; *A Boke off Recorde ... concerning ... Kirkbiekendall*, ed. Richard S. Ferguson (Cumbs. and Westmld. Antiqn. & Archaeol. Soc., 1892), 126; Phythian-Adams, *Desolation of a City*, 75; *Liverpool Town Books*, ed. J. A. Twemlow (1918), i. 14, 175, 349; S. H. Waters, *Wakefield in the Seventeenth Century* (Wakefield, 1933), 42; Geo. Fyler Townsend, *The Town and Borough of Leominster* (n.d.), 236; Tickell, *History of Kingston upon Hull*, 277; Willan, *Elizabethan Manchester*, 46, 51.
17. *The Diary of Ralph Josselin*, ed. Alan Macfarlane (Brit. Acad., 1976), 352–3, 192, 399, 431, 629; Walter Pope, *The Life of Seth Lord Bishop of Salisbury*, ed. J. B. Bamborough (Luttrell Soc., 1961), 146–7.
18. Robert Willis, *The Architectural History of the University of Cambridge*, ed. John Willis Clark (Cambridge, 1886), iii. 520.
19. David Loggan, *Cantabrigia Illustrata* (1690); *id.*, *Oxonia Illustrata* (1675).
20. Pepys, *Diary*, ix. 234; Ellis, iv(3). 133; *Diary of the Rev. John Ward*, ed. Charles Severn (1839), 112; Dean B. Lyman, *The Great Tom Fuller* (Berkeley, Calif., 1935), 26. On dogs in general see George R. Jesse, *Researches into the History of*

the British Dog (1866); Edward C. Ash, *Dogs; their History and Development* (1927); Clifford L. B. Hubbard, *An Introduction to the Literature of British Dogs* (Ponterwyd, 1949); Brian Vesey-Fitzgerald, *The Domestic Dog* (1957).

21. *The Travels of Peter Mundy*, ed. Sir Richard Carnac Temple (Hakluyt Soc., 1907–25), iv. 11; Edward Hughes, *North Country Life in the Eighteenth Century* (1952–65), i. 30–31, 389n; *OED*, 'turnspit'.
22. Richard Welford, *History of Newcastle and Gateshead*, iii (1887), 99, 108–9.
23. Pepys, *Diary*, viii. 339; *Thomas Jefferson's Farm Book*, ed. Edwin Morris Betts (Amer. Philos. Soc., Princeton, 1953), 140.
24. *The Retrospective Rev.*, i (1853), 413. Cf. Evelyn Hardy, 'Life on a Suffolk Manor in the 16th and 17th Centuries', *The Suffolk Rev.*, 3 (1968), 232; William Hamilton, 'Bonny Heck', in *The Penguin Book of Animal Verse*, ed. George MacBeth (Harmondsworth, 1965), 132–5. Dog skins were also used by tanners; *Chesterfield Wills and Inventories, 1521–1603*, ed. J. M. Bestall and D. V. Fowkes (Derbyshire Rec. Soc., 1977), 197.
25. Willson, *James VI and I*, 184, 186–7; *The Letters of John Chamberlain*, ed. Norman Egbert McLure (Amer. Philos. Soc., Philadelphia, 1939), i. 469; *The History of the King's Works*, ed. H. M. Colvin (1963–), iii(i). 125; Charlotte Fell Smith, *Mary Rich, Countess of Warwick* (1901), 29; M. A. Gibb, *Buckingham* (1939), 92.
26. Thomas Fuller, *The Worthies of England*, ed. John Freeman (1952), 421; *CSPD, 1611–18*, 434.
27. John Bowle, *Charles the First* (1975), 107; Evelyn, *Diary*, iii. 331; *Catalogue of the Pamphlets ... collected by George Thomason 1640–1661* (1908), i. 229, 237, 242, 243; C. V. Wedgwood, *The Trial of Charles I* (1964), 165.
28. Pepys, *Diary*, viii. 421 and n.; R. D. Middleton, *Dr Routh* (1938), chap. x. Cf. Antonia Fraser, *King Charles II* (1979), 291–2; Edmund Ludlow, *A Voyce from the Watch Tower*, ed. A. B. Worden (Camden ser., 1978), 183.
29. Evelyn, *Diary*, iii. 412; *Burnet's History of My Own Time*, ed. Osmund Airy (Oxford, 1897–1900), ii. 326–8; *Letters and the Second Diary of Samuel Pepys*, ed. R. G. Howarth (1933), 134–5.
30. Tilley, *Dictionary of the Proverbs*, 253. On the gentry's obsession with dogs see, e.g., Clement Ellis, *The Gentile Sinner* (4th edn, Oxford, 1668), 70–71; Samuel Butler, *Characters and Passages from Note-Books*, ed. A. R. Waller (Cambridge, 1908), 40.
31. E.g. *The Lismore Papers*, ed. Alexander B. Grosart (1886–8), 1st ser., ii. 305; iv. 205; 2nd ser., iv. 9; *The Lisle Letters*, ed. Muriel St Clare Byrne (1981), index, 'animals: dogs'; *The Memoirs of Ann Lady Fanshawe* (1907), 170.
32. Andrew Willet, *Hexapla in Leviticum* (1631), 414; *Memoirs of the Life of Mr Ambrose Barnes*, ed. W. H. D. Longstaffe (Surtees Soc., 1867), 33.
33. *The Poems of John Collop*, ed. Conrad Hilberry (Madison, Wisc., 1962), 71; William Cobbett, *Rural Rides*, ed. George Woodcock (Harmondsworth, 1967), 320; Byng, iv. 48.
34. [William Turner], *A New Booke of Spirituall Physik* (1555), fol. 67; Richard

Bernard, *Ruths Recompence* (1628), 199; Maurice Cranston, *John Locke* (1957), 426.

35. Edward Bury, *The Husbandmans Companion* (1677), 268 (and 311). Cf. *Letters and Papers of Henry VIII*, ed. J. S. Brewer *et al.* (1862–1932), xiv(2), no. 810.
36. *Manners and Meals in Olden Times*, ed. Frederick J. Furnivall (EETS, 1888), i. 182, 283; ii. 32–3; *The Household of Edward IV*, ed. A. R. Myers (Manchester, 1959), 65, 120–21, 169, 172; *The Berkeley Manuscripts*, ed. Sir John Maclean (Gloucester, 1883–5), ii. 367.
37. James Howell, *Familiar Letters*, ed. Joseph Jacobs (1890), 106–7.
38. John Hutchins, *The History and Antiquities of the County of Dorset* (1861–70), iii. 154.
39. Thomas Shadwell, *The Lancashire Witches* (1682), Act iii; *Of Building. Roger North's Writings on Architecture*, ed. Howard Colvin and John Newman (Oxford, 1981), 127, 129; Bingley, *Quadrupeds*, 123–4; George Roberts, *The Social History of the People of the Southern Counties* (1856), 30–31; John Cordy Jeaffreson, *A Book about the Table* (1875), i. 256; *Procs. of the Soc. of Antiquaries*, ii (1853), 75.
40. *Letters of a Grandmother, 1732–1735*, ed. Gladys Scott Thomson (1943), 59.
41. *The Works of John Bunyan*, ed. George Offor (1856), iii. 677; Adam Smith, *An Inquiry into the Nature and Causes of the Wealth of Nations*, ed. R. H. Campbell and A. S. Skinner (Oxford, 1976), i. 243.
42. Kent Archives Office, New Romney Borough Collection.
43. Middlesex R.O., Calendar of Sessions Records, 1639–44 (typescript), 48; F. P. Wilson, *The Plague in Shakespeare's England* (1963 edn), 38–40; *Poor Relief in Elizabethan Ipswich*, ed. John Webb (Suffolk Recs. Soc., 1966), 116–17.
44. G. R. Elton, *Reform and Renewal* (Cambridge, 1973), 127–8.
45. J. S. Bromley in *Britain and the Netherlands*, vi, ed. A. C. Duke and C. A. Tamse (The Hague, 1978), 179; *Connoisseur*, 64 (17 Apr. 1755); 36 Geo. III, c. 124 (1796); 38 Geo. III, c. 41 (1798); Stephen Dowell, *A History of Taxation and Taxes in England* (1884), iii. 292–304; *A Series of Letters of the First Earl of Malmesbury*, ed. Earl of Malmesbury (1870), i. 342.
46. G. Clark, *An Address to Both Houses of Parliament: containing Reasons for a Tax upon Dogs* (1791), 6, 13.
47. Revelation, xxii. 15; Thomas Brightman, *The Revelation of St John Illustrated* (4th edn, 1644), 888.
48. Beryl Rowland, *Blind Beasts* (Kent State U.P., 1971), 161; Caroline F. E. Spurgeon, *Shakespeare's Imagery* (Cambridge, 1935), 195–9; Tilley, *Dictionary of the Proverbs*, 163, 74, 168; F. Edward Hulme, *Proverb Lore* (1902), 164.
49. Mouffet, *Theater of Insects*, 1093.
50. Francis Rollenson, *Sermons preached before his Maiestie* (1611), 59–60; John Weemse, *An Exposition of the Second Table of the Morall Law* (1636), 163; George Foster, *The Pouring Forth of the Seventh and Last Viall* (1650), 21.
51. Thomas Brooks, *The Crown & Glory of Christianity* (1662), 54. See Ronald Paulson, *Popular and Polite Art in the Age of Hogarth* (1979), chap. 5.

52. [Benjamin Buckler?], *A Philosophical Dialogue concerning Decency* (1751), 21.
53. Sigmund Freud, *Civilization and its Discontents*, trans. Joan Rivière (1957), 67n.
54. *Macbeth*, iii. 1; Joshua Poole, *The English Parnassus* (1677 edn), 70; Edward, 2nd Duke of York, *The Master of Game*, ed. Wm. and F. Baillie-Grohman (1904), 42, 44.
55. *The Merry Travellers ... by the Author of the Cavalcade* (2nd edn, 1724), 20.
56. 13 Ric. II, st. 1, c. 13 (1389–90); 23 Car. II, c. 25 (1670 and 1671); *Liverpool Town Books*, i. 349 (my italics). Cf. *Munimenta Gildhallae Londoniensis*, ed. Henry Thomas Riley (Rolls Ser., 1859–62), i. xlii.
57. Pliny, *Nat. Hist.*, viii. 40; Giraldus Cambrensis, *The Itinerary through Wales* (EL, 1908), 63–5 (chap. 7); Stith Thompson, *Motif-Index of Folk-Literature* (rev. edn, Copenhagen, 1955–8), i. 426, 445; J. R. Porter and W. M. S. Russell, *Animals in Folklore* (Folklore Soc., 1978), 161–2.
58. In his *Le Saint Lévrier. Guinefort, guérisseur d'enfants depuis le XIIIe siècle* (Paris, 1979), Jean-Claude Schmitt reveals that the cult was still in existence in the later nineteenth century. His book provides a brilliant analysis of the myth and the elements which gave it its potency.
59. Roy Strong, *The English Icon* (1969), 290; E. K. Chambers, *Sir Henry Lee* (Oxford, 1936), 83.
60. *The Works of Thomas Nashe*, ed. R. B. McKerrow (1904–10), iii. 254–6; Robert Chester, *Love's Martyr* (New Shakspere Soc., 1878), 110; [Nicholas Cox], *The Gentleman's Recreation* (1677; reprint, East Ardsley, 1973), i. 27. For praise of the dog's fidelity see, e.g., *All the Workes of John Taylor the Water Poet* (1630), 225–31; Topsell, sig. A5; Johannes Jonstonus, *An History of the Wonderful Things of Nature*, trans. John Rowland (1657), 213–15.
61. *The Poems of Sir John Davies*, ed. Robert Krueger (Oxford, 1975), 136–7; Worlidge, *Systema Agriculturae*, 151.
62. Joseph Caryl, *An Exposition ... upon ... Job* (1643–66), viii. 17; Timothy Nourse, *A Discourse upon the Nature and Faculties of Man* (1686), 31.
63. E.g. John Bulwer, *Anthropometamorphosis* (1653), 486, 503. See in general the Hon. Mrs Neville Lytton, *Toy Dogs* (1911).
64. John Caius, *Of Englishe Dogges*, trans. Abraham Fleming (1880 edn), 21. Cf. *Lisle Letters*, iii. 60; Smith, *Mary Rich*, 328.
65. Samson Price, *The Two Twins of Birth and Death* (1624), 8–9; *Lisle Letters*, ii. 331; vi. 38; Bartholomew Batty, *The Christian Mans Closet*, trans. William Lowth (1581), fol. 32^{v}; *The Pilgrimage of Man* (1612), sig. A4.
66. Pepys, *Diary*, vi. 290, 293.
67. Jesse, *Researches*, i. 206–7. For some discussion of the British dog population in the 19th century see John K. Walton, 'Mad Dogs and Englishmen', *Jnl Social Hist.*, 13 (1979).
68. Goldsmith, iii. 272; Bewick, *Quadrupeds*, iv.
69. William Camden, *Britannia*, ed. Richard Gough (2nd edn, 1806), i. 168; *Travels of Peter Mundy*, iv. 50. Cf. *Lisle Letters*, iv. 460; v. 252–3.
70. *Records of the Virginia Company*, ed. Susan Myra Kingsbury (Washington,

1906–36), iii. 170; Peter Beckford, *Thoughts on Hunting* (n.d.), 3 (letter 1).

71. Bewick, *Quadrupeds*, 334; *OED*, 'bull-dog'.
72. Paulson, *Popular and Polite Art*, 57.
73. Mona Gooden, *The Poet's Cat* (1946), 23–4; Nikolaus Pevsner, *South and West Somerset* (The Buildings of England, Harmondsworth, 1958), 267. For other medieval cats, Charles Homer Haskins, *The Renaissance of the Twelfth Century* (Cambridge, Mass., 1927), 335–6; Thomas Wright, *Essays on Archaeological Subjects*, ii (1861), 118; Colin Platt, *Medieval Southampton* (1973), 104; P. D. A. Harvey, *A Medieval Oxfordshire Village* (1965), 63n. See in general [Augustin Paradis de Moncrif], *Les Chats* (Rotterdam, 1728); Brian Vesey-Fitzgerald, *The Domestic Cat* (1969), part 1.
74. Topsell, 106; George Lyman Kittredge, *Witchcraft in Old and New England* (New York, 1956), 178–9, 497n; John Swan, *Speculum Mundi* (2nd edn, Cambridge, 1644), 457; Jeremy Collier, *Essays upon Several Moral Subjects* (2nd edn, 1697), i. 218–19.
75. Horman, *Vulgaria*, fol. 54; Samuel Lysons, *The Model Merchant of the Middle Ages* (1860), 42.
76. Strong, *English Icon*, 261; Bodl., MS. Top. Gen. c 25, fol. 208v (cf. *Archaeologia*, 37 (1857), 197).
77. Ralph Thoresby, *Ducatus Leodiensis* (1715), 11, 615; Daniel Defoe, *Journal of the Plague Year* (New York, 1960 edn), 123.
78. A. Gibbons, *Ely Episcopal Records* (Lincoln, 1881), 88; D. R. Guttery, *The Great Civil War in Midland Parishes* (Birmingham, n.d.), 38.
79. *Correspondence of the Family of Hatton*, ed. Edward Maunde Thompson (Camden Soc., 1878), i. 157; John Brand, *Observations on ... Popular Antiquities*, ed. Sir Henry Ellis (new edn, 1849–55), iii. 39, 43; Shakespeare, *Much Ado about Nothing*, i. 1, line 210; *The Guardian*, 61 (1713).
80. R. T. Gunther, *Early British Botanists* (Oxford, 1922), 350.
81. Stuart Piggott, *William Stukeley* (Oxford, 1950), 152–3; *The Collected Poems of Christopher Smart*, ed. Norman Callan (1949), i. 313, 60.
82. Bingley, *Quadrupeds*, 143–8; Henry Mayhew, *London Labour and the London Poor* (1861), i. 181.
83. Eileen Power, *Medieval English Nunneries* (Cambridge, 1922), 305–7; Platt, *Medieval Southampton*, 104; Antonia Gransden, *Historical Writing in England* (1974), 499.
84. Topsell, 658 (cf. *The Diary of the Lady Anne Clifford* (1923), 53; *Locke Corr.*, i. 206; *Bodl. Lib. Record*, ix (1978), 375–7). For monkeys: *Lisle Letters*, ii. 317; iii. 548; Lady Newton, *The House of Lyme* (1917), 81; *id.*, *Lyme Letters* (1925), 194; Evelyn, *Diary*, iv. 349–50; J. P. Turbervill, *Ewenny Priory* (1901), 63–4; tortoises: Goldsmith, vi. 387, 392, 393; otters: Izaak Walton, *The Compleat Angler* (World's Classics, 1935), 60; *Miscellaneous Writings of Sir Thomas Browne*, 406; Bewick, *Quadrupeds*, 489; Morton, *Northants.*, 444.
85. *Dialect Dict.*, 'Anthony-pig'; 'cade', 'cosset', 'sock'; 'tiddle'.
86. William Borlase, *The Natural History of Cornwall* (Oxford, 1758), 289; William Cowper, 'The Task', iii. lines 334–9.

87. Robert Smith, *The Universal Directory for Taking Alive and Destroying Rats* (1768), 169; Bingley, *Quadrupeds*, 270.
88. Bingley, *Quadrupeds*, 39.
89. Pennant, *Zoology*, iii. 495–9 (following Gilbert White, *The Natural History and Antiquities of Selborne* (1788), letter xvii to Pennant); Goldsmith, vii. 95; W. Youatt, *The Obligation and Extent of Humanity to Brutes* (1839), 191.
90. Thomas Cooper, *A Briefe Exposition of Such Chapters of the Olde Testament as usually are red in the Church* (1573), fol. 177; *The Diary of Sir Simonds D'Ewes*, ed. Elisabeth Bourcier (Paris, 1974), 98; Pepys, *Diary*, ix. 99.
91. *Turner on Birds*, ed. A. H. Evans (Cambridge, 1903), 195; Cox, *Gentleman's Recreation*, iii. 61, 73, 89; Ray, *Willoughby*, 262.
92. Much information in *A Natural History of English Song-Birds*, with figures by Eleazar Albin (1737); *The Bird-Fancier's Recreation* (3rd edn, 1735); Pennant, *Zoology*, ii. 315–26; Goldsmith, part iv; George Edwards, *Gleanings of Natural History* (1758–64); R. Campbell, *The London Tradesman* (1747; reprint, Newton Abbot, 1969), 245.
93. James Edmund Harting, *The Ornithology of Shakespeare* (1871), 144.
94. Pepys, *Diary*, vi. 8; *Personal Recollections ... of Mary Somerville* (1873), 66–7.
95. *Poems of William Browne of Tavistock*, ed. Gordon Goodwin (n.d.), i. 89; John Donne, *Complete Poetry and Selected Prose*, ed. John Hayward (1955), 103; William Wordsworth, 'Poems of the Fancy, xv: The Redbreast chasing the Butterfly'. Cf. James Thomson, 'Winter', line 246; *The Poetical Works of John Langhorne* (1804), ii. 105; John Oswald, *The Cry of Nature* (1791), 46–7.
96. *Year Books, 12 Henry VIII*, Trin. pl. 3, pp. 3–4 (my trans.). Cf. above, p. 56.
97. *The Reports of Sir George Croke*, ed. Sir Harbottle Grimston (1659–69), i. 125–6; ii. 262.
98. Michael Dalton, *The Countrey Justice* (1635), 265. Cf. Coke, *Institutes*, iii. chap. 47; *Les Reports de S[r] Creswell Levinz* (1702), iii. 336–7.
99. Caius, *English Dogges*, 16; *The Poems of Henry Carey*, ed. Frederick T. Wood (n.d.), 94–5.
100. BL, MS. Harley 610, fol. 69.
101. *Tudor Royal Proclamations*, ed. Paul L. Hughes and James F. Larkin (1964–9), i. 537 (no. 384). Cf. *Visitation Articles and Injunctions*, ed. Walter Howard Frere and William McClure Kennedy (Alcuin Club, 1910), ii. 318; W. P. M. Kennedy, *Elizabethan Episcopal Administration* (1924), ii. 68.
102. Ash, *Dogs*, i. 116–20; *OED*, 'dog-whipper'; Gereth Spriggs, 'A Dog in the Pew', *Country Life* (12 Feb. 1976); Alfred Suckling, *The History and Antiquities of the County of Suffolk* (1846–8), i. 154; Edward Peacock, *A Glossary of Words used in the Wapentakes of Manley and Corringham* (EDS, 1877), 88.
103. Ian Green, 'Career Prospects and Clerical Conformity in the Early Stuart Church', *Past & Present*, 90 (1981), 114n; *John Lucas's History of Warton Parish*, ed. J. Rawlinson Ford and J. A. Fuller-Maitland (Kendal, 1931), 13; *2nd Report of the Commissioners appointed to inquire into the Rules ... for Regulating ... Public Worship* (1868), 564, 579.

104. J. Addy, 'Ecclesiastical Discipline in the County of York, 1559–1714' (M.A. thesis, Univ. of Leeds, 1960), appendix B, 31; Cambridge Univ. Lib., Ely Diocesan Records, B 2/12, fol. 35v. Cf. *Diocese of Norwich. Bishop Redman's Visitation, 1597*, ed. J. F. Williams (Norfolk Rec. Soc., 1946), 127.
105. *A Second and Most Exact Relation of those Sad and Lamentable Accidents ... in ... Wydecombe neere the Dartmoores* (1638), 9–10, 23; Nehemiah Wallington, *Historical Notices of Events*, ed. R. Webb (1869), i. 46–8.
106. *Connoisseur*, 89 (9 Oct. 1755).
107. Lévi-Strauss, *Savage Mind*, 205–7. Cf. Xenophon, *Cynegeticus*, vii; Columella, *De Re Rustica*, vii. 12.
108. Beckford, *Thoughts on Hunting*, 50–7 (letter v). Cf. Shakespeare, *Taming of the Shrew*, induction, i; Walton, *Compleat Angler*, i. 2; Cox, *Gentleman's Recreation*, i. 19–20; *The Letters of Daniel Eaton*, ed. Joan Wake and Deborah Champion Webster (Northants. Rec. Soc., 1971), 64, 80; Henry Fielding, *The Adventures of Joseph Andrews* (1742), iii. 6.
109. *Borthwick Institute: Subject Index*, 216–17.
110. C. M. Prior, *Early Records of the Thoroughbred Horse* (1924), 55, 84–5, 87–8, 93. Cf. Lévi-Strauss, *Savage Mind*, 206–7.
111. *Poems of Sir John Davies*, 149; William Young, *The History of Dulwich College* (1889), ii. 22, 24; John Taylor, *Bull, Beare, and Horse* (1638), sig. D8.
112. A. C. Edwards, *John Petre* (1975), 64; Thirsk, *Horses in Early Modern England*, 16.
113. Norreys Jephson O'Conor, *Godes Peace and the Queenes* (1934), 17. Cf. Smith, *Mary Rich*, 53.
114. Thomas Henry Taunton, *Portraits of Celebrated Racehorses* (1887–8), iv. 12.
115. *Berkeley Manuscripts*, ii. 363; William Addison, *Epping Forest* (1945), 62.
116. Charles Boutell, *Monumental Brasses and Slabs* (1847), 65.
117. Cf. John Taylor, *Wit and Mirth* (1630), 35 (in *Shakespeare Jest-Books*, ed. W. Carew Hazlitt (1864), iii).
118. Wakefield, *Instinct Displayed*, 54–6; *The Letters of Joseph Ritson* (1833), i. lxxii; *The Works of Jeremy Bentham*, ed. John Bowring (1843–59), xi. 80; Claude Lévi-Strauss in *Mélanges en l'honneur de Fernand Braudel* (Toulouse, 1973), ii. 332.
119. James Hart, *KΛINIKH, or the Diet of the Diseased* (1633), 84.
120. Topsell, 106; William Cowper, 'The Task', i. line 562; Bernard Mandeville, *The Fable of the Bees*, ed. F. B. Kaye (Oxford, 1924), i. 174.
121. *Come Hither*, ed. Walter de la Mare (new edn, 1928), 94–5; [Duncan], *Essays and Miscellanea* (Oxford, 1840), 219.
122. G. Eland, *At the Courts of Great Canfield* (1949), 92; Moryson, *Itinerary*, iv. 199.
123. Bacon, ii. 625; John Ashton, *Humour, Wit & Satire of the Seventeenth Century* (1883), 176–80.
124. Bingley, *Quadrupeds*, 435; Louis Simond, *Journal of a Tour and Residence in Great Britain* (2nd edn, Edinburgh, 1817), ii. 329.
125. A. S. Bicknell, 'Hippophagy', *Jnl of Soc. of Arts*, xvi (1867–8); Edward Smith, *Foods* (1873), 74–5; Frank Buckland, *Log-Book of a Fisherman and Zoologist* (n.d.), 61.

126. Ian Kershaw in *Past & Present*, 59 (1973), 9; *Agrarian History of England and Wales*, iv. 632; Hart, *Diet of Diseased*, 84; Wallington, *Historical Notices*, ii. 165–6.
127. James Ritchie, *The Influence of Man on Animal Life in Scotland* (Cambridge, 1920), 139; Robinson, *South Down Farm*, 62.
128. Kenneth Clark, *Animals and Men* (1977), 60.
129. Ray, *Observations*, 362; Pennant, *Zoology*, ii. 397. For the consumption of small birds in Tudor and Stuart times see, e.g., Thomas Dawson, *The Good Huswifes Iewell* (1596), sigs. A2, A3^{v}; *HMC, Portland*, ii. 274; Hart, *Diet of Diseased*, 80–81; Hannah Woolley, *The Gentlewomans Companion* (1675), 136; Cox, *Gentleman's Recreation*, ii. 61, 75; Ray, *Willoughby*, 188, 189, 191.
130. Ray, *Observations*, 361–2; Tobias Smollett, *Travels through France and Italy* (1766; 1907 edn), 174; *Personal Recollections of Mary Somerville*, 238; Muffett, *Healths Improvement*, 104. For Elphinstone's reaction, cf. George, Lord Lyttelton, *Dialogues of the Dead* (1760; 1795 edn), 55.
131. Ronald Paulson, *Emblem and Expression* (1975), 242.
132. Sir John Harington, *Nugae Antiquae*, ed. Thomas Park (1804), i. 380–84.
133. Prior, *Early Records of the Thoroughbred Horse*, 100–101; *HMC, Portland*, ii. 306. For others see, e.g., *The Diary of John Hervey, First Earl of Bristol*, ed. S. A. H. (Wells, 1894), opp. 52; Thomson, *Letters of a Grandmother*, 127; Taunton, *Portraits of Celebrated Racehorses*.
134. Byng, iii. 139.
135. Raymond Carr, *English Fox Hunting* (1976), 41.
136. See, e.g., A. Lytton Sells, *Animal Poetry in French and English Literature* (1957); *The Dog in British Poetry* (1893), ed. R. Maynard Leonard; *The Life of William Hutton ... by himself*, ed. Catherine Hutton (1816), 271–2.
137. Many are listed in Barbara Jones, *Follies and Grottoes* (2nd edn, 1974). For others see *The Travels through England of Dr Richard Pococke*, ed. James Joel Cartwright (Camden Soc., 1888–9), ii. 41; Nicholas Penny, *Church Monuments in Romantic England* (1977), 37, 150, 209 n40, 211 n34; William Cartwright, *The Poets of Yorkshire*, ed. John Holland (1845), 57; David Verey, *Gloucestershire: The Cotswolds* (Buildings of England, Harmondsworth, 1970), 117.
138. William H. Drummond, *The Rights of Animals* (1838), 94. For other legacies see, e.g., *The Flemings in Oxford*, ed. John Richard Magrath (Oxford Hist. Soc., 1904–24), iii. 220; *Notes & Queries*, 9th ser., iii (1899), 241–2; *Connoisseur*, 89 (1755); *Kirby's Wonderful and Eccentric Museum* (1820), iii. 126–7; v. 23, 27–9; Bingley, *Quadrupeds*, 143.
139. Sells, *Animal Poetry*, 96–100.
140. *The Wentworth Papers, 1705–1739*, ed. James J. Cartwright (1882), 40, 42, 45, 55, 214, 284.
141. Joseph Taylor, *The General Character of the Dog* (1804), 179. Cf. Mary Monk, *Marinda* (1716), 69; *Diary of John Hervey*, 86; Byron, 'Inscription on the Monument of a Newfoundland Dog'.
142. Edmund Burke, *Of the Sublime and Beautiful*, ii. 5, in *Works* (Bohn edn, 1854), i. 96; *Thraliana*, ed. Katharine C. Balderston (Oxford, 1942), i. 197.

143. Richard Capel, *Tentations* (1633), 357.
144. Pepys, *Diary*, vi. 2; Gabriel Towerson, *An Explication of the Decalogue* (1676), 410.
145. Figures drawn from the P.R.O. calendars of assize records (indictments) compiled by J. S. Cockburn (1975–80) and J. A. Sharpe, 'Crime in the County of Essex, 1620–1680' (Oxford D.Phil. thesis, 1978), 229–30.
146. Cf. John Berger, 'Vanishing Animals', *New Soc.*, 39 (31 March 1977), 664; Tom Forester, 'Animal Planning', *ibid.*, 32 (8 May 1975), 325.
147. Norman Ault, *New Light on Pope* (1949), chap. xxii; William Cowper, *Selected Letters* (EL, 1926), 263–72.
148. *Memoirs of the Life of Samuel Romilly*, ed. by his sons (2nd edn, 1840), ii. 240–41.
149. *The Poems of Andrew Marvell*, ed. Hugh Macdonald (1952), 17; Chaucer, 'The Canterbury Tales: General Prologue', lines 144–9.
150. Richard D. French, *Antivivisection and Medical Science in Victorian Society* (1975), 375. Cf. Frances Power Cobbe, *Italics* (1864), 443–4.

iii. The Narrowing Gap

1. Caius, *Englishe Dogges*, 31; William Ramesey, *Mans Dignity* (1661), 55; *Aubrey on Education*, ed. J. E. Stephens (1972), 133.
2. John Hayes, *Thomas Gainsborough* (1980), 77; Richard Dean, *An Essay on the Future Life of Brutes* (Manchester, 1767), ii. 71; *The Works of Mr Henry Needler* (2nd edn, 1728), 213; Erasmus Darwin, *Zoonomia* (1794–6), i. 169.
3. For individual stories see, e.g., Pliny, *Nat. Hist.*, viii. 3; Plutarch, *De Sollertia Animalium*; Aelian, *De Natura Animalium*; William Derham, *Physico-Theology* (1713), 203n; Dean, *Essay on Future Life of Brutes*, ii. xvi–xviii; *Kirby's Wonderful and Eccentric Museum*, ii. 5–6; and for anthologies, Bingley, *An. Biog.*; Taylor, *General Character of the Dog*; *id.*, *The Wonders of the Horse* (1808); Wakefield, *Instinct Displayed*; Capt. Thomas Brown, *Biographical Sketches and Authentic Anecdotes of Dogs* (1829); *id.*, *Biographical Sketches of Horses* (1830); F. O. Morris, *Records of Animal Sagacity* (1861); W. and R. Chambers, *Kindness to Animals illustrated by Stories* (1877); J. G. Wood, *Man and Beast Here and Hereafter* (8th edn, 1903).
4. Cit. *Dog in British Poetry*, ed. Leonard, 302–3.
5. Herschel Baker, *The Image of Man* (New York, 1961), 298; Arthur O. Lovejoy and George Boas, *Primitivism and Related Ideas in Antiquity* (reprint, New York, 1973), chap. 13; George Boas, *The Happy Beast in French Thought of the Seventeenth Century* (Baltimore, 1933).
6. Arthur Lake, *Sermons* (1629), 478; John Locke, *Two Treatises of Government*, ed. Peter Laslett (Cambridge, 1960), 200 (i. para. 58).
7. 'Philotheos Physiologus' [Thomas Tryon], *The Country-Man's Companion* (n.d. [1683]), sig. A2v–3. Cf. Ames, *Conscience*, iv. 198; Margaret Cavendish, Marchioness of Newcastle, *Orations of Divers Sorts* (1662), 195; Matthew Griffith, *Bethel* (1634), 296; Brooks, *Crown and Glory*, 42; Francis Rous, *Oile of Scorpions* (1623), 67–8.

8. M. J. Ingram, 'Ecclesiastical Justice in Wiltshire, 1600–1640' (Oxford D.Phil. thesis, 1976), 81. Cf. *HMC, Various Collections*, i. 132–3; J. A. F. Thomson, *The Later Lollards* (1965), 27, 82; G. B. Harrison in *Willobie his Avisa* (Edinburgh, 1966), 264; *Studies in Church History*, ii, ed. G. J. Cuming (1965), 255; *DNB*, 'North, Dudley, 4th Baron'. On the *libertins* see René Pintard, *Le Libertinage érudit* (Paris, 1943); J. S. Spink, *French Free-Thought from Gassendi to Voltaire* (1960).
9. Alexander Murray, 'Religion among the Poor in Thirteenth-Century France', *Traditio*, xxx (1974), 323; Emmanuel Le Roy Ladurie, *Montaillou*, trans. Barbara Bray (1978), 320–21; Carlo Ginzburg, *The Cheese and the Worms*, trans. John and Anne Tedeschi (1980), 69, 123. Cf. William Pemble, *Salomons Recantation* (1627), 37; William Darrell, *The Gentleman Instructed* (1738), 575–7.
10. Norman T. Burns, *Christian Mortalism from Tyndale to Milton* (Cambridge, Mass., 1972); Christopher Hill, *Milton and the English Revolution* (1977), chap. 25.
11. R. O[verton], *Mans Mortalitie* (Amsterdam, 1644), 6.
12. *The Prerogative of Man* (Oxford, 1645), 3.
13. Sir Kenelm Digby, *Two Treatises* (1645), ii. 86; *The Whole Works of Edward Reynolds* (1826), iv. 96; William Strachey, *The Historie of Travel into Virginia* (1612), ed. Louis B. Wright and Virginia Freund (Hakluyt Soc., 1953), 100.
14. Christopher Hill, 'John Reeve', in *Prophecy and Millenarianism*, ed. Ann Williams (1980), 321.
15. Aram Vartanian, *La Mettrie's L'Homme Machine* (Princeton, 1960), 162.
16. *The Works of ... Henry St John, Lord Viscount Bolingbroke* (1809), viii. 348; *Letters from Mrs Elizabeth Carter to Mrs Montagu*, ed. Montagu Pennington (1817), i. 115; M. F. Ashley Montagu, *Edward Tyson* (Philadelphia 1943), 288.
17. *The Family Memoirs of the Rev. William Stukeley*, ed. W. C. Lukis (Surtees Soc., 1882–7), i. 100; *The Parliamentary History of England*, xvii (1813), 782; James Raine, *A Memoir of the Rev. John Hodgson* (1857), i. 363–4.
18. W. H[owell], *The Spirit of Prophecy* (1679), 266.
19. Aristotle, *Hist. An.*, 588[b]; Sir Matthew Hale, *The Primitive Origination of Mankind* (1677), 49–50.
20. E.g. William Attersoll, *A Commentarie upon ... Numbers* (1618), 900.
21. Arthur O. Lovejoy, *The Great Chain of Being* (1936; New York, 1960), 227, 231; Winthrop D. Jordan, *White over Black* (Baltimore, 1968), 228; *The Works of Symon Patrick*, ed. Alexander Taylor, ix (Oxford, 1858), 273.
22. Thomas Ball, *The Life of the Renowned Doctor Preston*, ed. E. W. Harcourt (1885), 21–5. For a later disputation on a similar topic, Charles Webster, *The Great Instauration* (1975), 135.
23. John Locke, *An Essay concerning Human Understanding*, ed. Peter Nidditch (Oxford, 1975), 160 (ii. xi. 11).
24. *Paradise Lost*, viii. line 374.
25. Hale, *Primitive Origination*, 16; Richard Meggott, *A Sermon preached at White-Hall* (1683), 10; Humphry Ditton, *A Discourse concerning the Resurrection of Jesus Christ* (1712), 517.

26. David Hume, *Essays Moral, Political, and Literary*, ed. T. H. Green and T. H. Grose (new imp., 1898), ii. 85–8; David Hartley, *Observations on Man* (4th edn, 1801), i. 409–10. Cf. Mandeville, *Fable of the Bees*, ii. 166; *Works of Bolingbroke*, viii. 231; [John Gregory], *A Comparative View of the State and Faculties of Man* (2nd edn, 1766), 13; Lawrence, *Horses*, i. 78; Robert M. Young, 'Animal Soul', in *The Encyclopedia of Philosophy*, ed. Paul Edwards, i (1967), 124.
27. *Hawkins' Voyages*, 195–6; Tho[mas] Robinson, *A Vindication of the ... Mosaick System of the Creation* (1709), 97.
28. E.g. *Spectator*, 120 (1711); *Rambler*, 41 (1750); [James Burnet, Lord Monboddo], *Antient Metaphysics* (1779–99), iii, appendix, chap. iii.
29. Hume, *Essays*, ed. Green and Grose, ii. 88; Matthew Prior, *Poems on Several Occasions*, ed. A. R. Waller (Cambridge, 1905), 270. Cf. Smellie, *Philosophy of Natural History*, i. 152; Charles White, *An Account of the Regular Gradation in Man* (1799), 65; George Warren, *A Disquisition on the Nature and Properties of Living Animals* (1828), 135.
30. E.g. Richard Berenger, *A New System of Horsemanship* (1754), 64–5.
31. Charles Stevens and John Liebault, *Maison Rustique*, trans. Richard Surflet, ed. Gervase Markham (1616), 321; William Cobbett, *A Year's Residence in America* (Abbey Classics, n.d.), 139.
32. *Maroccus Extaticus* (1595); *DNB*, 'Banks, –'; Francis Douce, *Illustrations of Shakespeare* (new edn, 1839), 131–2. For other 'learned' animals, see, e.g., *Archaeologia*, xxxvii (1857), 200; Pepys, *Diary*, ix. 297, 301; *VCH, Warws.*, vii. 221; John Ashton, *Social Life in the Reign of Queen Anne* (new edn, 1897), 197–8; John Hill, *An History of Animals* (1752), 317; E. C. Cawte, *Ritual Animal Disguise* (Folklore Soc., 1978), 41; R. M. Wiles, 'Crowd-pleasing Spectacles in Eighteenth-Century England', *Jnl of Popular Culture*, i (1967), 94; Frederick Cameron Sillar and Ruth Mary Meylier, *The Symbolic Pig* (1961), 61–2; Richard D. Altick, *The Shows of London* (1978), 40–41, 306–7.
33. *An Elizabethan in 1582*, ed. Elizabeth Story Donno (Hakluyt Soc., 1976), 112; Topsell, 10.
34. Bacon, ii. 666; Digby, *Two Treatises*, i. 407.
35. Bingley, *An. Biog.*, ii. 256–7; Sternberg, 160; Caius, *Englishe Dogges*, 16; BL, Add. 53,726, fol. 37; Smith, *Universal Directory*, iv.
36. Henry Swinburne, *A Briefe Treatise of Testaments and Last Wills* (1635), 69; William Cowper, 'Pairing Time Anticipated', note (citing Rousseau, *Émile*).
37. Hunter, *Hallamshire Glossary*, 55–6; Moor, *Suffolk Words*, 15; Wright, *Rustic Speech*, 310–11.
38. *Gilbert White's Journals*, ed. Walter Johnson (1931; reprint, Newton Abbot, 1970), 79; Robinson, *South Down Farm*, 59; H. Kirke Swann, *A Dictionary of English and Folk-Names of British Birds* (1913; reissue, Detroit, 1968), 44.
39. Pennant, *Zoology*, ii. 365; Daines Barrington, 'Experiments and Observations on the Singing of Birds', *Philosophical Trans.*, lxiii (1773).
40. Nathaniel Homes, *The Resurrection-Revealed Raised above Doubts and Difficulties* (1661), 244. Also John Bulwer, *Chirologia* (1644), 5; John Webster,

Academiarum Examen (1654), 31; Tryon, *Country-Mans Companion*, 59–60.

41. John Oswald, *The Cry of Nature* (1791), 118–19; Montaigne, *Essays*, trans. John Florio (1603; 1893 edn), ii. 145.
42. Digby, *Two Treatises*, i. 374; William Gilpin, *Remarks on Forest Scenery* (1791), ii. 117.
43. Locke, *Essay*, 33–5 (ii. xxvii. 8); Ray, *Willoughby*, 109.
44. Margaret Cavendish, Marchioness of Newcastle, *Philosophical Letters* (1664), 40–41, 43. Cf. *ibid.*, sigs. a1, b2[v]; 34–5, 147, 192; *Poems and Phancies* (2nd imp., 1664), 115, 124, 128–9, 283; *The Worlds Olio* (1655), 140–43.
45. Especially perhaps the anonymous *Theophrastus Redivivus* (1659); see Spink, *French Free-Thought*, 69; and Montaigne, *Essays*, ed. Albert Thibaudet (Pléiade edn, Paris, 1950), 479.
46. N. S. Sutherland in *TLS*, 26 Dec. 1975, 1542.
47. Cavendish, *Poems and Phancies*, 86.
48. George Abbot, *An Exposition upon the Prophet Ionah* (1600), 469–70; Goldsmith, iv. 259.
49. Lovejoy, *Great Chain of Being*, 231.
50. See Thomas H. Huxley, *Man's Place in Nature* (1894), chap. 1; H. W. Janson, *Apes and Ape Lore in the Middle Ages and the Renaissance* (1952), chap. xi; Robert M. Yerkes and Ada W. Yerkes, *The Great Apes* (1929), part 1; Robert Wokler, 'Tyson and Buffon on the Orang-Utan', *Studies on Voltaire*, cli-clv (1976).
51. Sir Thomas Browne, *Religio Medici* (1643), i. para. 36. Cf. Richard Baxter, *Of the Immortality of Mans Soul* (1682), 40–41.
52. Edward Tyson, *Orang-Outang, sive Homo Sylvestris* (1699); Prior, *Poems on Several Occasions*, 11.
53. Carolus Linnaeus, *Systema Naturae* (10th edn, Stockholm, 1758–9), i. 20–32. For earlier classifications of non-human animals as 'irrational' see Walter Charleton, *Onomasticon Zoicon* (1668), 1. Aristotle, however, had classified man among the animals, including him among those which were 'political' and 'tame'; *Hist. An.*, 488[a].
54. James Burnet, Lord Monboddo, *Of the Origin and Progress of Language* (Edinburgh, 1774–92), esp. i. 144–5, 187–8, 270, 409; *id.*, *Antient Metaphysics*, iii. 40–42, 359–77.
55. Rousseau, *Discours sur l'inégalité* (1755), note J.
56. See Maurice Mandelbaum, 'The Scientific Background of Evolutionary Theory in Biology', *JHI*, 18 (1957), 352–3.
57. Henry Grove, *Sermons* (1740), ii. 163. Cf. *Primitivism and Related Ideas in Antiquity*, 208f., 221, 229, 243–6, 371–2, 374–5.
58. Thomas Starkey, *A Dialogue between Reginald Pole and Thomas Lupset*, ed. Kathleen M. Burton (1948), 60; *Wilson's Arte of Rhetorique*, ed. G. H. Mair (Oxford, 1909), sig. Avi[v]. Cf. Quentin Skinner, *The Foundations of Modern Political Thought* (Cambridge, 1978), ii. 346; Harro Höpfl and Martyn P. Thompson in *American Hist. Rev.*, 84 (1979), 936n; Bernard W. Sheehan, *Savagism to Civility* (Cambridge, 1980), 69–71.

59. John Aubrey, *Brief Lives*, ed. Anthony Powell (1949), 1; *Works of Henry Needler*, 17.
60. *Oldenburg*, v. 101–2, 119–20; Monboddo, *Antient Metaphysics*, iii. 41, 45–7, 57, 367–77; vi. 137n, 299; *Origin and Progress of Language*, i. 186–7, 199–201; *The Wild Man Within*, ed. Edward Dudley and Maximillian E. Novak (Pittsburgh, 1972), 183ff.; Ashton, *Social Life in Reign of Anne*, 209–10; Lucien Malson, *Wolf Children*, and Jean Itard, *The Wild Boy of Aveyron* (Eng. trans, 1972); *The Banks Letters*, ed. Warren R. Dawson (1958), 188–9; Richard Carlile, *The Republican*, xii (1825), 221; Jean Ehrard, *L'Idée de la nature en France dans la première moitié du XVIIIe siècle* (Paris, 1963), 686.
61. Ramesey, *Mans Dignity*, 53–4; Digby, *Two Treatises*, 403–5.
62. Darwin, *Zoonomia*, i. 178–9, 164, 160–61.
63. Bingley, *An. Biog.*, i. 24–5.
64. Joseph Butler, *The Analogy of Religion*, ed. W. E. Gladstone (World's Classics, 1907), 36; Monboddo, *Origin and Progress of Language*, i. 148–9; Tyson, *Orang-Outang*, 51–2; Vartanian, *La Mettrie's L'Homme Machine*, 160, 214; John C. Greene, *The Death of Adam* (Ames, Iowa, 1959), 177–9.
65. Pepys, *Diary*, ii. 160.
66. Monboddo, *Antient Metaphysics*, iii. 40.
67. Darwin, *Zoonomia*, i. 162–3; Bingley, *Quadrupeds*, 452–4; John Stuart Mill, *A System of Logic*, ed. J. M. Robson (1973–4), 859.
68. E. P. Evans, *Evolutional Ethics and Animal Psychology* (1897), 9–11.
69. Bulwer, *Anthropometamorphosis*, sig. B3; 450. Cf. Spink, *French Free-Thought*, 40.
70. John Hall of Richmond, *Of Government and Obedience* (1654), 444–6; *Locke Corr.*, ii. 344.
71. Monboddo, *Origin and Progress of Language*, i. 183.
72. See Francis C. Haber, *The Age of the World* (Baltimore, 1959), 264; Glyn Daniel, *The Idea of Prehistory* (1962).
73. Karl Marx, *Early Writings*, trans. Rodney Livingstone and Gregor Benton (Harmondsworth, 1975), 175.
74. Nourse, *Discourse upon the Nature of Man*, 359. Cf. [Samuel Gott], *Nova Solyma*, ed. Walter Begley (1902), i. 90; William Penn, *The Peace of Europe* (EL, n.d.), 36; Gregory, *Comparative View of the State and Faculties of Man*, 17–18; Robert M. Young in *Past & Present*, 43 (1969), 113; W. Lawrence, *Lectures on Physiology, Zoology, and the Natural History of Man* (1819), 393–5; Herbert Spencer, *The Study of Sociology* (14th edn, 1888), 369.
75. Stith Thompson, *Motif Index of Folk Literature* (rev. edn, Copenhagen, 1955–8), i. 461–9.
76. Margaret T. Hodgen, *Early Anthropology in the Sixteenth and Seventeenth Centuries* (Philadelphia, 1964), 19; Bulwer, *Anthropometamorphosis*, 410; John Block Friedman, *The Monstrous Races in Medieval Art and Thought* (1981), 11–21. Cf. S. Baring-Gould, *Curious Myths of the Middle Ages* (new edn, 1897), 145–9; James Orchard Halliwell, *A Dictionary of Archaic and Provincial Words* (5th edn, 1845), 'Long-tails'.

77. Monboddo, *Origin and Progress of Language*, i. 262n; [Samuel Gott], *The Divine History of the Genesis of the World* (1670), 420.
78. *DNB*, 'Siward'; W. G. Hoskins and H. P. R. Finberg, *Devonshire Studies* (1952), 108.
79. Richard Bernheimer, *Wild Men in the Middle Ages* (Cambridge, Mass., 1952); G. C. Druce, 'Some Abnormal and Composite Human Forms in English Church Architecture', *Archaeol. Jnl*, 72 (1915), 168–9; Lillywhite, *London Signs*, 247–50; David M. Bergeron, *English Civic Pageantry 1558–1642* (1971), 56, 70–71; Timothy Husband, *The Wild Man* (New York, 1981).
80. Linnaeus, *Systema Naturae* (10th edn), i. 20; 'A Description of Cleveland', *The Topographer & Genealogist*, ii (1853), 416; *Banks Letters*, 563, 601, 625. Cf. Altick, *Shows of London*, 302–3.
81. E.g. [Gott], *Nova Solyma*, ii. 16.
82. William Gouge, *Of Domesticall Duties* (3rd edn, 1634), 185; Ramesey, *Mans Dignity*, 97. Cf. Coke, *Institutes*, iii. chap. x; Andrew Willet, *Hexapla in Exodum* (1608), 505; *id.*, *Hexapla in Leviticum* (1631), 434; Bulwer, *Anthropometamorphosis*, 434–5, 447; Locke, *Essay*, 451 (iii. vi. 23).
83. Vavasor Powell, *God the Father Glorified* (1649), 85; W. A. Leighton, 'Early Chronicles of Shrewsbury', *Shropshire Archaeol. & Nat. Hist. Soc.*, iii. (1880), 283.
84. William Turner, *A Compleat History of the Most Remarkable Providences* (1697), pt ii, chap. 27; *Life and Times of Anthony Wood*, ii. 378.
85. Edward Tyson in *Philosophical Trans.*, 21 (1699), 431–5.
86. For orthodox monogenism see, e.g., Sir Christopher Heydon, *A Defence of Iudiciall Astrologie* (Cambridge, 1603), 529; Bacon, ii. 473; Hale, *Primitive Origination, passim*; Bulwer, *Anthropometamorphosis*, 467–8; Hodgen, *Early Anthropology*, 213–14.
87. Sir Charles Linné, *A General System of Nature*, ed. William Turton (1806), i. 9. Cf. James Cowles Prichard, *Researches into the Physical History of Man* (1813), ed. George W. Stocking (Chicago, 1973), liv–lv; David Brion Davis, *The Problem of Slavery in Western Culture* (Ithaca, N.Y., 1966), chap. 15; Jordan, *White over Black*, chap. 1; Léon Poliakov, *The Aryan Myth*, trans. Edmund Howard (1974), 160–75.
88. Tyson, *Orang-Outang*, 'Epistle Dedicatory'.
89. *Petty Papers*, ii. 31; Jordan, *White over Black*, 253; Henry Home of Kames, *Sketches of the History of Man* (new edn, Glasgow, 1817), i.
90. Edward Long, *The History of Jamaica* (1774; reprint, 1970), ii. 371; White, *Account of the Regular Gradation*, 80. On the growth of polygenism see Hodgen, *Early Anthropology*, 415–26; Poliakov, *Aryan Myth*, 175–82; Davis, *Problem of Slavery*, chap. 15.

iv. Animal Souls

1. Howell, *Spirit of Prophecy*, 266; above, p. 32; William Lambe, *Additional Reports on the Effects of a Peculiar Regimen* (1815), 227.

2. E.g., *The Theological, Philosophical and Miscellaneous Works of the Rev. William Jones* (1801), iii. 71–2; Thomas Adams, *A Commentary ... upon the ... Second Epistle ... by ... St Peter* (1633), 971.
3. *Letters and Correspondence of John Henry Newman*, ed. Anne Mozley (1891), ii. 291. Cf. Genesis, ix. 10.
4. Sternberg, 186; A. R. Wright, *British Calendar Customs* (Folk-Lore Soc., 1936–40), ii. 74–5; Peacock, *Glossary of Manley and Corringham*, 57; Thompson, *Motif-Index of Folk-Literature*, i. 407–11; Obelkevich, *Religion and Rural Society*, 263–4, 269, 310; Ransome, *Sacred Bee*, 229.
5. Pliny, *Nat. Hist.*, x. 41; *Lucans Pharsalia*, trans. Thomas May (4th edn, 1650), continuation, 41; *The Theological Works of ... Henry More* (1708), 34.
6. *Collected Poems of Christopher Smart*, i. 24; G. G[oodman], *The Creatures praysing God* (1624), 24. Cf. John T. McNeill, *The History and Character of Calvinism* (New York, 1967), 232; George H. Williams, *Wilderness and Paradise in Christian Thought* (New York, 1962), 103; Rowland Watkyns, *Flamma sine Fumo*, ed. Paul C. Davies (Cardiff, 1968), 16; Bury, *Husbandmans Companion*, 289–90; L. Tyerman, *Wesley's Designated Successor* (1882), 454; [John Keble], 'Redbreast in Church', *Lyra Innocentium* (Oxford, 1846), 285–7.
7. Alex. Clogie, *Vox Corvi* (1694); *The Diary of Ralph Thoresby*, ed. Joseph Hunter (1830), i. 264; [William Prynne], *A New Discovery of the Prelates Tyranny* (1641), sig. *3v (between 90 and 91). Cf. Toussin Bridoul, *The School of the Eucharist established upon the Miraculous Respects and Acknowledgements, which Beasts, Birds, and Insects ... have rendered to the Holy Sacrament of the Altar* (Eng. trans., 1687).
8. Halliwell, *Dictionary of Archaic Words*, 'yeth-hounds'; Obelkevich, *Religion and Rural Society*, 272; *The Folk-Lore Jnl*, v (1887), 182; Swainson, *Birds*, 98; Swann, *Dictionary of Folk-Names of Birds*, 167; Edward A. Armstrong, *The Folklore of Birds* (1958), 211–13.
9. Henry More, *An Antidote against Atheism* (2nd edn, 1655), 370 (though cf. 353–4); Gottfried Wilhelm Leibniz, *Philosophical Papers and Letters*, ed. Leroy E. Loemker (Dordrecht, 1969), 589; Richard Burthogge, *An Essay upon Reason* (1694), 236–7; *id.*, *The Soul of the World* (1699).
10. Laurence Clarkson, *Look About You* (1659), 98. Cf. John Reeve and Lodowicke Muggleton, *A Transcendent Spiritual Treatise* (1822 edn), 50.
11. Aquinas, *Summa Theologica*, iii. supp. q. xci, art. 5; *id.*, *Summa contra Gentiles*, ii. 82; Peter Martyr, *Most Learned and Fruitfull Commentaries upon the Epistle to the Romanes* (1568), fol. 219. For exceptions see D. S. Wallace-Hadrill, *The Greek Patristic View of Nature* (1968), 115–17; *The Dictionary ... of Mr Peter Bayle* (2nd edn, 1734–8), 'Sennertus, Daniel'.
12. Andrew Willet, *Hexapla ... upon the ... Epistle ... to the Romanes* (1611), 370; *Dictionary of Bayle*, 'Sennertus, Daniel'.
13. E.g. Martyr, *Commentaries upon Romanes*, fol. 219; Willet, *op. cit.*, 372. Cf. Thomas More, *Utopia* (EL, 1951), 121.
14. Thomas Wilson, *A Commentarie upon the Most Divine Epistle of S. Paul to the Romanes* (1614), 588; Elnathan Parr, *A Plaine Exposition upon the ... Epistle of*

Saint Paul to the Romanes (1620), 88–9; Homes, *Resurrection-Revealed*, 250–52; Ray, *Wisdom*, 39.

15. E.g. Abbot, *Exposition upon Ionah*, 466–7; Thomas Draxe, *The Earnest of our Inheritance* (1613), 5–10; Wilson, *Commentarie upon Romanes*, 588–9; Goodman, *Creatures praysing God*, 28–9; *The Workes of Mr Williā Cowper* (2nd edn, 1629), 116; William Gearing, *A Prospect of Heaven* (1673), 98–100.
16. Evelyn, *Diary*, iv. 106.
17. Thomas Horton, *Forty Six Sermons upon the whole Eighth Chapter of . . . Romans* (1674), 368–70; Parr, *Plain Exposition*, 89.
18. *The Writings of John Bradford*, ed. Aubrey Townsend (Parker Soc., Cambridge, 1848–53), i. 351–64.
19. Thomas Edwards, *Gangraena* (1646), iii. 36 (also i. 27). Cf. Burns, *Christian Mortalism*, 125; Overton, *Mans Mortallitie*, 15.
20. *The Prerogative of Man*, 41–2.
21. Horton, *Forty Six Sermons*, 370; Humphry Primatt, *A Dissertation upon the Duty of Mercy and Sin of Cruelty to Brute Animals* (1776), 42.
22. W. C[oward?], *The Just Scrutiny* (?1705], 97; Charles Leigh, *The Natural History of Lancashire, Cheshire, and the Peak* (1700), ii. 13; *Diary of Ralph Josselin*, 342.
23. Butler, *Analogy of Religion*, 36; Sells, *Animal Poetry*, xxiv; Hartley, *Observations on Man*, ii. 391; [Robert Wallace], *Various Prospects* (1761), 341.
24. Hildrop, *Free Thoughts upon the Brute-Creation*, ii. 56, 77; Lawrence, *Horses* (3rd edn, 1810), i. 85 (though in a MS. note he later declared that he was no longer convinced that the souls of either men or animals were immortal; Bodl. copy, shelfmark 18972 d 52); Matthew Henry, *A Commentary on the Holy Bible* (1710; new edn, n.d.), iii. 964; Morris, *Records of Animal Sagacity*, xi–xiii; John Wesley, *Sermons on Several Occasions* (9th edn, n.d.), no. lxv.
25. George Cheyne, *An Essay on Regimen* (1740), 86–7; Dean, *Essay on the Future Life of Brutes*, ii. 49.
26. *The Works of Augustus Toplady*, iii (1794), 465–6. For other believers in animal immortality, see, e.g., *The Love-Letters of Mary Hays*, ed. A. F. Wedd (1925), 120; *The Adventurer*, 37 (1753); *The Diaries of Thomas Wilson*, ed. C. L. S. Linnell (1964), 259. The Muggletonians were divided on the issue in the 1730s (*Procs. Literary and Philos. Soc. of Liverpool*, xxiv (1869–70), 234, and information from Prof. William Lamont).
27. Wakefield, *Instinct Displayed*, 192; Pope, 'An Essay on Man', i. lines 111–12; Joseph Spence, *Anecdotes*, ed. Samuel Weller Singer (1820), 294.
28. Cited by Henry S. Salt, *Animals' Rights* (1892), 19.
29. *The Dunlop Papers*, i. *Autobiography of John Dunlop*, ed. J. G. Dunlop (1932), 186.
30. Mandelbaum, 'Scientific Background of Evolutionary Theory', 353.
31. *Life of Frances Power Cobbe by herself* (1894), ii. 199; *Personal Recollections of Mary Somerville*, 349; Robert Southey, 'On the Death of a Favourite Old Spaniel'. For other 19th-century believers in animal salvation see, e.g., S. T. Coleridge, *Lay Sermons*, ed. R. J. White (1972), 183n; William H. Drummond, *The Rights of*

Animals (1838), 204; T. Forster, *Sati or Universal Immortality* (Bruges, 1843); [R. Armitage], *The Penscellwood Papers* (1846?), i; Edward Maitland, *Anna Kingsford* (3rd edn, 1913), ii. 311–12; M. H. Spielmann and G. S. Layard, *Kate Greenaway* (1905), 190; Wood, *Man and Beast*, chap. xvii; Morris, *Records of Animal Sagacity*, preface; T. S. Hawkins, *The Soul of an Animal* (n.d.), chap. 2.

32. Wood, *Man and Beast*, 19.
33. John Bird Sumner, *A Treatise on the Records of Creation* (1816), ii. 18; Harold E. Gruber, *Darwin on Man* (1974), 39.
34. Charles Darwin, *The Descent of Man* (1871), i. 35, 48–9, 68, 78, 105. Cf. the comments of Gertrude Himmelfarb, *Darwin and the Darwinian Revolution* (1959), 304–7; and Young, 'Animal Soul', in *Encyclopedia of Philosophy*.

CHAPTER IV COMPASSION FOR THE BRUTE CREATION

i. Cruelty

1. *The Miscellaneous Works of ... Edward Earl of Clarendon* (2nd edn, 1751), 347; *The Journal of William Beckford in Portugal and Spain*, ed. Boyd Alexander (1954), 154, 127; *Letters of Robert Southey*, ed. Maurice H. Fitzgerald (1912), 33.
2. Kenneth Woodbridge, *Landscape and Antiquity* (Oxford, 1970), 84–5.
3. Tobias Smollett, *Travels through France and Italy* (1907 edn), 174; Captain Jesse, *The Life of George Brummell* (rev. edn, 1886), ii.156; William S. Childe-Pemberton, *The Earl Bishop* (n.d.), i. 77.
4. See, e.g., *The Letters of Charles Dickens*, iv, ed. Kathleen Tillotson (Oxford, 1977), 272; Edward Maitland, *Anna Kingsford* (3rd edn, 1913), ii. 311–12.
5. John Houghton, *A Collection for Improvement of Husbandry and Trade*, v. 108 (24 Aug. 1694). On bear- and bull-baiting see E. K. Chambers, *The Elizabethan Stage* (Oxford, 1945 reprint), ii. 449–71, and Robert W. Malcolmson, *Popular Recreations in English Society, 1700–1850* (Cambridge, 1973), 45–6, 66–8. For baiting of other animals see, e.g., *HMC, Chequers*, 414; Charles Stevens and John Liebault, *Maison Rustique*, trans. Richard Surflet, ed. Gervase Markham (1616), 703; *Crime in England, 1550–1800*, ed. J. S. Cockburn (1977), 237; John Clare, 'Badger', *Poems*, ed. J. W. Tibble (1935), ii. 333–4 (badgers); *Lancashire Quarter Sessions Records*, ed. James Tait, i (Chetham Soc., 1917), 101 (apes); *Yorkshire Diaries*, ed. H. J. Morehouse (Surtees Soc., 1877), 307 (mules); Robert Surtees, *The History and Antiquities of the County Palatine of Durham* (1816–40), iv. 78n; *Notes & Queries*, 4th ser., xii (1873), 273 (horses).
6. *Materials for the History of Thomas Becket*, ed. James Craigie Robertson (Rolls Ser., 1875–85), iii. 9. For general accounts see *Games and Gamesters of the Restoration* (1930), 100–114; George Ryley Scott, *The History of Cockfighting* (n.d.); and Malcolmson, *Popular Recreations*, 49–50.
7. *HMC*, 14th rept., appendix, pt viii. 94.
8. Pepys, *Diary*, iv. 427–8; Laurence V. Ryan, *Roger Ascham* (1963), 229, 242; James Tyrrell, *A Brief Disquisition of the Law of Nature* (1692), 324.
9. *Shakespeare's Europe*, ed. Charles Hughes (2nd edn, New York, 1967), 477. For

general accounts see Henry L. Savage, 'Hunting in the Middle Ages', *Speculum*, viii (1933); Marcelle Thiébaux, 'The Mediaeval Chase', *ibid*., xlii (1967); J. W. Fortescue, 'Hunting', *Shakespeare's England* (Oxford, 1916); Raymond Carr, *English Fox Hunting* (1976); David C. Itzkowitz, *Peculiar Privilege* (Hassocks, 1977). On hare-coursing see *The Gentleman's Recreation* (1674), 39–47.

10. James Cleland, *'ΗΡΩ–ΠΑΙΔΕΙΑ, or The Institution of a Young Noble Man* (Oxford, 1607), 134.
11. *The Lisle Letters*, ed. Muriel St Clare Byrne (1981), vi. 177.
12. Gervase Markham, *Cavelarice* (1607), iii. 1; *Gentleman's Recreation*, 2.
13. William Dugdale, *Origines Juridiciales* (1666), 156; Evelyn Philip Shirley, *Some Account of English Deer Parks* (1867), 215–16; Roger North, *The Lives of Francis ... Dudley ... and ... John North*, ed. Augustus Jessopp (1890), ii. 39.
14. *HMC, Portland*, ii. 258–9; Arthur Young, *Travels in France and Italy* (EL, 1915), 62.
15. *The Essays of Montaigne*, trans. John Florio (1603; 1893 edn), ii. 119–20; Peter Beckford, *Thoughts on Hunting* (n.d.), 146 (letter xvi); William Cowper, *Selected Letters* (EL, 1926), 270. Cf. above, p. 29.
16. *Camden Miscellany*, xvi (Camden ser., 1936), iii. 90.
17. See, e.g., Izaak Walton, *The Compleat Angler* (1653), i. 3; i. 8.
18. William Hinde, *A Faithfull Remonstrance of the Holy Life and Happy Death of Iohn Bruen* (1641), 33; *Gentleman's Recreation*, 11–12.
19. *Captain Cox, his Ballades and Books; or, Robert Laneham's Letter*, ed. Frederick J. Furnivall (Ballad Soc., 1871), 13–14, 17.
20. John Nichols, *The Progresses and Public Processions of Queen Elizabeth* (1823), iii. 91.
21. Martin Lluellyn, 'Cock-Throwing', in *Seventeenth-Century Lyrics*, ed. Norman Ault (1928), 191. See John Brand, *Observations on ... Popular Antiquities*, ed. Sir Henry Ellis (new edn, 1849–55), i. 72–82; Jeffrey N. Boss in *Notes & Records of the Royal Society*, 32 (1977), 145.
22. Fabian Philipps, *Tenenda non Tollenda* (1660), sig. A4^{v}.
23. E.g. G. R. Owst, *The Destructorium Viciorum of Alexander Carpenter* (1952), 25–6; *Minor Poets of the Caroline Period*, ed. George Saintsbury (Oxford, 1921), iii. 342; *Diary of Thomas Isham*, trans. Norman Marlow, ed. Sir Gyles Isham (Farnborough, 1971), 119; James Orchard Halliwell, *A Dictionary of Archaic and Provincial Words* (5th edn, 1845), 'praaling', 'tail-piping'; Robert Holland, *A Glossary of Words used in the County of Chester* (EDS, 1884–6), 331; Ellis, vii(2). 104–5; *The Memoirs of James Stephen*, ed. Merle M. Bevington (1954), 62, 90–91; H. C. Maxwell Lyte, *A History of Eton College* (1875), 27.
24. *The Remaining Medical Works of ... Thomas Willis*, trans. S. P[ordage] (1681), 93.
25. *The Works of ... Robert Boyle* (1744), v. 439.
26. *The Portledge Papers*, ed. Russell J. Kerr and Ida Coffin Duncan (1928), 262.
27. L. Tyerman, *Wesley's Designated Successor* (1882), 260. Cf. Jonathan Swift, *Gulliver's Travels* (1726), ii. 1; W. A. L. Vincent, *The State of School Education, 1640–1660* (1950), 36; Francis Coventry, *The History of Pompey the Little*, ed.

Robert Adams Day (1974), 49; Lawrence, *Horses*, i. 137; Joseph Hunter, *The Hallamshire Glossary* (1829), 78.

28. James Edmund Harting, *The Ornithology of Shakespeare* (1871), 237; *The Art of Angling* (appended to *The Country Man's Recreation* (1654)), 11; Morton, *Northants.*, 422.
29. John Sykes, *Local Records* (new edn, Newcastle, 1833), i. 221; Halliwell, *Dictionary*, 'mumble-a-sparrow'.
30. BL, Add. MS. 32, 512, fol. 20.
31. Swift, *Gulliver's Travels*, ii. 1; William Edward Hartpole Lecky, *History of European Morals* (1913 edn), i. 134.
32. Raymond Firth, *Elements of Social Organization* (1951), 199–200.
33. *Diary of Thomas Isham, passim.*
34. Edmund Waller, 'The Battle of the Summer Islands'; Matthew Prior, *Poems on Several Occasions*, ed. A. R. Waller (Cambridge, 1905), 134.
35. William Blake, 'Auguries of Innocence'.
36. On the tracts for children see F. J. Harvey Darton, *Children's Books in England* (Cambridge, 1958), chap. x, and Samuel F. Pickering, *John Locke and Children's Books in Eighteenth-Century England* (Knoxville, 1981), 12, 14–39.
37. Much relevant poetical writing is cited in Dix Harwood, *Love for Animals and How it Developed in Great Britain* (New York, 1928) and Dagobert De Levie, *The Modern Idea of the Prevention of Cruelty to Animals and its Reflection in English Poetry* (New York, 1947).
38. 3 Geo. IV, c. 71 (1822); 3 & 4 Gul. IV, c. 19 (1833); 5 & 6 Gul. IV, c. 59 (1835); 12 & 13 Vic., c. 92 (1849); 17 & 18 Vic., c. 60 (1854); 39 & 40 Vic., c. 77 (1876). The legal position at the end of the nineteenth century is summarized in Percy M. Burton and Guy H. Guillum Scott, *The Law relating to the Prevention of Cruelty to Animals* (1906). For accounts of the agitation see E. S. Turner, *All Heaven in a Rage* (1964); Malcolmson, *Popular Recreations*, appendix; Brian Harrison, 'Animals and the State in Nineteenth-Century England', *Eng. Hist. Rev.*, lxxxviii (1973); James Turner, *Reckoning with the Beast* (1980).
39. Edward G. Fairholme and Wellesley Pain, *A Century of Work for Animals* (1924), 95.
40. Lawrence, *Horses*, i. 135; William Edward Hartpole Lecky, *The History of the Rise and Influence of the Spirit of Rationalism* (1910 edn), i. 303.

ii. New Arguments

1. *The Whole Works of . . . Jeremy Taylor*, ed. Reginald Heber, rev. Charles Page Eden (1847–54), ix. 359.
2. Lawrence, *Horses*, i. 131.
3. Samuel Richardson, *Clarissa* (3rd edn, 1750–51), iv. 342 (iv. letter 56); John Porter, *Kingsclere*, ed. Byron Webber (1896), 303.
4. *Memoirs of the Life of Sir Samuel Romilly*, ed. by his sons (2nd edn, 1840), ii. 293–4. Cf. M. Pillet, *L'Angleterre* (Paris, 1815), 270.
5. Aquinas, *Summa contra Gentiles*, iii. 113.

6. Other relevant texts were Numbers, xxii. 28; Deuteronomy, xxii. 6–7; Isaiah, i. 11; Jonah, iv. 11.
7. E.g. David Dickson, *An Exposition of all St Paul's Epistles* (1659), 53; William Burkitt, *Expository Notes with Practical Observations on the New Testament* (1703), on 1 Cor., ix. 9; William Young, *The History of Dulwich College* (1889), i. 221.
8. E.g. [William Alley], *The Poore Mans Library* (1571), fol. 142v; Michael Jermin, *Paraphrasticall Meditations* (1638), 245; James Usher, *A Body of Divinitie* (1645), 275; *Whole Works of Jeremy Taylor*, ix. 357.
9. John Calvin, *Commentaries on the Four Last Books of Moses*, ed. Charles William Bingham (Edinburgh, 1852–5), iii. 56–7; George Estie, *A Most Sweete and Comfortable Exposition upon the Tenne Commaundements* (1602), sigs. M8v–N1; James Durham, *A Practical Exposition of the X. Commandements* (1675), 210.
10. George Hughes, *An Analytical Exposition of ... Genesis and ... Exodus* (1672), 942.
11. [John Day], *Day's Descant on Davids Psalmes* (1620), 210.
12. Notable classical texts include Plutarch, *Lives: Marcus Cato*, v. 3; *Moralia: De Sollertia Animalium*; *De Esu Carnium*; Cicero, *De Republica*, iii. 11; *Epist. ad Familiares*, vii. 1; Marcus Aurelius Antoninus, *Meditations*, vi. 23; Porphyry, *De Abstinentia*; Oppian, *Halieutica*, iv. 530–61; v. 519–88.
13. Harwood, *Love for Animals*, 74; Peter Singer, *Animal Liberation* (1977 edn), 197, 202; John Passmore, 'The Treatment of Animals', *JHI*, 36 (1975), 215–16. For similar views see Lecky, *History of European Morals*, ii. 166; Edward Westermarck, *The Origin and Development of Moral Ideas* (1908), ii. 506–9.
14. Chaucer, 'The Manciple's Tale', lines 170–71 (presumably echoing Boethius, *Philosophiae Consolationis*, iii. 2); *Secular Lyrics of the XIVth and XVth Centuries*, ed. Rossell Hope Robbins (Oxford, 1952), 107–10; *The Penguin Book of Animal Verse*, ed. George MacBeth (Harmondsworth, 1965), 32–5.
15. *Beasts and Saints*, translations by Helen Waddell (1934), *passim*; Rosalind Hill, *Both Small and Great Beasts* (n.d.), 5–8.
16. Robert Charles Hope, *The Legendary Lore of the Holy Wells of England* (1893), 30; *Beasts and Saints*, 87–8; Alexander Murray, *Reason and Society in the Middle Ages* (1978), 379–80.
17. *Henry the Sixth. A Reprint of John Blacman's Memoir*, ed. M. R. James (Cambridge, 1919), 18; Bernard Lord Manning, *The People's Faith in the Time of Wyclif* (Cambridge, 1919), 115. In *The Writing of History in the Middle Ages*, ed. R. H. C. Davis and J. M. Wallace-Hadrill (Oxford, 1981), 439n, Roger Lovatt points out, however, that Henry VI is known to have hunted on occasions.
18. *Dives and Pauper*, ed. Priscilla Heath Barnum (EETS, 1976–), i(2). 35–6 (text modernized).
19. William Cowper, 'The Task', vi. lines 581–6 (my italics).
20. *The Workes of ... William Perkins*, ii (Cambridge, 1617), 141.
21. *The Sermons of M. John Calvin upon ... Deuteronomie*, trans. Arthur Golding (1583), 877, 776, 560–62, 774.

22. *The Poems of Sir Philip Sidney*, ed. William A. Ringler, Jr (Oxford, 1962), 103 (spelling modernized). The modern editor assumes that this is a fable in which the beasts represent the lower classes (*ibid.*, 412–13).
23. Hughes, *Analytical Exposition of Genesis*, 10; *The Works, Moral and Religious, of Sir Matthew Hale*, ed. T. Thirlwall (1805), ii. 274. Prof. John Passmore (*Man's Responsibility for Nature* (1974), 30–31) regards Hale's views as exceptional; to my mind, he underrates the strength of the tradition of Christian stewardship in early Protestantism. On its biblical origins see John Black, *The Dominion of Man* (Edinburgh, 1970), chap. 4.
24. 'Philotheos Physiologus' [Thomas Tryon], *The Country-man's Companion* (n.d. [1683]), sig. A2.
25. David Hartley, *Observations on Man* (4th edn, 1801), i. 415; Humphry Primatt, *A Dissertation on the Duty of Mercy and Sin of Cruelty to Brute Animals* (1776), iv.
26. *Calvin upon Deuteronomie*, 768–9; [Thomas Wilcox], *A Short ... Commentarie written on ... the Proverbes* (1589), fol. 38; Peter Muffett, *A Commentarie upon the ... Proverbes* (1592), 103. Cf. George Abbot, *An Exposition upon the Prophet Ionah* (1600), 630–31; John Rawlinson, *Mercy to a Beast* (Oxford, 1612), 41; Henry Ainsworth, *Annotations upon the Five Bookes of Moses* (1639), v. 90.
27. [Robert Wilkinson], *A Sermon preached at North-Hampton* (1607), sig. C3. Cf. John Collinges, *Several Discourses concerning the actual Providence of God* (1678), 128–9; Matthew Henry, *A Commentary on the Holy Bible* (1710; new edn, n.d.), i. 7.
28. Lancelot Andrewes, *The Pattern of Catechistical Doctrine* (1650), 404; Jeremiah Burroughs, *Four Books on the Eleventh of Matthew* (1659), ii. 284; Richard Steele, *The Husbandmans Calling* (1670), 208–10; Durham, *Exposition of the X. Commandements*, 341–2.
29. Edward Elton, *Gods Holy Minde* (1648), 103, 109; *The Workes of ... Gervase Babington* (1637), i. 8; John Dod and Robert Cleaver, *A Plaine and Familiar Exposition of the Ten Commandements* (18th edn, 1632), 197.
30. Topsell, *Beastes*, 621.
31. 'The Rationalist' [William Baker], *Peregrinations of the Mind* (1770), 180; *The Theological, Philosophical and Miscellaneous Works of the Rev. William Jones* (1801), iii. 103–4.
32. George Wither, *Hallelujah, or Britain's Second Remembrancer* (1641; 1847 edn), pt i, hymn xxii.
33. Thomas Draxe, *The Earnest of our Inheritance* (1613), 26–7; Thomas Wilson, *A Commentarie upon the Most Divine Epistle of S. Paul to the Romanes* (1614), 586. Also G. G[oodman], *The Creatures praysing God* (1624), 29; George Walker, *God made Visible* (1641), 160; Benjamin Parker, *A Second Volume of Philosophical Meditations* (1735), 47.
34. *The Select Works of Robert Crowley*, ed. J. M. Cowper (EETS, 1872), 16.
35. Thomas Babington Macaulay, *The History of England* (1905 edn), i. 144.
36. *Workes of Perkins*, ii. 141; John Spencer, *A Discourse of Divers Petitions* (1641), 58–9; *Philip Stubbes's Anatomy of Abuses in England*, ed. Frederick J. Furnivall

(New Shakspere Soc., 1877–9), 177–8; Hinde, *Faithfull Remonstrance*, 31–2; Henry Bedel, *A Sermon exhorting to Pitie the Poore* (1573), sig. Ej[v]; [John Dod and Robert Cleaver], *A Plaine and Familiar Exposition of the Eleventh and Twelfth Chapters of the Proverbes* (1607), 140–42; Thomas Beard, *The Theatre of Gods Iudgements* (3rd edn, 1631), 212; Elton, *Gods Holy Minde*, 108; Osmund Lake, *A Probe Theologicall* (1612), 95; Francis Rous, *Oile of Scorpions* (1623), 106; *Robert Dover and the Cotswold Games*, ed. Christopher Whitfield (1962), 223.

37. *Acts and Ordinances of the Interregnum*, ed. C. H. Firth and R. S. Rait (1911), ii. 861.
38. Robert Bolton, *Some Generall Directions for a Comfortable Walking with God* (1625), 155–7.
39. G. F[ox], *To the Parliament of the Commonwealth of England: Fifty Nine Particulars* (1659), 12; *The Journal of George Fox*, ed. Norman Penney (Cambridge, 1911), i. 177; Tryon, *Country-Mans Companion*, 2; Hinde, *Faithfull Remonstrance*, 36; Richard Ligon, *A True & Exact History of the Island of Barbados* (1657), 105.
40. *Examen Legum Angliae* (1656), 121.
41. *Records of Maidstone* (Maidstone, 1926), 131; Rupert H. Morris, *Chester in the Plantagenet and Tudor Reigns* (Chester, n.d.), 333–4; John Latimer, *The Annals of Bristol in the Seventeenth Century* (Bristol, 1900), 260; *Quarter Sessions Records for the County of Somerset*, i, ed. E. H. Bates (Somerset Rec. Soc., 1907), 7; Richard J. Murrell and Robert East, *Extracts from Records in . . . Portsmouth* (Portsmouth, 1884), 515–16.
42. Leslie Hotson, *The Commonwealth and Restoration Stage* (1928; reprint, New York, 1962), 59–70.
43. *The Harleian Miscellany* (1808–11), vii. 67, 69.
44. Evelyn, *Diary*, v. 317.
45. *The Poems of Sir John Davies*, ed. Robert Krueger (Oxford, 1975), 149.
46. *Essays of Montaigne*, trans. Florio, ii. 126; Pepys, *Diary*, vii. 246; Evelyn, *Diary*, iii. 549; Richard Butcher, *The Survey and Antiquity of Stamford* (1717), 76.
47. *Tatler*, 134 (1710); Oswald Dykes, *English Proverbs with Moral Reflexions* (1709), 231; David Ogg, *England in the Reigns of James II and William III* (Oxford, 1955), 519.
48. Malcolmson, *Popular Recreations*, 118–20; G. A. Cranfield, *The Development of the Provincial Newspaper* (Oxford, 1962), 267; John Money, *Experience and Identity* (Manchester, 1977), 57; R. M. Wiles, *Freshest Advices* ([Columbus], 1965), 346.
49. *Correspondence of the Reverend Joseph Greene*, ed. Levi Fox (Dugdale Soc., 1965), 42.
50. Malcolmson, *Popular Recreations*, 120–22; *A History of Northumberland* (1893–1935), iii. 217; A. S. Mahood, 'Some Notes on Blundell's School', *Trans. Devonshire Assoc.*, 84 (1952), 67–8; *VCH, City of York*, 246; *Miscellanies . . . of the Rev. Thomas Wilson*, ed. F. R. Raines (Chetham Soc., 1857), xviii.
51. *VCH, Warws.*, vii. 221; Malcolmson, *op. cit.*, 124–6; *The Later Records relating*

to North Westmorland, ed. John F. Curwen (Cumbs. and Westld. Antiqn. and Archaeol. Soc., 1932), 149.

52. E.g. Lawrence, *Horses*, ii. 12; William Chafin, *A Second Edition of the Anecdotes and History of Cranbourn Chase* (1818), 52–3.
53. *Boswell's London Journal 1762–1763*, ed. Frederick A. Pottle (1950), 87; Pepys, *Diary*, iv. 427–8; [John Dunton], *Athenian Sport* (1707), xix; W. Bennett, *A History of the Burnley Grammar School* (Burnley, 1940), 27.
54. Reprinted in *Games and Gamesters of the Restoration*, 112–14. For other attributions see *First-Line Index of English Poetry 1500–1800*, ed. Margaret Crum (Oxford, 1969), i. 277.
55. 'The Parish Register', in *The Life and Poetical Works of George Crabbe by his Son* (new edn, 1854), 135.
56. Malcolmson, *Popular Recreations*, 135.
57. Edmund Gibson, *Codex Juris Ecclesiastici Anglicani* (1713), i. 183; David Wilkins, *Concilia* (1737), iii. 721; Thomas Fuller, *The Church History of Britain* (new edn, 1837), iii. 286–8.
58. Denis Brailsford, *Sport and Society* (1969), 111; John Brayne, *The New Earth* (1653), 93–4; *The Practical Works of . . . Richard Baxter* (1707), i. 367; *Autobiography of Thomas Raymond and Memoirs of the Family of Guise*, ed. G. Davies (Camden ser., 1917), 118; Arthur Dent, *The Plaine-Mans Path-Way to Heaven* (21st edn, 1631), 170; Ruth Spalding, *The Improbable Puritan* (1975), 46; *A Cavalier's Note Book*, ed. T. Ellison Gibson (1880), 192.
59. *HMC, Laing*, i. 99–100. Cf. D. Harris Willson, *King James VI and I* (1956), 179–82.
60. Hinde, *Faithfull Remonstrance*, 32–42; *Extracts from the Diary of the Rev. Robert Meeke*, ed. Henry James Morehouse (1874), 13–14. On the lawfulness of hunting see also Wilson, *Commentarie upon Romanes*, 591; Andrew Willet, *Hexapla in Leviticum* (1631), 413–14; William Prynne, *Histrio-Mastix* (1633), 966; Lake, *Probe Theologicall*, 94; John Downame, *A Guide to Godlynesse* (n.d.), 266–7.
61. *The Memoirs of Edmund Ludlow*, ed. C. H. Firth (Oxford, 1894), i. 1; *The Journal of Nicholas Assheton*, ed. F. R. Raines (Chetham Soc., 1848), *passim*; BL, Add. MS. 37, 345, fols. 244–8.
62. John of Salisbury, *Policraticus*, ed. Clement C. J. Webb (Oxford, 1909), i. 21–35; Thomas More, *Utopia* (EL, 1951), 89, 128; Robert P. Adams, *The Better Part of Valor* (Seattle, 1962), 15, 46–7, 99, 145–6; V. Norskov Olsen, *John Foxe and the Elizabethan Church* (1973), 198–9.
63. *DNB*, 'Warham, William'; John Strype, *The Life and Acts of Matthew Parker* (1711), 504; *The Zurich Letters*, ed. Hastings Robinson (Parker Soc., Cambridge, 1842–5), ii. 86; *Stubbes's Anatomy of Abuses*, 181–2; Christopher Hooke, *A Sermon preached in Paules Church* (1603), sigs. B2^{v}–3.
64. Hinde, *Faithfull Remonstrance*, 47.
65. [Nicholas Cox], *The Gentleman's Recreation* (1677; East Ardsley, 1973), i. 3 (plagiarizing *Gratii Falisci Cynegeticon*, trans. Christopher Wafse (1654), 'A Preface to the Reader'); Thomas Tryon, *Wisdom's Dictates* (1691), 129.
66. *HMC, Salisbury*, xvii. 108.

67. Edward Bury, *The Husbandmans Companion* (1677), 222–3.
68. Walker, *God made Visible*, 160; J. B. Williams, *Memoirs of . . . Sir Matthew Hale* (1835), 203.
69. William C. Braithwaite, *The Second Period of Quakerism* (1919), 513, 565; *Tatler*, 134 (1710).
70. *Poems of William Browne of Tavistock*, ed. Gordon Goodwin (n.d.), i. 95.
71. J. P. Hore, *The History of the Royal Buckhounds* (1893), 397; *Gilbert White's Journals*, ed. Walter Johnson (1931; reprint, Newton Abbot, 1970), 301.
72. William Blane, *Cynegetica* (1788), 112. Cf. [Soame Jenyns], *Disquisitions on Several Subjects* (2nd edn, 1782), 23–4; Charles Leigh, *The Natural History of Lancashire, Cheshire, and the Peak* (Oxford, 1700), ii. 3; Luke Booker, *Poems* (Wolverhampton, 1785), i. 36; ii. 83; Lawrence, *Horses*, ii. 13–14; *The Hare, or Hunting Incompatible with Humanity* (Dublin, 1800); Nathaniel Bloomfield, *An Essay on War* (1803), xvii; Harwood, *Love of Animals*, 284–9; De Levie, *Modern Idea of Prevention of Cruelty*, 81–4.
73. William Somervile, *The Chace* (1735; 4th edn, 1749), bk iii; Richard Jago, *Edge-Hill* (1767), 51–2.
74. [Francis Mundy], *Needwood Forest* (Lichfield, 1776), 35, 36.
75. Rawlinson, *Mercy to a Beast*, 42, 45; James Thomson, 'The Seasons: Autumn' (1730), lines 401–501.
76. Itzkowitz, *Peculiar Privilege*, 142–3.
77. William Harrison, *The Description of England*, ed. Georges Edelen (Ithaca, N.Y., 1968), 325–6; *Suffolk in the XVIIth Century*, ed. Lord Francis Hervey (1902), 34; *Letters and Papers of Henry VIII*, xiv(2). no. 810.
78. J. W[orlidge], *Systema Agriculturae* (1669), 192; John R. Wise, *The New Forest* (1863), 230; *The Letters of Daniel Eaton*, ed. Joan Wake and Deborah Champion Webster (Northants. Rec. Soc., 1971), xlvi n6; Ralph Arnold, *A Yeoman of Kent* (1949), 134; Charles Brears, *Lincolnshire in the 17th and 18th Centuries* (1940), 76.
79. E.g. *Spectator*, 116 (1711); *Constable of Everingham Estate Correspondence*, ed. Peter Roebuck (Yorks. Archaeol. Soc., 1976), 26; Albinia Lucy Cust, *Chronicles of Erthig on the Dyke* (1914), i. 241–2.
80. *Pembroke Papers (1780–1794)*, ed. Lord Herbert (1950), 327–8; Ellis, iv(3), 124–5; Itzkowitz, *Peculiar Privilege*, 117.
81. E.g. T. N. Brushfield, 'On the Destruction of "Vermin" in Rural Parishes', *Trans. Devonshire Assoc.*, 29 (1897), 308–9. Cf. P. B. Munsche, *Gentlemen and Poachers* (Cambridge, 1981), 47.
82. Charles Durnford and Edward Hyde East, *Term Reports in the Court of King's Bench* (new edn, 1817), i. 334–8.
83. E.g. Primatt, *Dissertation on the Duty of Mercy*, 75–6; James Everett, *The Village Blacksmith or . . . a Memoir of the Life of Samuel Hick* (n.d.), 49; 'Josiah Frampton' [? = William Gilpin], *Three Dialogues on the Amusements of Clergymen* (2nd edn, 1797), 32–3, 45, 76–7; Itzkowitz, *Peculiar Privilege*, 139, 230 n24.

iii. The Dethronement of Man

1. Clarence J. Glacken, *Traces on the Rhodian Shore* (Berkeley and Los Angeles, 1967), 49–50, 52, 69, 179–80, 185–6; Passmore, *Man's Responsibility for Nature*, 14, 16; Henry Chadwick, *Origen contra Celsum* (Cambridge, 1953), x, 242–8; *id.* in *Jnl Theological Studs.*, xlviii (1947), 36–7.
2. *Select Works of Porphyry*, trans. Thomas Taylor (1823), 115–16.
3. *Essays of Montaigne*, trans. Florio, ii. 126; Alan M. Boase, *The Fortunes of Montaigne* (1935), 204; J. S. Spink, *French Free-Thought from Gassendi to Voltaire* (1960), 40.
4. *The Writings of John Bradford*, ed. Aubrey Townsend (Parker Soc., Cambridge, 1848–53), i. 361–2.
5. Thomas Edwards, *Gangraena* (1646), i. 20.
6. *The Morning Exercises at Cripplegate* (5th edn by James Nichols), iii (1854), 319; Henry More, *An Antidote against Atheism* (2nd edn, 1655), 124–5, 371.
7. Ray, *Wisdom*, 127–8. Cf. Joseph Caryl, *An Exposition ... of Job* (1643–66), xi. 208–10; William King, *An Essay on the Origin of Evil* (Eng. trans., 1731), 106.
8. *Oldenburg*, i. 277. Cf. Kevin Sharpe, *Sir Robert Cotton* (Oxford, 1979), 46.
9. Ray, *Wisdom*, 49; Bodl., Ashmole MS. 192, ii. 28; Bernard Tocanne, *L'Idée de la nature en France dans la seconde moitié du XVIIème siècle* (1978), 52–4; Oxford Univ. Archives, Qb 18, fol. 177^{v}; Bd 19, fol. 204^{v}.
10. Arthur O. Lovejoy, *The Great Chain of Being* (1936; New York, 1960), 188; Tocanne, *op. cit.*, 57; Gottfried Wilhelm Leibniz, *Philosophical Papers and Letters*, ed. Leroy E. Loemker (2nd edn, Dordrecht, 1969), 316.
11. Richard Parr, *The Life of ... James Ussher* (1686), 492–4; John Locke, *An Essay concerning Human Understanding* (EL, 1947), 269 (iv. 3. 24).
12. Clifford Dobell, *Antony van Leeuwenhoek and his 'Little Animals'* (1932; reprint, New York, 1958), 174, 243.
13. M. Adanson, *Familles des plantes* (Paris, 1763), i. cxvj; Wilma George, 'Sources and Background to Discoveries of New Animals', *History of Science*, 18 (1980).
14. René Descartes, *The Principles of Philosophy*, iii (in *A Discourse of Method* (EL, 1912), 212); Thomas Gray, 'Elegy written in a Country Church-yard' (1751).
15. Arthur Raistrick, *Quakers in Science and Industry* (1950), 261.
16. Charles Lyell, *Principles of Geology* (1830–33), i. 154, 80; Roy Porter, *The Making of Geology* (Cambridge, 1977), 66–7 and chap. 7; John C. Greene, *The Death of Adam* (Ames, Iowa, 1959), chap. 4; *Fore-Runners of Darwin, 1745–1859*, ed. Bentley Glass *et al.* (Baltimore, 1959), chap. 9.
17. Ray, *Wisdom*, 101; Cecil Schneer, 'The Rise of Historical Geology in the Seventeenth Century', *Isis*, 45 (1954); Martin J. Rudwick, *The Meaning of Fossils* (2nd edn, New York, 1976), 65, 83, 105, 112; Margaret 'Espinasse, *Robert Hooke* (1956), 76. Cf. *Philosophical Letters*, ed. W. Derham (1718), 350.
18. Roy S. Porter, 'Philosophy and Politics of a Geologist', *JHI*, 39 (1978), esp. 443.
19. *Quarterly Rev.*, 108 (1860), 258; D. R. Oldroyd, *Darwinian Impacts* (Milton Keynes, 1980), 151.
20. *The Works of ... Henry St John, Lord Viscount Bolingbroke* (1809), viii. 184.

21. [Henry] Baker, *The Universe* (n.d.), 33; Prior, *Poems on Several Occasions*, 278. See also Alexander Pope, 'Essay on Man'; Bernard de Mandeville, *The Fable of the Bees* (Oxford, 1924), i. 179–80; Lovejoy, *Great Chain*, 207.
22. [Edward Bancroft], *An Essay on the Natural History of Guiana* (1769), 223–4; G. Gregory, *The Economy of Nature* (1796), iii. 103.
23. *Works of Robert Boyle*, iv. 515.
24. For this paragraph see Glacken, *Traces on the Rhodian Shore*, chap. 11.
25. 'Philotheos Physiologus' [Thomas Tryon], *Friendly Advice to the Gentlemen-Planters of the East and West Indies* (1684), 183; Margaret Cavendish, Marchioness of Newcastle, *Poems and Phancies* (2nd imp., 1664), 137, 86–90; Tryon, *Country-Mans Companion*, 171.
26. Thomas Hobbes, *The Elements of Law*, ed. Ferdinand Tönnies (2nd edn, 1969), 130–31; Hobbes, *EW*, ii. 113–14; v. 166, 185–6, 187–8; *id.*, *Thomas White's De Mundo examined*, trans. Harold Whitmore Jones (1976), 444–5; above, p. 166.
27. John Bramhall, *Castigations of Mr Hobbes* (1658), 190.
28. Tryon, *Country-Mans Companion*, sig. A2^{v}.
29. John Flavell, *Husbandry Spiritualized* (1669), 210; Benjamin Parker, *A Survey of the Six Days Work of the Creation* (1745), 195; *The Collected Poems of Christopher Smart*, ed. Norman Callan (1949), i. 227.
30. Burns, 'To a Mouse'; Blake, 'The Fly'; Coleridge, 'To a Young Ass'.

iv. New Sensibilities

1. John Dyer, 'The Fleece', ii. lines 22–3; *Collected Poems of Christopher Smart*, i. 251, 252; *Child-Rearing. Historical Sources*, ed. Philip C. Greven (Itasca, Illinois, 1973), 66; James Granger, *An Apology for the Brute Creation* (1772), 17; Allen, *Naturalist*, 145–7. Cf. Cowper, 'The Task', vi. 560, 562–3.
2. Laurence Sterne, *The Life and Opinions of Tristram Shandy* (1759–67), bk ii, chap. 12; *The Diary of Sylas Neville*, ed. Basil Cozens-Hardy (1950), 25; [William Melmoth], *The Letters of Sir Thomas Fitzosborne* (1742; 1776 edn), letter 16; *The Works of August Toplady* (new edn, 1853), 526; Chafin, *Anecdotes and History of Cranbourn Chase*, 63–4.
3. *Essays of Montaigne*, trans. Florio, ii. 119; Margaret Cavendish, Duchess of Newcastle, *The Life of William Cavendish*, ed. C. H. Firth (2nd edn, n.d.), 175.
4. Shakespeare, *Measure for Measure*, iii. 1. line 79; *Titus Andronicus*, iii. 2. lines 54–65; and references cited in Caroline Spurgeon, *Shakespeare's Imagery* (Cambridge, 1935), 104–9; *John Stubbs's Gaping Gulf*, ed. Lloyd E. Berry (Charlottesville, 1968), 28.
5. *Truth's Innocency and Simplicity shining through ... Thomas Taylor* (1697), 128–9; Francis Rous, *Archaeologiae Atticae* (8th edn, Oxford, 1675), 271; Frederick J. Powicke, 'The Reverend Richard Baxter's Last Treatise', *Bull. John Rylands Lib.*, x (1926), 197.
6. *Deeds against Nature, and Monsters by Kinde* (1614), sig. A4; *The Works of Gerrard Winstanley*, ed. George H. Sabine (Ithaca, N.Y., 1941), 329; Pepys, *Diary*,

ix. 203; *Oldenburg*, ix. 187; Wallace Shugg, 'Humanitarian Attitudes in the Early Animal Experiments of the Royal Society', *Ann. Sci.*, 24 (1968), 231–2.

7. MS. note in copy in present writer's possession of William Turner, *A Compleat History of the Most Remarkable Providences* (1697), iii. 129.
8. *Collected Poems of Christopher Smart*, ii. 990–91.
9. R. S. Crane, 'Suggestions toward a Genealogy of the "Man of Feeling"', *ELH. A Journal of English Literary History*, i (1934).
10. John Hall of Richmond, *Of Government and Obedience* (1654), 300; Samuel Parker, *A Demonstration of the Divine Authority of the Law of Nature* (1681), 54–5; Cecil A. Moore, *Backgrounds of English Literature, 1700–1760* (Minneapolis, 1953), chap. 1; [William Wollaston], *The Religion of Nature* (5th edn, 1731), 139; *The Adventurer*, 37 (1753).
11. 'The Task', vi. line 345.
12. Richard Cumberland, *A Treatise of the Laws of Nature*, trans. John Maxwell (1727), 302; David Hume, *An Enquiry concerning the Principles of Morals*, iii. 1. 152 (in *Enquiries*, ed. L. A. Selby-Bigge (2nd edn, Oxford, 1902), 190–91); Philip Doddridge, *A Course of Lectures on the Principal Subjects in Pneumatology, Ethics, and Divinity* (1763), 130; *Collected Poems of Christopher Smart*, i. 313, 290.
13. Primatt, *Dissertation*, 7; *The Political Writings of Jean Jacques Rousseau*, ed. C. E. Vaughan (reprint, Oxford, 1962), i. 138.
14. Jeremy Bentham, *An Introduction to the Principles of Morals and Legislation*, ed. Wilfrid Harrison (Oxford, 1948), 412n; and a similar passage in Bentham MSS. in Univ. Coll. London cit. by Amnon Goldworth in *The New York Rev. of Books*, xx (20 Sep. 1973), 42. It is possible that Bentham subsequently reverted to the argument that cruelty to animals was bad because it led to cruelty to men; see David Baumgardt, *Bentham and the Ethics of Today* (Princeton, N.J., 1952), 338–9, 362–3.
15. *The Journal of John Wesley*, ed. Nehemiah Curnock (1909–16), iv. 176.
16. Granger, *Apology for Brute Creation*, 8; John Oswald, *The Cry of Nature* (1791), ii; Lawrence, *Horses*, i. 5; Sir Richard Phillips, *Golden Rules of Social Philosophy* (1826), 42; W. Youatt, *The Obligation and Extent of Humanity to Brutes* (1839), 195; R. Fletcher, *A Few Notes on Cruelty to Animals* (1846), 4; Wordsworth, 'Hart Leap Well'.
17. Thomas Sheridan, *A Course of Lectures on Elocution* (1762), 104.
18. *A Memoir of Thomas Bewick written by himself*, ed. Ian Bain (1979 edn), 6, 178.
19. See authorities cited by Walton, *Compleat Angler*, i. 1.
20. De Levie, *Modern Idea of Prevention of Cruelty*, 90–91; Lawrence D. Stewart, *John Scott of Amwell* (Berkeley and Los Angeles, 1956), 12–13.
21. *The Letters of Charles Lamb*, ed. Alfred Ainger (1904), i. 106; Byron, 'Don Juan', canto xiii, stanza 106. Cf. John Wolcot (in *Come Hither*, ed. Walter de la Mare (new edn, 1928), 510); Jenyns, *Disquisitions*, 24; John Armstrong, 'The Art of Preserving Health', iii. lines 97–8; Primatt, *Dissertation*, 270; Anne Radcliffe, *The Mysteries of Udolpho* (1794; EL, 1931), 8 (i. 1); Thomas Young, *An Essay on*

Humanity to Animals (1798), 89–91; Henry Crowe, *Zoophilos* (1819), 25–9; James Jennings, *Ornithologia* (1828), xxiii, 6.

22. Above, p. 21; *Johnson on Shakespeare*, ed. Walter Raleigh (1908), 181. Cf. *The Idler*, 17 (1758).
23. *The Banks Letters*, ed. Warren R. Dawson (1958), 506. Cf. *Gentleman's Mag.*, x (1740), 194; Pennant, *Zoology*, i. 165; [Robert Wallace], *Various Prospects* (1761), 17; [T. J. Mathias], *The Pursuits of Literature* (5th edn, 1798), 340–42; Joseph Spence, *Anecdotes*, ed. Samuel Weller Singer (1820), 293.
24. Daniel Defoe, *The Complete English Tradesman* (1841 edn), ii. 228–9; A Prebendary of York, *An Enquiry about the Lawfulness of Eating Blood* (1733), 10; Priscilla Wakefield, *Instinct Displayed* (1811), 289; Young, *Essay on Humanity to Animals*, chap. vi; William Cobbett, *Cottage Economy* (1926 edn), 126.
25. 21 Geo. III c. 67 (1781); 26 Geo. III c. 71 (1786); Lawrence, *Horses*, i. 155–8; *Memoirs of Sir Samuel Romilly*, ii. 109–11.
26. E.g. Baker, *Peregrinations of the Mind*, 181; James Burgh, *An Account of the First Settlement . . . of the Cessares* (1764), 74–5; Cowper, 'The Task', vi. lines 432–3; *Works of Toplady*, 443; Primatt, *Dissertation*, 289; Lawrence, *Horses*, i. 123; Byng, iv. 151; *The Sporting Magazine*, iv (July 1794), 199.
27. E.g. L. Brettle, *A History of Queen Elizabeth's Grammar School for Boys, Mansfield* (Mansfield, n.d. [1961]), 43; W. T. Carless, *A Short History of Hereford School* (Hereford, 1914), 74.
28. E.g. *Bedfordshire County Records. Notes and Extracts from the . . . Quarter Sessions Rolls from 1714 to 1832* (Bedford, n.d.), i. 60; *The Sporting Magazine*, iv (July 1794), 188.
29. William Whewell, *Lectures on the History of Moral Philosophy* (1852), 223–5; John Stuart Mill, *Essays on Ethics, Religion and Society*, ed. J. M. Robson (1969), 185–7.
30. *Essays*, trans. Florio, ii. 126. Cf. Spink, *French Free-Thought*, 56, 61.
31. Nathaniel Homes, *The Resurrection-Revealed raised above Doubts and Difficulties* (1661), 244; J. W[orlidge], *Systema Horti-Culturae* (1677), 283; *The Autobiography of Edward, Lord Herbert of Cherbury*, ed. Sidney Lee (2nd edn, n.d.), 31.
32. *A Philosophical Survey of Nature* (1763), 29; George Bell, 'On the Physiology of Plants', in A. Hunter, *Georgical Essays* (York, 1803–4), i. 542; Thomas Percival, 'Speculations on the Perceptive Power of Vegetables', *Memoirs of the Literary & Philosophical Soc. of Manchester*, ii (1785), 115. Cf. [John Dunton], *The Athenian Oracle* (1703–4), ii. 448–51; James Perchard Tupper, *An Essay on the Probability of Sensation in Vegetables* (1811); Philip C. Ritterbush, *Overtures to Biology* (1964), 151; Charles Webster, 'The Recognition of Plant Sensitivity by English Botanists in the Seventeenth Century', *Isis*, 57 (1966).
33. Wordsworth, 'Lines written in Early Spring'; *The Prelude*, ed. Jonathan Wordsworth *et al.* (1979), 493.
34. Francis Hutcheson, *A System of Moral Philosophy* (1755), i. 314; Thomas Percival, *A Father's Instructions* (5th edn, 1781), 26–7; Bentham, *Introduction to Principles of Morals*, 412n; Lawrence, *Horses*, i. 123. Cf. Baker, *Peregrinations*,

182. Thomas Taylor's *A Vindication of the Rights of Brutes* (1792) cannot certainly be aligned with this school of thought. It is a curious work, in part satirical, in part serious, in part mildly pornographic.

35. Tryon, quoted above, p. 155; [James Burnet, Lord Monboddo], *Antient Metaphysics* (1779–99), ii. 103.
36. Mandeville, *Fable of the Bees*, i. 180; P. B. Shelley, *A Vindication of Natural Diet* (1813; 1884 edn), 13; William Lambe, *Additional Reports on the Effects of a Peculiar Regimen* (1815), 238.
37. E.g. Henry S. Salt, *Animals' Rights* (1892), 15.
38. Obadiah Walker, *Of Education* (1673), 91.
39. As does John Passmore, who writes that 'by the end of the eighteenth century . . . *even* Christians began to argue . . . that callousness toward animal suffering was morally wrong'; *Man's Responsibility for Nature*, 186 (my italics).

v. New Conditions

1. Cf. the comments of John Berger, 'Vanishing Animals', *New Soc.*, 39 (31 March 1977).
2. For figures see F. M. L. Thompson, *Victorian England: The Horse-Drawn Society* (1970); *id.*, in *Econ. Hist. Rev.*, 2nd ser., xxix (1976), 80; and Henry Edmunds in *Jnl Royal Agricultural Soc. Eng.*, 138 (1977), 53.
3. Seamus Heaney, *Death of a Naturalist* (1966), 23.
4. *Autobiography of Thomas Raymond*, 118; Malcolmson, *Popular Recreations*, esp. 160–63.
5. Pepys, *Diary*, vi. 295.
6. Sir Thomas Elyot, *The Boke named the Governour*, ed. H. H. S. Croft (1883), i. 189 (paraphrasing Xenophon); Somervile, *The Chace*; R. S. Surtees, *Handley Cross* (1843), chap. 48.
7. Sir Thomas Cockaine, *A Short Treatise of Hunting* (1591), sig. A3[v]; Hinde, *Faithfull Remonstrance*, 41.
8. G. W[ilson], *The Commendation of Cockes* (1607), *passim*; Ray, *Willoughby*, 155; *The Travels of Peter Mundy*, ed. Sir Richard Carnac Temple (Hakluyt Soc., 1907–25), iv. 50; William Machrie, *An Essay upon the Royal Recreation and Art of Cocking* (Edinburgh, 1705), dedication. There are marked analogies with the Balinese cock-fight immortalized by Clifford Geertz, 'Deep Play', *Daedalus* (Winter 1972).
9. *VCH, Hants.*, ii. 313; George Chapman, *A Treatise on Education* (Edinburgh, 1773), 227.
10. *The Diary of Abraham de la Pryme*, ed. Charles Jackson (Surtees Soc., 1870), 33.
11. *The Letters of the Earl of Chesterfield*, ed. Charles Strachey (1901), i. 324; ii. 176 (cf. Voltaire, *Anti-Machiavel* (Eng. trans., 1741), 153n–164n; William Shenstone, *Essays on Men and Manners* (1794), 88); *The Guardian*, 61 (21 May 1713).
12. Thomas Paine, *The Rights of Man* (1792), ii. 65; Carr, *English Fox Hunting*, 127.
13. Byng, iv. 10; Peter Borsay, 'The English Urban Renaissance', *Social History*, 5 (1977), 583; Carr, *op. cit.*, 59–61, 129–32.

14. E. P. Thompson, *Whigs and Hunters* (1975), 40, 64, 255–6; 37 Hen. VIII, c. 6 (1545); 22 & 23 Car. II, c. 7 (1670 & 1671); 9 Geo I, c. 22 (1722).
15. *Examen Legum Angliae*, 54, 56.
16. Cowper, *Selected Letters*, 273–9; Primatt, *Dissertation*, 11, 27–8; *The Works of Anna Laetitia Barbauld*, ed. Lucy Aikin (1825), i. 173–9; Granger, *Apology*, 20.
17. *Memoir of Thomas Bewick*, 136; George Nicholson, *On the Conduct of Man to Inferior Animals* (Stourport, n.d.), 53–63.
18. Oswald, *Cry of Nature*, ii; *The Complete Writings of Thomas Paine*, ed. Philip S. Foner (New York, 1945), i. 512; Mary Wollstonecraft, *The Rights of Woman* (1792; EL, 1929), 190.
19. *The Parliamentary History of England*, xxxvi (1820), 833–4, 839.
20. *Cruelties of Civilization*, ed. Henry S. Salt (n.d.), vii.
21. *Olla Podrida*, 6 (1787). Cf. Wollstonecraft, *Rights of Woman*, 191.
22. Maitland, *Anna Kingsford*, i. 46.
23. Malcolmson, *Popular Recreations*, 138, 152–7; Brian Harrison, 'Religion and Recreation in Nineteenth-Century England', *Past & Present*, 38 (1967), 117–18.
24. Lyte, *History of Eton College*, 283, 302–3.
25. Baker, *Peregrinations*, 225; *Works of Rev. William Jones*, iii. 103. Cf. Pennant, *Zoology*, ii. 314 (and Goldsmith, v. 163).
26. Fletcher, *Notes on Cruelty to Animals*, 55, 69; Lawrence, *Horses*, i. 148; *id.*, *A General Treatise on Cattle* (2nd edn, 1809), 283; 5 & 6 Gul. IV, c. 59 (1835).
27. Harrison, 'Religion and Recreation', 116–19; Lewis Gompertz, *Fragments in Defence of Animals* (1852), 86–7.
28. *The Later Letters of John Stuart Mill*, ed. Francis E. Mineka and Dwight N. Lindley (1972), iii. 1423–4; Mill, *Principles of Political Economy*, ed. J. M. Robson (1965), ii. 952 (v. xi. 9).
29. Youatt, *Humanity to Brutes*, 141; Fletcher, *Notes on Cruelty*, 23.
30. Cf. David Brion Davis, *The Problem of Slavery in the Age of Revolution, 1770–1823* (1975), 348–50, 354, 366, 373–8, 451, 466–7.
31. Lawrence, *Horses*, i. 135; *Cruelty to Animals. The Speech of Lord Erskine* (1809), 27; Malcolmson, *Popular Recreations*, 135–6.
32. Davis, *The Problem of Slavery in the Age of Revolution*, as summarized by George M. Fredrickson in *New York Rev. of Books*, xxii (16 Oct. 1975). Cf. Karl Marx, *Capital*, trans. Eden and Cedar Paul (EL, 1930), i. 234.
33. Gompertz, *Fragments*, 34; Margaret T. Hodgen, *Early Anthropology in the Sixteenth and Seventeenth Centuries* (Philadelphia, 1964), 366–7; William Cobbett, *Rural Rides* (Harmondsworth, 1967), 62; J. G. Wood, *Man and Beast Here and Hereafter* (8th edn, 1903), 228, 427, 429.
34. Charles Darwin, *The Descent of Man* (1871; 2nd edn, 1891), ii. 440; Frances Power Cobbe, 'The Ethics of Zoophily', *Contemporary Rev.*, lxviii (1895), 504.
35. Percival, *Father's Instructions*, 64; Darwin, *op. cit.*, i. 180, 188.
36. Topsell, 106; Fuller, cit. Emma Phipson, *The Animal-Lore of Shakespeare's Time* (1883), 136.
37. *Killing for Sport*, ed. Henry S. Salt (1914), xii.

38. Topsell, 621; John Locke, *Two Treatises of Government*, ed. Peter Laslett (Cambridge, 1960), 346 (ii. para. 93).
39. Markham, *Cavelarice*, i. 77, 82, 85; ii. 15, 63–4, 90, 96–7, 106; v. 46. Cf. Thomas de Grey, *The Compleat Horse-Man* (3rd edn, 1656), 47–8; Joan Thirsk, *Horses in Early Modern England* (Reading, 1978), 18.
40. Stevens and Liebault, *Maison Rustique*, 90, 123; John Laurence, *A New System of Agriculture* (1726), 130.
41. Alfred Saunders, *Our Horses* (1886), ix. 89; *Cruelty to Animals. The Speech of Lord Erskine*, 4.
42. Thomas Muffett, *Healths Improvement*, enlarged by Christopher Bennet (1655), 43–4; George Cheyne, *An Essay of Health* (6th edn, Dublin, 1725), 14–15. See also Ray, *Willoughby*, 155; John Arbuthnot, *An Essay concerning the Nature of Aliments* (Dublin, 1731), 39.
43. Muffett, *op. cit.*, 42.
44. *The Laws and Liberties of Massachusetts*, ed. Max Farrand (Cambridge, Mass., 1929), 16 (my italics). For prosecutions under this law, Stephen Foster, *Their Solitary Way* (1971), 23n.
45. *Stuart Royal Proclamations*, ed. James F. Larkin and Paul L. Hughes (Oxford, 1973), nos. 36, 100; *Tudor and Stuart Proclamations*, ed. Robert Steele (Oxford, 1910), i. no. 1521.
46. Richard Bagwell, *Ireland under the Stuarts*, i (1909), 124–5; Aidan Clarke, *The Old English in Ireland* (1966), 240; T. C. Barnard, *Cromwellian Ireland* (1975), 76.
47. [Sir Peter Pett], *The Happy Future State of England* (1688), 22.
48. *Cruelty to Animals. The Speech of Lord Erskine*, 18–21.
49. Williams, *Memoirs of Sir Matthew Hale*, 203; Spence, *Anecdotes*, 328; Stuart Piggott, *William Stukeley* (Oxford, 1950), 144–5.
50. E.g. Pennant, *Zoology*, i. 12; William Gilpin, *Remarks on Forest Scenery* (1791), ii. 255–64; *Monthly Magazine*, 39 (March 1815), 115.
51. E.g. Worlidge, *Systema Agriculturae*, 172–4; Joseph Warder, *The True Amazons* (3rd edn, 1716); John Thorley, *ΜΕΛΙΣΣΗΛΟΓΙΑ or, the Female Monarchy* (1744), 4, 177; John Mills, *An Essay on the Management of Bees* (1766); Stephen White, *Collateral Bee-Boxes* (1756); John Keys, *The Practical Bee-Master* (n.d.), chap. viii; Thomas Nutt, *Humanity to Honey Bees* (1832).
52. *Boswell's Life of Johnson*, ed. George Birbeck Hill, rev. by L. F. Powell (Oxford, 1934–64), iv. 197; *The Letters and Prose Writings of William Cowper*, ed. James King and Charles Ryskamp (Oxford, 1979–), ii. 68–9.
53. Bodl., MS. Eng. Misc. d. 73, fol. 47^{v}.

CHAPTER V TREES AND FLOWERS

i. The Wild Wood

1. This paragraph is based on H. C. Darby, 'The Clearing of the English Woodlands', *Geography*, xxxvi (1951); *id.*, 'The Clearing of the Woodland in Europe', in *Man's Role in Changing the Face of the Earth*, ed. William L.

Thomas, i. (1956); *A New Historical Geography of England*, ed. H. C. Darby (Cambridge, 1973); Oliver Rackham, *Trees and Woodland in the British Landscape* (1976); *id.*, *Ancient Woodland* (1980); and the volumes in the *History of the English Landscape* series, ed. W. G. Hoskins. Much of the earlier part of the story inevitably remains speculative and controversial.

2. George Hammersley, 'The Crown Woods and their Exploitation in the Sixteenth and Seventeenth Centuries', *Bulletin Institute of Hist. Research*, 30 (1957); *id.*, 'The Charcoal Iron Industry and its Fuel', *Econ. Hist. Rev.*, 2nd ser., xxvi (1973); Michael W. Flinn, 'Timber and the Advance of Technology', *Ann. Sci.*, 15 (1959); *id.*, 'Consommation du bois et développement sidérurgique en Angleterre', *Actes du Colloque sur la Forêt, Besançon 21–22 Oct. 1966* (Paris, 1967).
3. *The Works of Michael Drayton*, ed. J. William Hebel (Oxford, 1961), v. 426 (and 465); George Owen of Henllys, *The Description of Penbrokshire*, ed. Henry Owen (Cymmrodorion Rec. Soc., 1892), 87.
4. *Northamptonshire Past & Present*, iii (1960–65), 277; *Glamorgan County History*, iv, ed. Glanmor Williams (Cardiff, 1974), 6; T. R. Potter, *The History and Antiquities of Charnwood Forest* (1842), 5; *The Diary of Ralph Thoresby*, ed. Joseph Hunter (1830), i. 362; William Gilpin, *Observations on the Western Parts of England* (1798), 331.
5. In addition to the *History of the English Landscape* volumes, see E. C. K. Gonner, *Common Land and Inclosure* (1912), 268–9, and *Camden's Britannia*, ed. Edmund Gibson (1695), col. 267; *Studies of Field Systems*, ed. Alan R. H. Baker and Robin A. Butlin (Cambridge, 1973), 364–5; David Roden, 'Woodland and its Management in the Medieval Chilterns', *Forestry*, 41 (1968); *VCH, Leics.*, ii. 267, 269.
6. *Seventeenth-Century Economic Documents*, ed. Joan Thirsk and J. P. Cooper (Oxford, 1972), 779; Roger Fisher, *Heart of Oak* (2nd edn, 1764), 29; Byng, iii. 229; *New Historical Geography of England*, 424; William Somerville, 'Forestry in some of its Economic Aspects', *Jnl Royal Statistical Soc.*, lxxii (1909), 42; *Man's Role in Changing the Face of the Earth*, 416.
7. Roderick Nash, *Wilderness and the American Mind* (rev. edn, New Haven, 1973), x–xi.
8. Joshua Poole, *The English Parnassus* (1657; 1677 edn), 90.
9. Nathaniel Morton, *New Englands Memoriall* (1669), ed. Howard J. Hall (New York, 1937), 13; *Bradford's History of Plymouth Plantation*, ed. William T. Davis (New York, 1908), 96; Peter N. Carroll, *Puritanism and the Wilderness* (1969), 193–4, 219–20 and *passim*; Alan Heimert, 'Puritanism, the Wilderness and the Frontier', *New Eng. Qtly*, xxvi (1953); Hans Huth, *Nature and the American* (Berkeley and Los Angeles, 1957), 5.
10. Heimert, *art. cit.*, 369n.
11. John Manwood, *A Treatise and Discourse of the Lawes of the Forrest* (1598), fols. 11v–12. Cf. J. N.[orden], *The Surveyors Dialogue* (1607), 233–4.
12. William Browne, *Britannia's Pastorals* (1613–16), i. 1. line 510.
13. Giraldus Cambrensis, *Topographia Hibernica*, ed. James F. Dimock (Rolls Ser., 1867), 151; Glyn Daniel, *The Idea of Prehistory* (1962), 17–18; John Stow, *A Survey of . . . London*, enlarged by John Strype (1720), i. 7.

14. D. B. Quinn, *The Elizabethans and the Irish* (Ithaca, N.Y., 1966), 136; John Locke, *Two Treatises of Government*, ed. Peter Laslett (Cambridge, 1960), 201 (i. para. 58); *The Writings and Speeches of Edmund Burke*, v, ed. P. J. Marshall (Oxford, 1981), 389; Samuel Sorbière, ep. ded. to Thomas Hobbes, *Elemens philosophiques du bon citoyen* (Paris, 1651), sig. *viij^{v}.
15. Philip A. J. Pettit, *The Royal Forests of Northamptonshire* (Northants. Rec. Soc., 1968), 16, 133, 163; *The Agrarian History of England and Wales*, iv, ed. Joan Thirsk (Cambridge, 1967), 111–12; *The Berkeley Manuscripts*, ed. Sir John Maclean (Gloucester, 1883–5), iii. 328; W. H. Stevenson and H. E. Salter, *The Early History of St John's College Oxford* (Oxford, 1939), 249–50; Cyril E. Hart, *The Free Miners of the Royal Forest of Dean* (Gloucester, 1953), 174–5, 197.
16. Buchanan Sharp, *In Contempt of All Authority* (1980); E. P. Thompson, *Whigs and Hunters* (1975); Douglas Hay, 'Poaching and the Game Laws on Cannock Chase', in *Albion's Fatal Tree*, ed. Hay *et al.* (1975); Charles Vancouver, *General View of the Agriculture of Hampshire* (1813), 496.
17. Robert W. Malcolmson in *An Ungovernable People*, ed. John Brewer and John Styles (1980), 86–7, 92; William Camden, *Britannia*, ed. Richard Gough (2nd edn, 1806), i. 111.
18. William Worcestre, *Itineraries*, ed. John H. Harvey (Oxford, 1969), 209; *A Memoir of . . . William Gilpin* (1851), 156; John Rous, *Historia Regum Angliae*, ed. Thomas Hearne (2nd edn, Oxford, 1745), 123–4; cf. John Vowell *alias* Hoker, *An Account of the Sieges of Exeter* (Exeter, 1911), 222.
19. E.g. [Francis Trigge], *To the Kings Most Excellent Maiestie* (1604), sig. B1.
20. *Cyvile and Uncyvile Life* (1579), sig. Fii^{v}.
21. Thomas Churchyard, *The Worthines of Wales* (1587; 1776 edn), 51–2; Walter Blith, *The English Improver Improved* (1653), 83.
22. Samuel Hartlib, *The Compleat Husband-Man* (1659), 13; Timothy Nourse, *Campania Foelix* (1700), 26–8, 60; Ellis, v(2). 57; viii. 175; John Kennedy, *A Treatise upon Planting* (Dublin, 1784), 157–8; Gonner, *Common Land and Inclosure*, 312–13.
23. John Houghton, *A Collection of Letters for the Improvement of Husbandry & Trade* (1681–3), ii. 74; Morton, *Northants.*, 12–13; John Dyer, 'The Fleece', i. lines 115–16.
24. Robert Greenhalgh Albion, *Forests and Sea Power* (Cambridge, Mass., 1926), 119.
25. Psalms, lxxiv. 5; *Camden Miscellany*, iii (Camden Soc., 1855), iii. 9; D. A. Hamer, 'Gladstone: the making of a political myth', *Victorian Studies*, 22 (1978–9), 37–8.

ii. Tree-Planting

1. H. G. Richardson, 'Some Remarks on British Forest History', *Trans. Scottish Arboricultural Soc.*, 35 (1921); Roden, 'Woodland and its Management'; Rackham, *Trees and Woodland*, *passim*; *id.*, *Ancient Woodland*, 111–60.
2. *English Historical Documents c. 500–1042*, ed. Dorothy Whitelocke (2nd edn, 1979), 403–4; Warren O. Ault, *Open-Field Husbandry and the Village Community*

(Trans. Amer. Philos. Soc., 1965), appendix, nos. 166, 180, 192, 210, 212, 215; Roden, *art. cit.*, 67; Charles R. Young, *The Royal Forests of Medieval England* (Leicester, 1979), 123, 170.

3. See Albion, *Forests and Sea Power*, 122–3; Hammersley, 'Crown Woods and their Exploitation'.
4. [John Fitzherbert], *Here begynneth ... the Boke of Surveyeing* (1523), fol. xlvi^{v}; Norden, *Surveyors Dialogue*, 116, 214; [Arthur Standish], *The Commons Complaint* (1611), 2; *id.*, *New Directions ... for the Planting of Timber* (1613); Gervase Markham, *The English Husbandman* (1635), bk ii, pt 2; Leonard Meager, *The Mystery of Husbandry* (1697), 130–37; Charles Webster, *The Great Instauration* (1975), 477, 480.
5. *Proceedings in Parliament* 1610, ed. Elizabeth Read Foster (1966), i. 51; John Smith, *Englands Improvement Revived* (1670), 8–9.
6. John Evelyn, *Sylva* (3rd edn, 1679), 'To the King'; Michael Hunter, *Science and Society in Restoration England* (Cambridge, 1981), 93.
7. Lindsay Sharp, 'Timber, Science, and Economic Reform in the Seventeenth Century', *Forestry*, xlviii (1975); Evelyn, *Sylva*, 9, 13, 26, 53, 115; Evelyn/Hunter, 87, 176, 282, 352–3, 580–81.
8. *VCH, Berks.*, ii. 349.
9. 19 & 20 Car. II, c. 8 (1668); 9 Gul. III, c. 33 (1697–8).
10. R. H. Richens, 'Studies on *Ulmus*. vii. Essex Elms', *Forestry*, 40 (1967), 201–2; John Steane, *The Northamptonshire Landscape* (1974), 83; John Harvey, *Early Nurserymen* (1974), 22–3; Rackham, *Trees and Woodland*, 96. More evidence for medieval planting is provided by John Harvey, *Mediaeval Gardens* (1981), 13–16.
11. William Dugdale, *Monasticon Anglicanum*, ed. John Caley *et al.* (1846), v. 43. Cf. Teresa McClean, *Medieval English Gardens* (1981), 20, 55, 57–8; Markham, *English Husbandman*, ii. 40.
12. Fitzherbert, *Boke of Surveyeing*, fol. xlvi; *id.*, *The Book of Husbandry*, ed. Walter W. Skeat (EDS, 1882), 82.
13. Oliver Rackham, *Hayley Wood* (Cambridge, 1975), 36; Ault, *Open-Field Husbandry*, appendix, no. 215; *Agrarian History of England*, iv. 459, 677; John H. Harvey in *The Local Historian*, 13 (1978), 233.
14. Evelyn, *Sylva*, 37; *The Works of ... Robert Boyle* (1744), v. 454.
15. Hugh Prince, *Parks in England* (Shalfleet Manor, I.O.W., 1967), 13–14; Peter Roebuck, *Yorkshire Baronets 1640–1760* (Oxford, 1980), 38, 241, 245, 326; Adam Smith, *An Inquiry into the Nature and Causes of the Wealth of Nations*, ed. R. H. Campbell and A. S. Skinner (Oxford, 1976), i. 183; Pepys, *Diary*, viii. 201.
16. Sharp, *In Contempt of All Authority*, 135–6 and *passim*; Rackham, *Hayley Wood*, 10.
17. Evelyn, *Sylva*, 108.
18. Standish, *New Directions*, 4; Nourse, *Campania Foelix*, 59. Cf. J. A. Sharpe, 'Crime in the County of Essex, 1620–1680' (Oxford D. Phil. thesis, 1979), 303; Pettit, *Royal Forests*, 125, 162–3; George Roberts, *The Social History of the People of the Southern Counties* (1856), 357–8; Frank Emery, *The Oxfordshire Landscape* (1974), 114; Ellis, ii(3).153; R. W. Bushaway, 'Grovely, Grovely and all

Grovely. Custom, Crime and Conflict in the English Woodland', *History Today*, 31 (1981).

19. Manwood, *Lawes of the Forrest*, fol. 1.
20. Richardson, 'Remarks on British Forest History', 162.
21. George Hammersley, 'The Revival of Forest Laws under Charles I', *History*, xlv (1960). Cf. Christopher Taylor, *The Making of the English Landscape. Dorset* (1970), 132; *Agrarian History of England*, iv. 98.
22. 'J. Norden's Survey of Medieval Coppices in the New Forest A.D. 1609', *Hants. Field Club*, x (1927), 105.
23. See Evelyn Philip Shirley, *Some Account of English Deer Parks* (1867); Prince, *Parks in England*; Maurice Beresford, *History on the Ground* (1971 edn), chap. 7; L. M. Cantor and J. Hatherly, 'The Medieval Parks of England', *Geography*, 64 (1979).
24. Houghton, *Collection of Letters*, ii. 73–4.
25. 'Date Stones in Althorp Park, Northamptonshire', *Northants Notes & Queries*, new ser., iii (1909).
26. Manwood, *Lawes of the Forrest*, fol. 33^{v}.
27. [Dudley, Lord North], *Observations and Advices Oeconomical* (1669), 104; A. G. Dickens, *The Register or Chronicle of Butley Priory* (Winchester, 1951), 54.
28. Lionel M. Munby, *The Hertfordshire Landscape* (1977), 151.
29. D. M. Palliser, *The Staffordshire Landscape* (1976), 108. For others see Trevor Rowley, *The Shropshire Landscape* (1972), 124; Steane, *Northamptonshire Landscape*, 192–3; Peter Bigmore, *The Bedfordshire and Huntingdonshire Landscape* (1979), 143–5; M. W. Beresford and J. K. St Joseph, *Medieval England* (2nd edn, Cambridge, 1979), 58–60; J. D. Chambers, *Nottinghamshire in the Eighteenth Century* (2nd edn, 1966), 15–16.
30. Stephen Switzer, *Ichnographia Rustica* (2nd edn, 1742), i. 273.
31. McClean, *Medieval English Gardens*, 20, 57–8; Harvey, *Mediaeval Gardens*, 4, 16–17; *The Itinerary of John Leland*, ed. Lucy Toulmin Smith (1964 edn), iv. 7.
32. Ra[lph] Austen, *A Dialogue ... between the Husbandman, and Fruit-Trees* (Oxford, 1676), 59. Cf. *Agrarian History of England and Wales*, iv. 195–6.
33. Blith, *English Improver Improved*, 166; Meager, *Mystery of Husbandry*, 108; James Anderson, *Essays relating to Agriculture* (3rd edn, Edinburgh, 1784), i. 54–5, 83–6, 96, 101–2. Cf. *Considerations concerning Common Fields* (1654), 11; *Progress Notes of Warden Woodward for the Wiltshire Estates of New College, Oxford*, ed. R. L. Rickard (Wilts. Rec. Soc., 1957), 4.
34. E.g. Kalm, 129.
35. *Works of Drayton*, iv. 524; J. W. F. Hill, 'Sir George Heneage's Estate Book, 1625', *Lincs. Architectural & Archaeol. Soc.*, new ser., i. (1938), 76.
36. Pettit, *Royal Forests*, 57; William Richard Fisher, *The Forest of Essex* (1887), 239; Owen Manning and William Bray, *The History and Antiquities of the County of Surrey* (1804–14), ii. 605.
37. *The Statutes of the Realm* (1810–24), i. 221; 'The Canterbury Tales: General Prologue', lines 606–7.
38. William Harrison, *The Description of England*, ed. Georges Edelen (Ithaca, N.Y., 1968), 275; I[ohn] B[eale], *Herefordshire Orchards* (1657), 35–6; Mr Salmon,

The Foreigner's Companion through the Universities of Cambridge and Oxford (1748), i. 2; Parkinson, *Paradisi*, 607.

39. Sir Edward Coke, *Institutes of the Laws of England*, i. chap. 7; Markham, *English Husbandman*, 21; North, *Observations*, 102.

40. Walter Blith, *The English Improver* (1649), 123; Christopher Merrett, *Pinax Rerum Naturalium Britannicarum* (1667), 1; *The Berkeley Manuscripts*, ed. Sir John Maclean (Gloucester, 1883–5), iii. 4.

41. Robert Willis, *The Architectural History of the University of Cambridge*, ed. John Willis Clark (Cambridge, 1886), i. 123, 4, 82, 241; ii. 5, 352, 635; *The Works of Nicholas Ridley*, ed. Henry Christmas (Parker Soc., Cambridge, 1841), 407.

42. Willis, *op. cit.*, i. 187n, 216n, 567–8; ii. 180–81, 322–3, 408, 639–42. Cf. J. H. Gray, *The Queens' College* (1899), 123.

43. E. G. W. Bill, *Christ Church Meadow* (Oxford, 1965), 21.

44. John Speed, *The Theatre of the Empire of Great Britaine* (1611); W. G. Hoskins, *The Making of the English Landscape* (1955), 109. Cf. *HMC, Portland*, ii. 269.

45. F. M. Stenton, *Norman London* (Hist. Assoc., 1934), 27; A. D. Webster, *London Trees* (1920), 167–8; [Richard Johnson], *The Pleasant Walkes of Moore-Fields* (1607), sig. A3v; Norman G. Brett-James, *The Growth of Stuart London* (1935), 452–7; *The Letters of John Chamberlain*, ed. Norman Egbert McClure (Philadelphia, 1939), i. 235; Brigid Mary Urswick Boardman, 'The Gardens of the London Livery Companies', *Jnl of Garden History*, 2 (1982).

46. Stow, *Survey*, ed. Strype, *passim*; Kalm, 85–7.

47. F. J. Fisher, 'The Development of London as a Centre of Conspicuous Consumption', *Trans. Royal Hist. Soc.*, 4th ser., xxx (1948), 47; Pepys, *Diary*, vii. 213; William Bond, *Buckingham House* (included in BL copy of *The Altar of Love* (1727)).

48. Walter Pope, *The Life of Seth Lord Bishop of Salisbury*, ed. J. B. Bamborough (Luttrell Soc., Oxford, 1961), 49; *Reliquiae Hearnianae*, ed. Philip Bliss (2nd edn, 1869), ii. 170; Salmon, *Foreigner's Companion*, ii. 67.

49. Peter Borsay, 'The English Urban Renaissance', *Social History*, 5 (1977), 583. Numerous examples are given in Dr Borsay's unpublished Ph.D. thesis, 'The English Urban Renaissance: Landscape and Leisure in the Provincial Town, *c.* 1660–1770' (Univ. of Lancaster, 1981), chap. 8, appendix 2.

50. Joseph Spence, *Anecdotes, Observations, and Characters*, ed. Samuel Weller Singer (1820), 424–5; Howard Erskine-Hill, *The Social Milieu of Alexander Pope* (1975), 27–8.

51. *The Journeys of Celia Fiennes*, ed. Christopher Morris (1947), 23, 227, 238, 271; Daniel Defoe, *A Tour through England and Wales* (EL, 1928), i. 162; *Traherne's Poems of Felicity*, ed. H. I. Bell (Oxford, 1910), 45–6.

52. John Evelyn, *Fumifugium* (1661; 1772 edn, reissued Oxford, 1930), pt iii; Evelyn, *Diary*, ii. 46, 47, 67, 139.

53. Thomas Fairchild, *The City Gardener* (1722), 12–14; Batty Langley, *A Sure Method of Improving Estates by Plantations* (1728), 143. Jonas Hanway propounded 'the idea of a *rural city*' in 1767; *Letters on the Importance of the Rising Generation* (1767), ii. 130–31.

54. John Norden, *Speculi Britanniae pars altera; or a Delineation of Northamptonshire* (1720), 50; Lucy Hutchinson, *Memoirs of the Life of Colonel Hutchinson* (EL, n.d.), 19, 290. For others see Frank Emery in *Jnl of Hist. Geog.*, 2 (1976), 37; John Nassau Simpkinson, *The Washingtons* (1860), appendix (A) 4, lx; Northants. R.O., Finch-Hatton 2475; Edward Earl of Clarendon, *The History of the Rebellion*, ed. W. Dunn Macray (Oxford, 1888), v. 156; John Aubrey, *Brief Lives*, ed. Anthony Powell (1949), 194–8; [John Coker], *A Survey of Dorsetshire* (1732), 124; Beale, *Herefordshire Orchards*, 37–9; and references cited from Evelyn, *Sylva*, in note 7 above.
55. Michael Jermin, *A Commentary, upon the Whole Booke of Ecclesiastes* (1639), 36.
56. Roger North, *The Lives of ... Francis ... Dudley ... and ... John North*, ed. Augustus Jessopp (1890), i. 171. Cf. Horace Walpole, *Anecdotes of Painting in England*, ed. Ralph N. Wornum (n.d.), iii. 96; Mark Girouard, *Life in the English Country House* (1978), 145.
57. Stephen Switzer, *Ichnographia Rustica* (1718), i. xxxviii; Walpole, *op. cit.*, iii. 85.
58. Evelyn, *Diary*, iii. 180; *CSPV, 1617–19*, 250. For others see Shirley, *English Deer Parks*, 65; Plot, *Oxon.*, 158; Evelyn, *Diary*, iv. 200, 594; *Camden Miscellany*, xvi (Camden ser., 1936), iii. 18; Moses Cook, *The Manner of Raising, Ordering and Improving Forest-Trees* (3rd edn, 1724), 30–31; *Locke Corr.*, ii. 613, 684, 685.
59. Leonard Knyff and Joannes Kip, *Britannia Illustrata* (1709–16); Marvell, 'Upon Appleton House', lines 619–20; *Journeys of Celia Fiennes*, 228. Cf. John Harris, *The Artist and the Country House* (1979), 172–3 and *passim*.
60. *The Letters of William and Dorothy Wordsworth*, ed. Ernest de Selincourt, i (2nd edn, rev. Chester L. Shaver, Oxford, 1967), 625.
61. J. W[orlidge], *Systema Agriculturae* (1669), 72; William Marshall, *Planting and Rural Ornament* (2nd edn, 1796), i. 283–4; Uvedale Price, *Essays on the Picturesque* (1810), iii. 113–14.
62. H[umphry] Repton, *Observations on the Theory and Practice of Landscape Gardening* (1803), 107n; *id.*, *Sketches and Hints on Landscape Gardening* (n.d. [1794]), 81.
63. North, *Observations*, 102; Erskine-Hill, *Social Milieu of Pope*, 59. Cf. Worlidge, *Systema Agriculturae*, 212; Plot, *Staffs.*, 41.
64. *The Landscape Gardening and Landscape Architecture of the late Humphry Repton*, new edn by J. C. Loudon (1840), 347.
65. *Spectator*, 414 (25 June 1712); Spence, *Anecdotes*, 460; Charles Marshall, *An Introduction to the Knowledge and Practice of Gardening* (1796), 154; [Henry Home, Lord Kames], *Elements of Criticism* (6th edn, 1785), ii. 440; Price, *Essays on Picturesque*, i. 259; Edmund Bartell, *Hints for Picturesque Improvements in Ornamented Cottages* (1804), 59n.
66. Evelyn, *Sylva*, 'To the Reader'; S. Arthur Strong, *A Catalogue of Letters and Other Historical Documents ... at Welbeck* (1903), 209; Robert Turner, *ΒΟΤΑΝΟΛΟΓΙΑ. The British Physician* (1664), 21; William Hanbury, *A Complete Body of Planting and Gardening* (1770), i. 54.

67. Evelyn, *Diary*, iii. 370n; iv. 409; David Green, *The Gardens and Parks at Hampton Court* (1974); *The History of the King's Works*, ed. H. M. Colvin (1963–), v. 171–2, 221–5, 329, 457–8.
68. John Cornforth, 'The Making of the Boughton Landscape', *Country Life* (11 March 1971); Steane, *Northamptonshire Landscape*, 242. For others see, e.g., Cook, *Manner of Raising Forest-Trees*, 30–32; Nourse, *Campania Foelix*, 109; [Sir Peter Pett], *The Happy Future State of England* (1688), 101; Vivian de Sola Pinto, *Peter Sterry* (Cambridge, 1934), 55; Evelyn, *Diary*, iv. 305–6; R. W. Ketton-Cremer, *Felbrigg* (1962), 61–2; *Locke Corr.*, i. 418, 434, 444; iii. 54, 513.
69. Defoe, *Tour*, i. 167.
70. Gilpin, *Observations on Western Parts of England*, 40; Byng, iii. 229; Price, *Essays on Picturesque*, i. 249; Humphry Repton, *Odd Whims and Miscellanies* (1804), ii. 155–6.
71. Dorothy Stroud, *Capability Brown* (new edn, 1975), 153, 157.
72. Louis Simond, *Journal of a Tour and Residence in Great Britain* (new edn, Edinburgh, 1817), i. 194 (also 258).
73. Emery, *Oxfordshire Landscape*, 130; Robert Newton, *The Northumberland Landscape* (1972), 119–21.
74. Derek Hudson and Kenneth W. Luckhurst, *The Royal Society of Arts* (1954), 87–8; Moelwyn Williams, *The Making of the South Wales Landscape* (1975), 145–6; Henrey, ii. 557–9; *New Historical Geography of England*, 426–8.
75. E.g. *Six North Country Diaries*, ed. J. C. Hodgson (Surtees Soc., 1910), 131n; Roebuck, *Yorkshire Baronets*, 38; Ellis, v(i). 7; v(2). 71; Nathaniel Kent, *Hints to Gentlemen of Landed Property* (1775), 202–3; Morton, *Northants.*, 486; Henrey, ii. 569; *Constable of Everingham Estate Correspondence*, ed. Peter Roebuck (Yorks. Archaeol. Soc., 1976), *passim*; William C. Braithwaite, *The Second Period of Quakerism* (1919), 424–5.
76. Sir John Davies, *Historical Tracts* (1786), 286 (but for Irish planting see Eileen McCracken, *The Irish Woods since Tudor Times* (Newton Abbot, 1971), 135–41); *Johnson's Journey to the Western Islands*, ed. R. W. Chapman (1930), 9 (cf. 126–7). For Scottish planting see R. N. Millman, *The Making of the Scottish Landscape* (1975), 141–9; Mark L. Anderson, *A History of Scottish Forestry*, ed. Charles J. Taylor (1967), i. chaps 10 and 11.
77. John Laurence, *A New System of Agriculture* (1726), 218, 245; [Arthur Young], *The Farmer's Letters* (2nd edn, 1768), 260; Albion, *Forests and Sea Power*, 99–100, 112–13; J. H. Plumb, *The Pursuit of Happiness* (New Haven, 1977), 26.
78. John Harvey, *Early Gardening Catalogues* (1972), 4–5; *id.*, *Early Nurserymen*; *id.* in *The Local Historian*, xi (1974), 235–6; Kalm, 316.
79. See in general William Aiton, *Hortus Kewensis* (1789); Laurence, *New System of Agriculture*, bk ii; James E. Gillespie, *The Influence of Overseas Expansion on England to 1700* (1920; reprint, New York, 1974), 97–8; P. J. Jarvis, 'North American Plants and Horticultural Innovation in England, 1550–1770', *Geog. Rev.*, 63 (1973); *id.*, 'A History of the Cedar of Lebanon in Britain', *Jnl Royal Horticultural Soc.*, xcix (1974); *id.*, 'Seventeenth-Century Cedars', *Garden History*, iv (1976); 'Plant Introductions to England', in *Change in the Countryside*,

ed. H. S. A. Fox and R. A. Butlin (1979); Alice M. Coats, *Garden Shrubs and their Histories* (1963).

80. John Rea, *Flora: seu De Florum Cultura* (2nd edn, 1676), 228.
81. Evelyn, *Diary*, v. 425–6; *The Collected Poems of Christopher Smart*, ed. Norman Callan (1949), i. 267.
82. J. C. Loudon, *Arboretum et Fruticetum Britannicum* (1838), i. 126.
83. *ibid.*, i. 12.

iii. The Worship of Trees

1. *Spectator*, 393 (31 May 1712); Hugh Bilbrough in *Garden History*, i (1973), 10.
2. Archibald Alison, *Essays on the Nature and Principles of Taste* (Edinburgh, 1790), 19; Hesther Lynch Piozzi, *Anecdotes of Samuel Johnson* (Cambridge, 1932), 170.
3. L. Tyerman, *The Life of the Rev. George Whitefield* (1876), i. 22.
4. William Gilpin, *Remarks on Forest Scenery* (1791), i. 1; Spence, *Anecdotes*, 11; William Shenstone, *Essays on Men and Manners* (1794 edn), 69; Vicesimus Knox, *Lucubrations* (1787), no. 93.
5. E.g. Alexander Cozens, *The Shape, Skeleton and Foliage of Thirty-Two Species of Trees* (1771) (cited in *DNB*); Gilpin, *Forest Scenery*; Hayman Rooke, *Descriptions and Sketches of Some Remarkable Oaks* (1791); Edward Kennion, *An Essay on Trees in Landscape* (1815); Jacob George Strutt, *Sylva Britannica* (1822); Henry W. Burgess, *Eidodendron* (1827); *id.*, *Studies of Trees* (1837); James Grigor, *The Eastern Arboretum* (1841); Charles Empson, *The Cowthorpe Oak* (1842); John Ruskin, *Modern Painters* (EL, n.d.), ii. 113–37; v. 1–98.
6. Grigor, *op. cit.*, iv; Charles Millard, 'Images of Nature', in *Nature and the Victorian Imagination*, ed. U. C. Knoepflmacher and G. B. Tennyson (1977), 25.
7. William Marsden, *The History of Sumatra* (1783), 62–3.
8. *The Correspondence of Alexander Pope*, ed. George Sherburn (Oxford, 1956), iv. 323–4; above, p. 203; *Poems by Anne, Countess of Winchilsea*, ed. John Middleton Murry (1928), 88; William Cowper, 'The Poplar Field'; Wordsworth, 'The Excursion', vii: argument and lines 590–631; 'Poems of the Imagination: Yew-Trees' (and cf. 'Memorial of a Tour in Scotland, 1803: Sonnet composed at — Castle'); *The Poems of John Clare*, ed. J. W. Tibble (1935), 4–5; Campbell, 'The Beech-Tree's Petition'; Tennyson, 'Tithonus'; Hopkins, 'Binsey Poplars'; *Glamorgan County History*, iv. 565–6; [Francis Noel Clerk Mundy], *Needwood Forest* (Lichfield, n.d.); *The Oxford Dictionary of Quotations* (2nd edn, 1953), 358.
9. *Cowper, illustrated by a Series of Views* (1804), 44–5.
10. *Camden Miscellany, iii*, iii. 8; Gilpin, *Forest Scenery*, 305.
11. E.g. Charlotte Fell Smith, *Mary Rich, Countess of Warwick* (1901), 336; Cook, *Manner of Raising Forest-Trees*, 163–4; Albion, *Forests and Sea Power*, 112; *The Family Memoirs of the Rev. William Stukeley*, ed. W. C. Lukis (Surtees Soc., 1882–7), i. 89; [George Nicholson], *The Cambrian Traveller's Guide* (2nd edn, 1813), 373; Steane, *Northamptonshire Landscape*, 285; *New Historical Geography of England*, 274.

12. Walpole, *Anecdotes of Painting*, iii. 92.
13. *English Historical Documents, c. 500–1042*, 455, 475; Jean-Claude Schmitt, *Le Saint Lévrier* (Paris, 1979), 38–9; Robert Charles Hope, *The Legendary Lore of the Holy Wells* (1893), xxii.
14. Charles Phythian-Adams, *Local History and Folklore* (1975), 15–16, 19; J. G. Frazer, *The Magic Art* (1913), ii. chap. x; *EDS, Series B: Reprinted Glossaries*, ed. Walter W. Skeat (1874), 58.
15. John Milton, 'Il Penseroso', lines 133–8.
16. *Heresy Trials in the Diocese of Norwich*, ed. N. P. Tanner (Camden ser., 1977), 95; Stow, *Survey*, ed. Strype, i(ii). 66.
17. R. T. Gunther, *Early British Botanists* (Oxford, 1922), 265; F. Lewes de Granada, *Of Prayer and Meditation* (1601).
18. Smith, *Mary Rich*, 204, 227, 257, 306; Beale, *Herefordshire Orchards*, 48–9; Francis Hutcheson, *Enquiry into the Original of Our Ideas of Beauty and Virtue* (1725), 76.
19. Mrs Rowe, *Friendship in Death* (1736), cit. Maren-Sofie Røstvig, *The Happy Man* (Oxford, 1954–8), ii. 200.
20. Spence, *Anecdotes*, 12; Arthur O. Lovejoy, *Essays in the History of Ideas* (New York, 1955), 153–4; Sir James Hall, *Essay on the Origin, History, and Principles of Gothic Architecture* (1813), chap. iv.
21. A. L. Owen, *The Famous Druids* (Oxford, 1962), 9, 171; Evelyn/Hunter, 610n; Cowper, 'Yardley Oak', line 9.
22. Hugh Honour, *Romanticism* (1979), 157, 347; Price, *Essays on the Picturesque*, i. 247–8; *The Poetical Works of William Lisle Bowles*, ed. George Gilfillan (1855), i. 215; Stuart Piggott, *Ruins in a Landscape* (Edinburgh, 1976), 72–3.
23. Coleridge, 'Christabel', lines 280–81; 'Ancient Mariner', line 511; Wordsworth, 'Poems of Sentiment and Reflection: The Tables Turned'; *Poetical Works of William Cullen Bryant* (1881), 79; *The Works of Ralph Waldo Emerson* (1902), 548; Peter J. Schmitt, *Back to Nature* (New York, 1969), 144.
24. E.g. Manwood, *Lawes of the Forrest*, fols. 4, 9^{v}; Beresford, *History on the Ground*, 42, 51; Fisher, *Forest of Essex*, 401; *VCH, Hants.*, iii. 67; *Forestry*, 40 (1967), 199; *Rural Economy in Yorkshire in 1641*, ed. C. B. Robinson (Surtees Soc., 1857), 42; *The Field Book of Walsham-le-Willows 1577*, ed. Kenneth Melton Dodd (Suffolk Recs. Soc., 1974); Edward T. MacDermot, *The History of the Forest of Exmoor* (Taunton, 1911), 4, 349; Evelyn/Hunter, 622; George Laurence Gomme, *Primitive Folk-Moots* (1880), 209–11; H. S. Bennett, *Life on the English Manor* (Cambridge, 1960), 203; Kalm, 162; Vaughan Cornish, *Historic Thorn Trees in the British Isles* (n.d.).
25. Evelyn, *Diary*, iii. 69; Erskine-Hill, *Social Milieu of Pope*, 33–4; Ellis, viii. 44; *The Magazine of Natural Hist.*, iii (1830), 555.
26. Evelyn/Hunter, 502, 100n; Sidney J. Madge, *The Domesday of Crown Lands* (1938), 107–8; Plot, *Oxon.*, 159; Plot, *Staffs.*, 279.
27. Henrey, ii. 568; Smith, *Englands Improvement Revived*, 77. Cf. E. E. Evans-Pritchard, *Nuer Religion* (Oxford, 1956), 6.
28. Evelyn, *Diary*, ii. 205, 421.

29. Rooke, *Descriptions and Sketches*, 12; *VCH, Berks.*, ii. 351; Evelyn/Hunter, 496; Camden, *Britannia*, ed. Gough, i. 186; *Letters and the Second Diary of Samuel Pepys*, ed. R. G. Howarth (1933), 361; *The Times*, 18 Nov. 1978. Cf. others listed in [Samuel Hayes], *A Practical Treatise on Planting* (Dublin, 1794), 125–7; Joseph Taylor, *Arbores Mirabiles* (1812); Mary Roberts, *Ruins and Old Trees associated with Remarkable Events* (*c.* 1843); and their modern counterparts in *Historic Trees* (Colourmaster, n.d.).
30. Marvell, 'Upon the Hill and Grove at Bill-Borow'. Cf. James Turner, *The Politics of Landscape* (Oxford, 1979), 98.
31. *Correspondence of Edmund Burke* (Cambridge, 1958–78), ii, ed. L. S. Sutherland, 377; Smith, *Mary Rich*, 271.
32. Ronald Paulson, *Emblem and Expression* (1975), 154–5; *The Poetical Works of Mrs Hemans* (1897), 383.
33. William Hanbury, *An Essay on Planting* (1758), 18–19. On Pullen's tree see *Reliquiae Hearnianae*, ii. 238.
34. Coles, *Eden*, 568.
35. *The Life of William Hutton . . . by himself*, ed. Catherine Hutton (1816), 253.
36. *A Tour in the Lakes made in 1797 by William Gell*, ed. William Rollinson (Newcastle upon Tyne, 1968), 15, 50; Michel Devèze, *La Vie de la forêt française au XVI*[e] *siècle* (Paris, 1961), i. 35–6.
37. Evelyn, *Sylva*, sig. A4; Edmund Waller, 'On St James's Park'; Thompson, *Whigs and Hunters*, 119; A. S. Turberville, *A History of Welbeck Abbey* (1938–9), i. 146; *The Beauties of England* (2nd edn, 1764), 107.
38. *The Life and Times of Anthony Wood*, ed. Andrew Clark (Oxford Hist. Soc., 1891–1900), ii. 479; Pepys, *Diary*, viii. 269.
39. *Autobiography of Thomas Raymond, and Memoirs of the Family of Guise*, ed. G. Davies (Camden ser., 1917), 115; *Wordsworth's Guide to the Lakes* (5th edn, 1835), ed. Ernest de Selincourt (1977), 146.
40. Jeremy Bentham, *Auto-Icon, or Farther Uses of the Dead to the Living* (1842?).
41. Thompson, *Whigs and Hunters*, 27; *id.*, in *Social History*, 3 (1978), 159; Røstvig, *Happy Man*, i. 402; Cook, *Manner of Raising Forest-Trees*, 32; Sir William Chambers, *An Explanatory Discourse by Tan Chet-Qua* (1773; Augustan Reprint Soc., 1978), 139–40.
42. William Turner, *A New Herball* (London and Cologne, 1551–62), ii. fol. 68[v]; *A Glossary of Provincial Words used in Herefordshire* (1839), 55; *OED*, 'dotard'.
43. Abraham Cowley, *Several Discourses by way of Essays*, ed. Harry Christopher Minchin (1904), 27; Turner, *Politics of Landscape*, 97; Rowland Watkyns, *Flamma sine Fumo* (1662), ed. Paul C. Davies (Cardiff, 1968), 75.
44. Marshall, *Planting and Rural Ornament*, i. 278; *The Letters of Mrs Gaskell*, ed. J. A. V. Chapple and Arthur Pollard (Manchester, 1966), 30–31.
45. David M. Bergeron, *English Civic Pageantry 1558–1642* (1971), 56; Shenstone, *Essays*, 68.
46. A. C. Forbes, *English Estate Forestry* (1904), 235; Albion, *Forests and Sea Power*, 16, 25.

47. Laurence, *New System of Agriculture*, 294. Cf. Miles Hadfield, *Topiary and Ornamental Hedges* (1971), chaps. 2 and 3.
48. *Spectator*, 414 (25 June 1712); *Guardian*, 173 (29 Sep. 1713).
49. Cook, *Manner of Raising Forest-Trees*, 58–9; Rackham, *Trees and Woodland*, 170.
50. Arthur Young, *A Six Weeks Tour through the Southern Counties* (1769), 308; *VCH, Berks*., ii. 351.
51. Byng, ii. 22; Gilpin, *Forest Scenery*, ii. 79; James Grahame, *British Georgics* (2nd edn, Edinburgh, 1812), 253.
52. Evelyn/Hunter, 379n; J. C. Loudon, *An Encyclopaedia of Gardening* (new edn, n.d. [*c*. 1834]), 245. Cf. Nikolaus Pevsner, *Studies in Art, Architecture and Design* (1968), i. 100–101; Mary Wollstonecraft, *A Vindication of the Rights of Woman* (1792; EL, 1929), 118.
53. *OED*, 'rustic'. There are designs for 'primitive huts' and 'hermitages' using 'rude branches and roots of trees' in William Wrighte, *Grotesque Architecture* (1767).
54. Thomas Tryon, *Wisdom's Dictates* (1691), 63.
55. Walter Minchinton, 'Cider and Folklore', *Folk Life*, 13 (1975), 67–71; Peter Clark, *English Provincial Society from the Reformation to the Revolution* (Hassocks, 1977), 166; *Dialect Dict*., 'youling'; Austen, *Dialogue*, sig. *4.
56. Marchioness of Newcastle, *Poems and Phancies* (2nd imp., 1664), 82; John Aubrey, *Remaines of Gentilisme and Judaisme*, ed. James Britten (Folk-Lore Soc., 1881), 247; *John Constable's Discourses*, ed. R. B. Beckett (Suffolk Recs. Soc., 1970), 71.
57. William Morris, *On Art and Socialism* (1947), 111; *Scotland and Scotsmen in the Eighteenth Century*, ed. Alexander Allardyce (Edinburgh, 1888), ii. 104; Frederick A. Whiting, 'The Pine Tree of Monte Mario', *Country Life* (22 Apr. 1976).

iv. Flowers

1. John Stow, *A Survey of London*, ed. Charles Lethbridge Kingsford (Oxford, 1908), i. 334. On flowers in medieval decoration see Joan Evans, *Nature in Design* (1933), 86, 97 and *id*., *Pattern* (1976), i. 61–4; and on medieval flower gardens, Harvey, *Early Gardening Catalogues*, intro. and chap. i; Teresa McLean, *Medieval English Gardens* (1981), chap. 5.
2. *Thomas Tusser . . . His Good Points of Husbandry*, ed. Dorothy Hartley (1931), 152–3; *CSPV, 1527–33*, 288; Sir Hugh Plat, *The Garden of Eden* (5th edn, 1659), 44–50; *The Berkeley Manuscripts*, ii. 367; iii. 26; John Gerard, *The Herball* (1597), *passim*; Henrey, i. 73; *OED*, 'bough-pots'.
3. Gerard, *Herball*, ed. Johnson, 151, 43; Parkinson, *Paradisi*, 324. Cf. Henrey, i. 55–73.
4. Harvey, *Early Nurserymen*, and *id*., *Early Gardening Catalogues*.
5. Harvey, *Early Nurserymen*, 29, 160; Parkinson, *Paradisi*, 336; Switzer, *Ichnographia Rustica* (1718), i. 79.
6. *Clennenau Letters and Papers*, ed. T. Jones Pierce (Nat. Lib. Wales *Jnl*, supp. iv

(1947), 61). Cf. Edward Hughes, *North Country Life in the Eighteenth Century* (1952–65), i. 29; McClean, *Medieval English Gardens, passim.*

7. Other names can be found in the published gardening books of the period, in household and estate accounts (e.g. Northants. R.O., FH 2452) and in contemporary correspondence (e.g. *Works of Robert Boyle*, v. 499).
8. E.g. *Locke Corr.*, ii. 271–2; *The Garden Book of Sir Thomas Hanmer*, with intro. by Eleanour Sinclair Rohde (1933), *passim*; *Constable of Everingham Estate Correspondence*, 46.
9. Lady Newton, *The House of Lyme* (1917), 314–15.
10. Switzer, *Ichnographia Rustica* (1718), i. 80–81; David Green, *Gardener to Queen Anne* (1956), 55, 163.
11. Henrey, ii. 365; Harvey, *Early Nurserymen*, 6.
12. Henrey, i. 3, 77; ii. 3, 363–4, 369, 302.
13. See Wilfrid Blunt, *The Art of Botanical Illustration* (1950), chaps. 11–18.
14. Joan Thirsk, *Economic Policy and Projects* (Oxford, 1978), *passim*; *The Cambridge Economic History of Europe*, iv, ed. E. E. Rich and C. H. Wilson (Cambridge, 1967), 276–9.
15. Turner, *British Physician*, 30; William Hughes, *The Flower Garden* (3rd edn, 1683), 7.
16. Harvey, *Early Nurserymen*, 128.
17. See Aiton, *Hortus Kewensis*; Gillespie, *Influence of Overseas Expansion*, 91–7; W. van Dijk, 'Notes on Some Plant Introductions in the Seventeenth Century', *Jnl Royal Horticultural Soc.*, lxxi (1946); William T. Stearn, 'The Origin and Later Development of Cultivated Plants', *ibid.*, xc (1965); and articles by P. J. Jarvis cited in note 79 (p. 376 above).
18. Raven, *Naturalists*, 117; Turner, *New Herball*, ii. fol. 69; Mary Dewar, *Sir Thomas Smith* (1964), 109; Logan Pearsall Smith, *The Life and Letters of Sir Henry Wotton* (Oxford, 1907), i. 59; *DNB*, 'Sherard, William'; Gerard, *Herball*, ed. Johnson, 314.
19. Parkinson, *Paradisi*, 133; Northants R.O., F.H. 3451.
20. Mea Allen, *The Tradescants* (1964).
21. *An Elizabethan in 1582*, ed. Elizabeth Story Donno (Hakluyt Soc., 1976), 116–17.
22. Parkinson, *Paradisi*, 88.
23. Raven, *Naturalists*, 235–6; Lindsay Sharp, 'Walter Charleton's Early Life', *Ann. Sci.*, 30 (1973), 338; *Garden Book of Hanmer*, ix, xx; Sir Hans Sloane, *A Voyage to the Islands* (1707–25), i. sig. A2 and p. 83; Henrey, i. 144; Alice M. Coats, 'The Hon. and Rev. Henry Compton, Lord Bishop of London', *Garden History*, iv (1976); Charles Evelyn, *The Lady's Recreation* (1717), 2–3. The huge scale of the Duchess's activities is revealed by her catalogues of the plants grown at Badminton; BL, Sloane MSS. 4070–72.
24. *A Selection of the Correspondence of Linnaeus*, ed. Sir James Edward Smith (1821), i. 9–11; *DNB*, 'Collinson, Peter'. For other notable importers see Henrey, ii. 335.
25. Harrison, *Description of England*, 265.
26. Turner, *New Herball*, ii. fol. 84v. Some of the gardens mentioned by Turner are

discussed in Raven, *Naturalists*, chap. vi. For others see Gunther, *Early British Botanists*.

27. *Garden Book of Hanmer*, 37–42; Hanmer to Evelyn, 1 May 1668 (Clwyd R. O., D/CL/48).
28. Rea, *Flora*, 'To the Reader'; J. W[orlidge], *Systema Horti-Culturae* (1677), 5.
29. Hughes, *Flower Garden*, sig. A3; Henrey, i. 204.
30. *HMC, Portland*, ii. 289, 295.
31. *The Miscellaneous Writings of Sir Thomas Browne*, ed. Geoffrey Keynes (1946), 238; *The Travels through England of Dr Richard Pococke*, ed. James Joel Cartwright (Camden Soc., 1888–9), i. 6; George Smith, *Six Pastorals* (1770), 6; *The Letters of Sir William Jones*, ed. Garland Cannon (Oxford, 1970), i. 39; *Sophie in London in 1786*, trans. Clare Williams (1933), 86; Byng, ii. 174; Simond, *Journal of a Tour*, i. 274, 336, 343; Abbé Le Blanc, *Letters on the English and French Nations* (1747), i. 321; William Cobbett, *A Year's Residence in America* (1818; Abbey Classics, n.d.), 3.
32. Simond, *op. cit.*, ii. 298. Cf. Loudon, *Encyclopaedia*, 362.
33. Thomas Fuller, *The Worthies of England*, ed. John Freeman (1952), 419; Evelyn, *Diary*, iii. 595; J. H. Plumb, *The Commercialisation of Leisure in Eighteenth-Century England* (Reading, 1973), 8n; Hanbury, *Complete Body of Planting*, i. 286; M. Sturge Henderson, *Three Centuries in North Oxfordshire* (1902), 131–2; information concerning Bristol from Mr Jonathan Barry.
34. *Poems of Christopher Smart*, i. 328–9.
35. J. C. Loudon, *The Suburban Gardener and Villa Companion* (1838), 272; Hanbury, *Complete Body of Planting*, i. 286; Miles Hadfield, *A History of British Gardening* (3rd edn, 1979), 262.
36. Loudon, *Encyclopaedia*, 1036, 1227; *id.*, *Suburban Gardener*, 270–74; Coats, *Garden Shrubs*, 209.
37. Henrey, ii. 469, 408; Harvey, *Early Gardening Catalogues*, 41.
38. *The Minutes of the First Independent Church (now Bunyan Meeting) at Bedford*, ed. H. G. Tibbutt (Beds. Hist. Rec. Soc., 1976), 161; Bartell, *Hints for Picturesque Improvements*, 132. On flower-gardening by the Nottinghamshire working classes in the nineteenth century see William Howitt, *The Rural Life of England* (1838), ii. 305–11; S. Reynolds Hole, *A Book about Roses* (1869), 17–19.
39. E.g. Gerard, *Herball*, ed. Johnson, 600, 957; Worlidge, *Systema Horticulturae*, 1, 4; Rea, *Flora*, sig. A3.
40. Levinus Lemnius, *An Herbal for the Bible*, trans. Thomas Newton (1587), 224; William Whateley, *A Care-Cloth* (1624), 14; Gerard, *Herball*, ed. Johnson, 850.
41. Raven, *Naturalists*, 51; *Philip Stubbes's Anatomy of Abuses*, ed. Frederick J. Furnivall (New Shakspere Soc., 1877–9), 149; *The Diaries of Thomas Wilson*, ed. C.L.S. Linnell (1964), 241; Richard Bradley, *Dictionarium Botanicum* (1728), i. sig. $H2^v$. On St Barnabas Day roses, see Hilderic Friend in *The Church Treasury*, ed. William Andrews (1898), 235.
42. *CSPV, 1617–19*, 135; *OED*, 'rosemary'; Parkinson, *Paradisi*, 426; Hilderic Friend, *Flowers and Flower Lore* (3rd edn, 1886), 113–15, 125–6, 128.
43. Coles, *Simpling*, 64; *Family Memoirs of William Stukeley*, i. 57; Shakespeare,

Pericles, iv. i. lines 13–16; *Twelfth Night*, ii. iv. line 57; Mrs [Elizabeth] Stone, *God's Acre* (1858), 273–4.

44. Mary Martha Sherwood, *The Fairchild Family*, pt ii (1842), chap. 23; *Camden's Britannia*, ed. Gibson, 162; Evelyn/Hunter, 629; Simond, *Journal of a Tour*, ii. 358; Loudon, *Encyclopaedia*, 298; *Memoirs of the Life of Sir Samuel Romilly written by himself* (2nd edn, 1840), i. 342.
45. Nicholas Penny, *Church Monuments in Romantic England* (1977), 25–6, 31, 34; Stone, *God's Acre*, 115–16.
46. Henry Phillips, *Flora Historica* (1824), i. xxxi. Cf. Mary Douglas and Baron Isherwood, *The World of Goods* (1979), 58.
47. Marshall, *Introduction to the Knowledge and Practice of Gardening*, 357.
48. Parkinson, *Paradisi*, 8, 9, 281; Rea, *Flora*, 39.
49. *Garden Book of Hanmer*, 19; Gerard, *Herball*, 430; Samuel Gilbert, *The Florists Vade Mecum* (1682), sig. A8v; Rea, *Flora*, sig. B4; Hanbury, *Complete Body of Planting*, i. 307, 336–7.
50. Stephen Blake, *The Compleat Gardeners Practice* (1664), 76; Worlidge, *Systema Horti-Culturae*, 73–5; *id.*, *Systema Agriculturae*, 84.
51. William Boutcher, *A Treatise on Forest-Trees* (Edinburgh, 1775), 55; Grahame, *British Georgics*, 171–4.
52. Parkinson, *Paradisi*, 518, 513, 521, 526, 560; Henrey, ii. 478.
53. Plat, *Garden of Eden*, 51.
54. Henrey, ii. 392. Cf. Thirsk, *Economic Policy and Projects*, 106–7.
55. Parkinson, *Paradisi*, 111–32, 306–14, 199–214, 48–61; Evelyn, *Diary*, iii. 33; Bradley, *Dictionarium Botanicum*, sig. K1v; Richard Weston, *The Universal Botanist and Nurseryman* (1777), iv. 52–128 (2nd pagination).
56. Coats, *Garden Shrubs*, 290–91; Harvey, *Early Nurserymen*, 202.
57. Harvey, *Early Gardening Catalogues*, 80; Richard Bradley, *New Improvements of Planting and Gardening* (1717–18), ii. 98–9; Weston, *Universal Botanist*, iv. 99–107, 126–8; *Letters of Josiah Wedgwood*, ed. K. E. Farrer (1903–6; reprint, Didsbury [1973]), ii. 403, 413.
58. [J. L. Knapp], *The Journal of a Naturalist* (3rd edn, 1830), 67; G. W. Francis, *The Little English Flora* (1839), 25.
59. William Cobbett, *The English Gardener* (1833; 1980 edn), 39 (para. 57); William Mitford, *Principles of Design in Architecture* (2nd edn, 1824), 280. Cf. A. Hunter, *Georgical Essays*, vi (1804), 297; Loudon, *Encyclopaedia*, 1225–6; Phillips, *Flora Historica*, i. v–vi, xxix; Knapp, *op. cit.*, 16; George H. Ford, 'Felicitous Space', in *Nature and the Victorian Imagination*, 35.
60. *Poetical Works of William Lisle Bowles*, ii. 67.
61. [J. Percival], *The Morality of Cumberland and Westmoreland* (1865), 16, 36; *The Gardener's Chronicle*, 7 July 1894; Hole, *Book about Roses*, 25. For more on this theme see S. Martin Gaskell, 'Gardens for the Working Class', *Victorian Studies*, 23 (1980).
62. William Horman, *Vulgaria* (1519), fol. 245; Thomas More, *Utopia* (EL, 1951), 61.
63. *The Atlas of Historic Towns*, ed. M. D. Lobel (1969–), i (Nottingham), 6; ii (Bristol), 23; *The Parliamentary Survey of the Lands . . . of the Dean and Chapter of*

Worcester, ed. Thomas Cave and Rowland A. Wilson (Worcs. Hist. Soc., 1924), xix; *VCH, City of York*, 160; McLean, *Medieval English Gardens*, chap. 2; *Camden Miscellany*, xvi (Camden ser., 1936), iii. 57.

64. Worlidge, *Systema Horti-Culturae*, 4; Fairchild, *City Gardener*, 7.
65. Kalm, 34; Switzer, *Ichnographia* (1718), i. xxxix; Fairchild, *op. cit.*, 14.
66. 'The Task', iv. lines 765–9, 771–9; Charlotte Brontë, *Villette* (1853), chap. 12.
67. The work, respectively, of Samuel Collins (1717), an anonymous author (1728) and John Parkinson (1629).
68. Hughes, *Flower Garden*, sig. A2; James Shirley, 'The Garden', in *Poems &c* (1646), 136. In *The Garden of Eden* (1981) John Prest shows how the botanic gardens of the period embodied this aspiration.
69. Green, *Gardens and Parks at Hampton Court*, 9; Evelyn, *Diary*, iii. 324–5; L. Tyerman, *John Wesley's Designated Successor* (1882), 150; *OED*, 'paradise'; Anthony Wood, *Survey of the Antiquities of the City of Oxford*, ed. Andrew Clark (Oxford Hist. Soc., 1889–99), ii. 410. Cf. Stanley Stewart, *The Enclosed Garden* (1966).
70. Leonard Forster, 'Meditation in a Garden', *German Life and Letters*, xxxi (1977), 33. Cf. Røstvig, *Happy Man*, *passim*.
71. J. Crofts, 'Wordsworth and the Seventeenth Century', *Procs. Brit. Acad.*, xxvi (1940), 187; [Anthony Walker], *The Holy Life of Mrs Elizabeth Walker* (1690), 19.
72. William Turner, *A Compleat History of the Most Remarkable Providences* (1697), iv. 53; *The Diary of Ralph Josselin*, ed. Alan Macfarlane (Brit. Acad., 1976), 449.
73. Christopher Hussey, *The Picturesque* (1967 edn), 131–2.
74. Jermin, *Ecclesiastes*, 34–5.
75. Laurence, *New System of Agriculture*, sig. b1, 287; John Hacket, *Scrinia Reserata* (1693), ii. 29; Switzer, *Ichnographia* (1718), i. 44; Patrick Collinson, *Archbishop Grindal* (1979), 40; 'Joseph Frampton' [= William Gilpin], *Three Dialogues on the Amusements of Clergymen* (2nd edn, 1797), 180.
76. Robert Barclay, *An Apology for the True Christian Divinity* (1678), 388; Henrey, ii. 310–11; Arthur Raistrick, *Quakers in Science and Industry* (1950), chap. 8; *Works of Robert Boyle*, v. 499; *Bishop Fell and Nonconformity*, ed. Mary Clapinson (Oxon. Rec. Soc., 1980), 27.
77. I. A. Richmond, *Archaeology and the After-Life in Pagan and Christian Imagery* (1950), 25–7; *The Somersetshire Quarterly Meeting of the Society of Friends*, ed. Stephen C. Marland (Somerset Rec. Soc., 1978), 20; D. G. Stuart, 'The Burial-grounds of the Society of Friends in Staffordshire', *South Staffs. Archaeol. and Hist. Soc., Trans.*, xii (1970–71); *Sir William Temple upon the Gardens of Epicurus*, ed. Albert Forbes Sieveking (1908), 198; Evelyn/Hunter, 624–9; Vincent T. Harlow, *Christopher Codrington* (Oxford, 1928), 138.
78. *The Poetical Works of Wordsworth*, ed. Thomas Hutchinson, rev. Ernest de Selincourt (1904), 930; Loudon, *Encyclopaedia*, 341; Penny, *Church Monuments*, 53; *id.*, 'The Commercial Garden Necropolis of the Early Nineteenth Century', *Garden History*, ii (1974).
79. Louise Imogen Guiney, *Recusant Poets* (1938), 93.

80. Coles, *Simpling*, 121; John Laurence, *The Gentleman's Recreation* (1716), sig. A7v. Cf. Ronald A. Knox, 'Man and his Garden', *On Getting There* (1929), 176.
81. Turner, *New Herball*, i. sig. Aiijv.
82. Parkinson, *Paradisi*, 65; *Sir William Temple upon the Gardens of Epicurus*, 44. Cf. *Several Tracts written by Sir Matthew Hale* (1684), iii. 16.
83. Loudon, *Suburban Gardener*, 6; *English History from Essex Sources, 1550–1750*, ed. A. C. Edwards (Chelmsford, 1952), 35; Parkinson, *Paradisi*, 348, 389.
84. William Gurnall, *The Christian in Complete Armour* (2nd edn, 1656; reprint, Evansville, Indiana, 1958), 311.
85. Coles, *Simpling*, 120 ('96'); F. E. Manuel, *A Portrait of Isaac Newton* (Cambridge, Mass., 1968), 105.
86. Bacon, 'Of Gardens', *Essays* (World's Classics, 1937), 190. For the ubiquity of bowling greens in seventeenth-century England see, e.g., *HMC*, *Portland*, ii. 271; Borsay, 'The English Urban Renaissance', 311–13; Dennis Brailsford, *Sport and Society* (1969), 116–17.
87. Geoffrey Taylor, *The Victorian Flower Garden* (1952), 191; Loudon, *Encyclopaedia*, 552–3.
88. Victor Crittenden, 'Australia's First Gardening Books, 1835–1838', *The Push from the Bush* (Nedlands, W. Aus.), 4 (1979).
89. Byng, iii. 301 (though cf. Simond, *Journal of a Tour*, i. 274).
90. Loudon, *Encyclopaedia*, 293.
91. Cobbett, *Year's Residence in America*, 3. Cf. Robert Beverley, *The History and Present State of Virginia*, ed. Louis B. Wright (Chapel Hill, 1947), xxxiii.
92. D. H. Lawrence, *Selected Essays* (Harmondsworth, 1950), 119.
93. Cf. the comments of A. Croxton Smith, *Dogs since 1900* (1950), preface.

CHAPTER VI THE HUMAN DILEMMA

i. Town or Country?

1. G. M. Trevelyan, *English Social History* (3rd edn, 1946), 374; C. W. Chalklin, *The Provincial Towns of Georgian England* (1974), 3; Phyllis Deane and W. A. Cole, *British Economic Growth 1688–1959* (Cambridge, 1962), 7–11; Adna Ferrin Weber, *The Growth of Cities in the Nineteenth Century* (1899; Ithaca, N.Y., 1967), 144–5.
2. *Cyvile and Uncyvile Life* (1579), sig. Nivv; Ruth Kelso, *The Doctrine of the English Gentleman in the Sixteenth Century* (Univ. of Illinois Studs. in Lang. & Lit., xiv, 1929), 58–9; Alexander Murray, *Reason and Society in the Middle Ages* (Oxford, 1978), 237–8; above, p. 195.
3. E.g. *Pearl*, ed. E. V. Gordon (Oxford, 1953), 37; *Select Poetry*, ed. Edward Farr (Parker Soc., Cambridge, 1845), ii. 428; *Old English Ballads 1553–1625*, ed. Hyder E. Rollins (Cambridge, 1920), 153; William Gouge, *A Learned ... Commentary on the whole Epistle to the Hebrewes* (1655), iii. 27; William Gearing, *A Prospect of Heaven* (1673), 121; *The Works of ... William Bates* (1700), 479. Cf. Revelation, xxi. 2.

4. *The Itinerary of John Leland*, ed. Lucy Toulmin Smith (1964 edn), v. 39; i. 228; ii. 88, 96; *Glamorgan County History*, iv, ed. Glanmor Williams (Cardiff, 1974), 43; *The Journeys of Celia Fiennes*, ed. Christopher Morris (1947), 72.
5. *The Letters of William Shenstone*, ed. Marjorie Williams (Oxford, 1939), 126.
6. *Cal. Patent Rolls, 1281–92*, 207, 296; *Cal. Close Rolls, 1302–7*, 537.
7. J. U. Nef, *The Rise of the British Coal Industry* (1932), i. 157–8, 198; Margaret, Duchess of Newcastle, *The Life of William Cavendish*, ed. C. H. Firth (2nd edn, n.d.), 66; James Ward, 'Phoenix Park' (1717), cit. Maren-Sofie Røstvig, *The Happy Man* (Oxford, 1954–8), ii. 231; Jean-Paul Hutin, *La Ville et les écrivains anglais 1770–1820* (thèse, Lille and Paris, 1978), 461–3.
8. Peter Brimblecombe, 'Interest in Air Pollution among the early Fellows of the Royal Society', *Notes & Recs. of the Royal Soc.*, 32 (1978); Kalm, 138.
9. Timothy Nourse, *Campania Foelix* (1700), 350–52; Kalm, 40.
10. 12 Ric. II, c. 13 (1388); 27 Hen. VIII, c. 18 (1536); *Tudor Royal Proclamations*, ed. Paul L. Hughes and James F. Larkin (1964–9), iii. no. 722; *Stuart Royal Proclamations*, ed. Larkin and Hughes (Oxford, 1973–), nos. 75, 107, 112, 200, 226; *Acts of the Privy Council, 1627*, 433–4, 444–5; *ibid.,1627–8*, 20, 169–70; Sir George Clark, *A History of the Royal College of Physicians* (Oxford, 1964–6), i. 255; *HMC, House of Lords*, xi. 382–4.
11. *Diary of Thomas Burton*, ed. John Towill Rutt (1828), ii. 221; C. H. Collins Baker, *The Life and Circumstances of James Brydges, First Duke of Chandos* (Oxford, 1949), 288.
12. *HMC, Shrewsbury & Talbot*, ii. xviii; William Stukeley, *Itinerarium Curiosum* (1776), 65.
13. *Archaeologia*, iv (1777), 227n.
14. John Graunt, *Natural and Political Observations* (1665), 141–2.
15. *The London Magazine*, xii (1743), 43.
16. Kalm, 138–9; *Locke Corr.*, iii. 583; Røstvig, *Happy Man*, i. 62–3; Alexandre Beljame, *Men of Letters and the English Public in the Eighteenth Century*, ed. Bonamy Dobrée, trans. E. O. Lorimer (1948), 223; Abraham Cowley, 'The Garden', *Several Discourses*, ed. Harry Christopher Minchin (1904), 65.
17. *The Works of Michael Drayton*, ed. J. W. Hebel (Oxford, 1961), iv. 279; OED, 'air'; 'fresh'; Robert Burton, *The Anatomy of Melancholy* (EL, 1932), ii. 75.
18. Norris, cit. Røstvig, *Happy Man*, i. 372; *A Dialogue between Reginald Pole and Thomas Lupset*, ed. Kathleen M. Burton (1948), 27.
19. Røstvig, *Happy Man, passim.*
20. *The Political and Commercial Works of Charles Davenant*, ed. Sir Charles Whitworth (1781), ii. 181.
21. *The Parliamentary Diary of Narcissus Luttrell*, ed. Henry Horwitz (Oxford, 1972), 132, 395.
22. Starkey, *Dialogue*, 92, 161; *Cyvile and Uncyvile Life*, sig. Aiv^{v}.
23. F. J. Fisher, 'The Development of London as a Centre of Conspicuous Consumption', *Trans. Royal Hist. Soc.*, 4th ser., xxx (1948); Anthony Fletcher,

A County Community in Peace and War (1975); Peter Clark, *English Provincial Society from the Reformation to the Revolution* (Hassocks, 1977), 404; Lawrence Stone, 'The Residential Development of the West End', in *After the Reformation*, ed. Barbara C. Malament (Manchester, 1980), 174–8.

24. H. J. Dyos, 'Greater and Greater London', in *Britain and the Netherlands*, iv, ed. J. S. Bromley and E. H. Kossmann (The Hague, 1971), 110–11; Mark Girouard, *Life in the English Country House* (1978), 5–6; Stephen Switzer, *Ichnographia Rustica* (2nd edn, 1742), i. xxxix; *The Banks Letters*, ed. Warren R. Dawson (1958), xxv; Sir Lewis Namier, *England in the Age of the American Revolution* (2nd edn, 1961), 16.
25. *Horae Subsecivae* (1620), 135.
26. William Blane, *Cynegetica* (1788), 4.
27. Colin Platt, *The English Medieval Town* (1979 edn), 122–3; *OED*, 'summer house'.
28. Peter Clark in *The English Commonwealth 1547–1640*, ed. Clark *et al.* (Leicester, 1979), 174; John T. Evans, *Seventeenth-Century Norwich* (Oxford, 1979), 24; John Stow, *A Survey of London*, ed. Charles Lethbridge Kingsford (Oxford, 1908), ii. 78.
29. Daniel Defoe, *A Tour through England and Wales* (EL, 1928), i. 161; Norman G. Brett-James, *The Growth of Stuart London* (1935), 99; *The Autobiography of Sir John Bramston*, ed. Lord Braybrooke (Camden Soc., 1845), 104; *CSPV, 1617–19*, 246; John Summerson, 'The Classical Country House in 18th-Century England, iii', *Jnl Royal Soc. of Arts*, cvii (1959); K. J. Allison, *The East Riding of Yorkshire Landscape* (1976), 181–2; information concerning Bristol from Mr Jonathan Barry.
30. Pepys, *Diary*, viii. 339–40n; John Stow, *A Survey of the Cities of London and Westminster*, enlarged by John Strype (1720), i. 227.
31. Kalm, 35; *Connoisseur*, 33 (12 Sep. 1754); [Arthur Young], *Rural Oeconomy* (1770), 175.
32. Uvedale Price, *Essays on the Picturesque* (1810), i. 162; Edmund Bartell, *Hints for Picturesque Improvements in Ornamented Cottages* (1804), 5; William Gilpin, *Observations on the Western Parts of England* (1798), 308–11; T. D. W. Dearn, *Sketches in Architecture* (1807), 5; Joseph Burke, *English Art, 1714–1800* (Oxford, 1976), 378; John Summerson, *Georgian London* (rev. edn, Harmondsworth, 1978), 273–5. There are designs for 'summer retreats' in T. Rawlins, *Familiar Architecture* (1768).
33. Stow, *Survey*, ed. Kingsford, i. 98; Brett-James, *Growth of Stuart London*, 452–3.
34. Pepys, *Diary*, viii. 339; *The Travels of Peter Mundy*, ed. Sir Richard Carnac Temple (Hakluyt Soc., 1907–25), iv. 29; John Milton, 'Paradise Lost', ix. lines 445–51; Stone, 'Residential Development of West End', 177.
35. Kalm, 37–8, 96–7.
36. Røstvig, *Happy Man*, i. 307; *The Prose of John Clare*, ed. J. W. and Anne Tibble (1951), 32; *The Works of Mr Henry Needler*, ed. William Duncombe (2nd edn, 1728), 135; Vivian de Sola Pinto, *Peter Sterry* (Cambridge, 1934), 194; R. H. Whitelocke, *Memoirs ... of Bulstrode Whitelocke* (1860), 450.

37. 'The Task', line 749; Thomas Jackson, *A Treatise containing the Originall of Unbeliefe* (1625), 196; William Penn, *The Peace of Europe* (EL, n.d.), 46; D. H. Lawrence, *Selected Essays* (Harmondsworth, 1950), 119.
38. *HMC, Skrine*, 161; Pepys, *Diary*, viii. 338–9; Hugh Blair, *Lectures on Rhetoric and Belles Lettres* (8th edn, 1801), iii. 108.
39. See David R. Coffin, *The Villa in the Life of Renaissance Rome* (Princeton, N.J., 1979).
40. I owe the expression to Mr Peter Clark. For examples of the process see Evelyn/Hunter, 518; Chalklin, *Provincial Towns*, 66–7; *VCH, Oxon.*, iv. 89, 113; *The Atlas of Historic Towns*, ed. M. D. Lobel (1969–), i (Nottingham), 8; ii (Coventry), 12.
41. John Fletcher, 'The Faithful Shepherdess', To the Reader, in *The Works of Beaumont and Fletcher*, ed. Henry Weber (Edinburgh, 1812), iv. 13–14. Cf. Helen Cooper, *Pastoral* (Ipswich, 1977), 135.
42. John Evelyn, *Publick Employment and an Active Life* (1667), 13; I[ohn] B[eale], *Herefordshire Orchards* (1657), 39.
43. As is emphasized by Raymond Williams, *The Country and the City* (1973); James Turner, *The Politics of Landscape* (Oxford, 1979); and John Barrell, *The Dark Side of the Landscape* (Cambridge, 1980).
44. Margaret M. Fitzgerald, *First Follow Nature* (New York, 1947), 56.
45. Hulin, *La Ville et les écrivains anglais*, 268–70, 491–2, 500–502.
46. Røstvig, *Happy Man*, i. 174; also 22, 54, 60–61, 249–50, 267–8. Cf. David Underdown, *Pride's Purge* (Oxford, 1971), 56.
47. Røstvig, *op. cit.*, 235–7, 410; Whitelocke, *Memoirs of Bulstrode Whitelocke*, 449–50; *DNB*, 'Sir William Temple'.
48. 'On Love'.
49. *Locke Corr.*, iv. 305.
50. [Dudley, Lord North], *Observations and Advices Oeconomical* (1669), 111–12; *Locke Corr.*, i. 98; Michael Hunter, *Science and Society in Restoration England* (1981), 80; Turner, *Politics of Landscape*, 175–6; *The Family Memoirs of the Rev. William Stukeley*, ed. W. C. Lukis (Surtees Soc., 1882–7), i. 107–9; Girouard, *Life in the English Country House*, 5–6; *The Rambler*, 135 (2 July 1751).
51. Bernard S. Horne, *The Compleat Angler 1653–1967: a New Bibliography* (Pittsburgh, 1970); Edward A. Martin, *A Bibliography of Gilbert White* (1934).
52. *Selected Essays of William Hazlitt*, ed. Geoffrey Keynes (1946), 3–8.
53. *Traherne's Poems of Felicity*, ed. H. I. Bell (Oxford, 1910), 29; Ray, *Wisdom*, 117.
54. Ebenezer Howard, *Garden Cities of To-Morrow* (1902), 18; above, p. 206.

ii. Cultivation or Wilderness?

1. Walter Blith, *The English Improver Improved* (1653), chap. 19; 5 & 6 Edw. VI, c. 5 (1551–2); Starkey, *Dialogue*, 30, 76; *The Agrarian History of England and Wales*, iv, ed. Joan Thirsk (Cambridge, 1967), xxxiii; Eric Kerridge, *The Agricultural Revolution* (1967), *passim*.
2. John Houghton, *A Collection of Letters for the Improvement of Husbandry &*

Trade (1681–3), i. 14. Cf. Samuel Hartlib, *The Compleat Husband-Man* (1659), 41.

3. Sir Edward Coke, *Institutes of the Laws of England* (1794–1817 edn), i. sect. 117; *Agrarian History of England*, iv. xxxiv–xxxv; Joan Thirsk, *Economic Policy and Projects* (Oxford, 1978), 147–8; Eric Kerridge, *Agrarian Problems in the Sixteenth Century and After* (1969), 120–21.
4. *The Voyages and Colonising Enterprises of Sir Humphrey Gilbert*, ed. David Beers Quinn (Hakluyt Soc., 1940), ii. 468; *The Miscellaneous Works of ... Edward, Earl of Clarendon* (2nd edn, 1751), 195; *Seventeenth-Century Economic Documents*, ed. Joan Thirsk and J. P. Cooper (Oxford, 1972), 135–6; Nourse, *Campania Foelix*, 99. Cf. above, pp. 14–15.
5. *Seventeenth-Century Economic Documents*, 779; W. G. Hoskins, *The Making of the English Landscape* (1955), 138.
6. [Arthur Young], *Observations on the Present State of the Waste Lands of Great Britain* (1773), 38; *Letters from Mrs Elizabeth Carter to Mrs Montagu*, ed. Montagu Pennington (1817), ii. 59.
7. Walter Blith, *The English Improver* (1649), sig. A4^{v}; Young, *op. cit.*, 37; *id.*, *View of the Agriculture of Oxfordshire* (1809), 228.
8. *Itinerary of Leland*, iv. 19; i. 1. Cf. John Ruskin, *Modern Painters* (EL, n.d.), iii. 172–81 (pt iv, chap. xiii); Ernst Robert Curtius, *European Literature and the Latin Middle Ages*, trans. Willard R. Trask (New York, 1953), 185–6; D. S. Wallace-Hadrill, *The Greek Patristic View of Nature* (Manchester, 1968), 90–91.
9. Sir Richard Weston, *A Discourse of Husbandrie*, ed. Samuel Hartlib (2nd edn, 1652), 27; Nourse, *Campania Foelix*, 2; Thomas Traherne, *Christian Ethicks* (1675), 103; John Norden, *Speculum Britanniae Pars: An Historical and Chorographical Description of Middlesex and Hartfordshire* (1723), 2nd pagination, 11.
10. William Cobbett, *Rural Rides*, ed. G. D. H. and Margaret Cole (1930), ii. 623.
11. William Lambarde, *A Perambulation of Kent* (1826 edn), 223; William Camden, *Britannia*, ed. Richard Gough (2nd edn, 1806), i. 341.
12. Blith, *English Improver Improved*, 155; Stephen Blake, *The Compleat Gardeners Practice* (1664), sig. A3^{v}; Ra[lph] Austen, *A Dialogue ... the Husbandman, and Fruit-Trees* (Oxford, 1676), 13. Cf. Sir Thomas Browne, *The Garden of Cyrus* (in *Hydriotaphia* (1658)).
13. Parkinson, *Paradisi*, 536.
14. Henry More, *An Antidote against Atheism* (2nd edn, 1655), 93. Cf. Arthur O. Lovejoy, *Essays in the History of Ideas* (reprint, New York, 1955), 99–100.
15. *Of Building. Roger North's Writings on Architecture*, ed. Howard Colvin and John Newman (Oxford, 1981), 13; John Laurence, *A New System of Agriculture* (1726), 51.
16. Oliver Rackham, *Trees and Woodland in the British Landscape* (1976), 115; E. C. K. Gonner, *Common Land and Inclosure* (1912), 82n.
17. See John Barrell, *The Idea of Landscape and the Sense of Place 1730–1840* (Cambridge, 1972), esp. chap. 2.
18. John Laurence, *The Gentleman's Recreation* (1716), 19; Cobbett, *Rural Rides*,

ed. Cole, i. 118; iii. 181; Samuel Collins, *Paradise Retriev'd* (1717), 61 (an 'old trite pun', according to William Hanbury, *A Complete Body of Planting* (1770), i. preface); Hesther Lynch Piozzi, *Anecdotes of Samuel Johnson*, ed. S. C. Roberts (1932), 169.

19. William Gilpin, *Remarks on Forest Scenery* (1791), ii. 166; *Wordsworth's Guide to the Lakes* (5th edn, 1835), ed. Ernest de Selincourt (1977), 151.

20. Thomas More, *Utopia* (EL, 1951), 111; George Owen, *The Description of Penbrokeshire*, ed. Henry Owen (Cymmrodorion Rec. Ser., 1892), 44. Cf. John Block Friedman, *The Monstrous Races in Medieval Art and Thought* (1981), 149.

21. A. L. Rowse, *The England of Elizabeth* (1950), 70–71; *Journeys of Celia Fiennes*, 222, 198, 196.

22. Roger North, *The Lives of . . . Francis . . . Dudley . . . and . . . John North*, ed. Augustus Jessopp (1820), i. 181; *The Diary of Ralph Thoresby*, ed. Joseph Hunter (1830), i. 105–6, 267; *Johnson's Journey to the Western Isles of Scotland*, ed. R. W. Chapman (1930), 34; Morton, *Northants.*, 20.

23. Raven, *Naturalists*, 289; Frank Emery, *Edward Lhuyd* (Cardiff, 1971), 31–5; Coles, *Eden*, 614; *The Familiar Letters of James Howell*, ed. Joseph Jacobs (1890), 95.

24. Blair, *Lectures on Rhetoric*, i. 55.

25. There is an excellent account of this change in Marjorie Hope Nicolson, *Mountain Gloom and Mountain Glory* (1959; New York, 1963). See also her article, 'Literary Attitudes toward Mountains', in *Dictionary of the History of Ideas*, ed. Philip P. Wiener (New York, 1973–4), iii. The novelty of the eighteenth-century feeling for wild nature had already been emphasized by G. M. Trevelyan in his Rickman Godlee Lecture (above, p. 14).

26. Nicolson, *Mountain Gloom*, 105–10, 116–17 and chaps. 5 and 6.

27. Erasmus Warren, *Geologia* (1690), 144, 146; Anthony, Earl of Shaftesbury, *Characteristicks* (6th edn, 1737), ii. 388.

28. *HMC, Portland*, ii. 302; Frank Emery, 'A New Account of Snowdonia, 1693', *Nat. Lib. of Wales Jnl*, xviii (1974), 409.

29. Byng, i. 147; *Collected Letters of Samuel Taylor Coleridge*, ed. Earl Leslie Griggs (Oxford, 1956–71), i. 610. On the growth of this tourism see James Holloway and Lindsay Errington, *The Discovery of Scotland* (Edinburgh, 1978); Norman Nicholson, *The Lakers* (1955); J. H. Plumb, *Georgian Delights* (1980), 17–21, 128–9.

30. Archibald Alison, *Essays on the Nature and Principles of Taste* (4th edn, Edinburgh, 1815), ii. 443. Cf. C. A. Moore, 'The Return to Nature in the English Poetry of the Eighteenth Century', *Studs. in Philology*, xiv (1917), 290; Ernst Lee Tuveson, *The Imagination as a Means of Grace* (Berkeley and Los Angeles, 1960), 98; George H. Williams, *Wilderness and Paradise in Christian Thought* (New York, 1962).

31. *Collected Letters of Coleridge*, ed. Griggs, ii. 916; Kenneth Woodbridge, *Landscape and Antiquity* (Oxford, 1970), 164.

32. Basil Willey, *The Eighteenth Century Background* (1940), 64; Daniel Mornet, *Le*

Sentiment de la nature en France de J.-J. Rousseau à Bernardin de Saint-Pierre (Paris, 1907).

33. Richard Ford, *Gatherings from Spain* (1846), 18; David Robertson, 'Mid-Victorians amongst the Alps', *Nature and the Victorian Imagination*, ed. U. C. Knoepflmacher and G. B. Tennyson (1977), 120.
34. Christopher Hussey, *The Picturesque* (1967 edn), 100–101; Stuart Piggott, *Ruins in a Landscape* (Edinburgh, 1976), 115, 124; Kenneth Smith, *Early Prints of the Lake District* (Nelson, 1973), (1); C. M. L. Bouch, *Prelates and People of the Lake Counties* (Kendal, 1948), 348.
35. *A New Historical Geography of England*, ed. H. C. Darby (Cambridge, 1973), 403; Hoskins, *Making of the English Landscape*, 143; J. D. Chambers and G. E. Mingay, *The Agricultural Revolution* (1966), 77; Wordsworth, 'The Excursion', viii. lines 128–30.
36. Cobbett, *Rural Rides*, ed. Cole, i. 73–4; William Gilpin, *Observations ... made in the year 1772, on ... Cumberland and Westmorland* (2nd edn, 1788), i. 7–8.
37. For general accounts see Elizabeth Wheeler Manwaring, *Italian Landscape in Eighteenth Century England* (1925); Hussey, *The Picturesque*; *id.*, *English Gardens and Landscapes, 1700–1750* (1967); Edward Malins, *English Landscaping and Literature 1660–1840* (1966); Nikolaus Pevsner, *Studies in Art, Architecture and Design* (1968), i. chaps. iv–viii; Miles Hadfield, *A History of British Gardening* (3rd edn, 1979), chap. 5; *The Genius of the Place*, ed. John Dixon Hunt and Peter Willis (1975).
38. Louis Simond, *Journal of a Tour and Residence in Great Britain* (2nd edn, Edinburgh, 1817), i. 263.
39. J. C. Loudon, *An Encyclopaedia of Gardening* (n.d.), 425; *id.*, *The Suburban Gardener and Villa Companion* (1838), 162. On Loudon's debt to Archibald Alison for this theory, see T. H. D. Turner in *Jnl of Garden History*, 2 (1982), 178–9.
40. William Marsden, *The History of Sumatra* (1783), 112–13.
41. Claude Colleer Abbott, *The Life and Letters of George Darley* (Oxford, 1967 reprint), 267.
42. William Gilpin, *Observations ... made in the year 1776, on ... the High-Lands of Scotland* (2nd edn, 1789), ii. 143.
43. Archibald Alison, *Essays on the Nature and Principles of Taste* (Edinburgh, 1790), 312–13.
44. *Ibid.*, 71–2.
45. *John Constable's Discourses*, ed. R. B. Beckett (Suffolk Recs. Soc., 1970), 72. Cf. Eilert Ekwall, *The Concise Oxford Dictionary of English Place-Names* (4th edn, Oxford, 1960), xxviii; Girouard, *Life in the English Country House*, 78; John Norden, *Speculi Britanniae Pars Altera; or a Delineation of Northamptonshire* (1720), 24.
46. See E. H. Gombrich, 'The Renaissance Theory of Art and the Rise of Landscape', *Norm and Form* (2nd edn, 1971); Manwaring, *Italian Landscape*; Henry V. S. Ogden and Margaret S. Ogden, *English Taste in Landscape in the Seventeenth Century* (Ann Arbor, 1955).

47. Edward Waterhouse, *Fortescutus Illustratus* (1663), 373; *Journeys of Celia Fiennes*, 353; John Ray, *The Wisdom of God Manifested* (2nd edn, 1692), i. 203. Other examples in Turner, *Politics of Landscape*, 9.
48. Hannah Woolley, *The Queen-Like Closet* (5th edn, 1684), ii. 53; Luke Herrmann, *British Landscape Painting of the Eighteenth Century* (1973); Martin Hardie, *Water-Colour Painting in Britain*, i, ed. Dudley Snelgrove *et al.* (2nd edn, 1967), chap. ix; *Scenery of Great Britain and Ireland in Aquatint and Lithography 1770–1860 from the Library of J. R. Abbey* (1977 edn); *Beauty, Horror and Immensity. Picturesque Landscape in Britain, 1750–1850*, exhibition catalogue by Peter Bicknell (Cambridge, 1981).
49. Holloway and Errington, *Discovery of Scotland*, chap. 4; Hussey, *Picturesque*, 126 (West was evidently following W. Hutchinson, *An Excursion to the Lakes* (1776), 191).
50. *Gilbert White's Journals*, ed. Walter Johnson (1931; reprint, Newton Abbot, 1970), 143, 112–13, 131.
51. *John Constable's Correspondence*, ed. R. B. Beckett (Suffolk Recs. Soc., 1962–8), vi. 98.
52. William Gilpin, *Observations on the River Wye* (2nd edn, 1789), 30–31; *id.*, *Observations on Several Parts of the Counties of Cambridge, Norfolk, Suffolk, and Essex* (1809), 176.
53. Cit. Kenneth Clark, *Landscape into Art* (new edn, 1976), 68.
54. Gilpin, *Observations on Cumberland and Westmoreland*, i. 127.
55. 'The Prelude', xii. lines 111–21; 'The Excursion', viii. lines 151–5; Pevsner, *Studies*, i. 120.
56. *Wordsworth's Guide to the Lakes*, 160, 150–55.
57. Robert Southey, *Letters from England*, ed. Jack Simmons (1951), 165. Similar comments in *Memoirs and Correspondence of Francis Horner*, ed. Leonard Horner (1843), i. 119–20.
58. Wordsworth, 'Liberty', line 32. Cf. the comments of Havelock Ellis, 'The Love of Wild Nature', *Contemporary Rev.*, 95 (1909), and Roderick Nash, *Wilderness and the American Mind* (rev. edn, 1973).
59. Hobbes, *EW*, ii. 2. Cf. Janette Dillon, *Shakespeare and the Solitary Man* (1981), part 1.
60. Hugh Honour, *Romanticism* (1979), 256–8; Røstvig, *Happy Man*, *passim*. Cf. Samuel Cradock, *Knowledge and Practice* (1659), sig. A7.
61. Holloway and Errington, *Discovery of Scotland*, 106–7.
62. John Stuart Mill, *Principles of Political Economy* (1848; 1965 edn), iv. 6. 2.
63. Cit. Peter J. Schmitt, *Back to Nature* (New York, 1969), 67. On the move to preserve wild scenery see Donald Fleming, 'Roots of the New Conservation Movement', *Perspectives in American History*, vi (1972); Nash, *Wilderness and the American Mind*; Hans Huth, *Nature and the American* (Berkeley and Los Angeles, 1957).

iii. Conquest or Conservation?

1. Parkinson, *Paradisi*, 324, 224, 242, 249.
2. [George Cornewall Lewis], *A Glossary of Provincial Words used in Herefordshire* (1839), 40; Rackham, *Trees and Woodland*, 32.
3. Shakespeare, *Henry V*, v. ii. lines 45–7; Blith, *English Improver Improved*, 125–6; *id.*, *English Improver*, 60, 139, 153–6; Hartlib, *Compleat Husband-Man*, 8, 41.
4. Ellis, i(2). 79; i(3). 16–20; ii(2). 36–50; iv(2). 39; vii. 25; viii. 4, 293–304; *Of Building. Roger North's Writings on Architecture*, 14.
5. Coles, *Eden*, 394; Northants. R.O., FH 2455. Cf. Bacon, 'Of Gardens'; Gerard, *Herball*, ed. Johnson, 479; Christopher Merrett, *Pinax Rerum Naturalium Britannicarum* (1667), 3; *Spectator*, 477 (6 Sep. 1712).
6. Robert Sharrock, *The History of the Propagation & Improvement of Vegetables* (1660), 144; Henry Peacham, *The Gentlemans Exercise* (1612), 57–8; Wilfrid Blunt, *The Art of Botanical Illustration* (1950), 130.
7. *The First and Seconde Partes of the Herbal of William Turner ... corrected and enlarged with the Thirde Parte* (Cologne, 1568), i. sig. *iij.
8. On the growth of plant-hunting see R. T. Gunther, *Early British Botanists* (Oxford, 1922); Raven, *Naturalists*; H. Wallis Kew and H. E. Powell, *Thomas Johnson Botanist and Royalist* (1932); Allen, *Naturalist*, chaps. 1 and 2; *A Seventeenth Century Flora of Cumbria*, ed. E. Jean Whittaker (Surtees Soc., 1981), xxvii–xxxviii. There is a facsimile reprint with trans. and intro. by J. S. L. Gilmour of Thomas Johnson, *Botanical Journeys in Kent and Hampstead* (Pittsburgh, 1972).
9. Charles Cardale Babington, *Flora of Cambridgeshire* (1860), intro.; Henrey, ii; [Richard Gough], *Anecdotes of British Topography* (1768), 95–6; Samuel Pegge, *Curialia Miscellanea* (1818), lxxxiii.
10. *The Sloane Herbarium*, comp. James Britten, ed. J. E. Dandy (1958).
11. Alice M. Coats, *Garden Shrubs and their Histories* (1963), 346–7; G. W. Francis, *The Little English Flora* (1839), 111–12.
12. Hanbury, *Complete Body of Planting*, i. 509, 420, 824; William Marshall, cit. Coats, *Garden Shrubs and their Histories*, 312; [Henry Home, Lord Kames], *Elements of Criticism* (6th edn, 1785), i. 302; Cowper, 'The Task', i. lines 526–30; *The Poems of John Clare*, ed. J. W. Tibble (1935), i. 235, 534–5; ii. 126, 148, 281, 315, 325.
13. Loudon, cit. A. A. Tait, *The Landscape Garden in Scotland* (Edinburgh, 1980), 241; James Bolton, *Filices Britannicae* (1785), iii; David Elliston Allen, *The Victorian Fern Craze* (1969); Francis George Heath, *The Fern Paradise* (3rd edn, 1876), 295.
14. *The Works of John Ruskin*, ed. E. T. Cook and Alexander Wedderburn (1903–12), xxv. 439; i. 156.
15. Tennyson, 'Amphion'; Hopkins, 'Inversnaid'; John Sheail, *Nature in Trust* (1976), 39–41.
16. X. de Planhol, 'Le Chien de berger', *Bulletin de l'association de géographes français*, 370 (March 1969); Hartlib, *Compleat Husband-Man*, 74 ('78');

Camerarius, *The Walking Librarie*, trans. John Molle (1621), 98–9; Waterhouse, *Fortescutus Illustratus*, 379–80; E. Estyn Evans, *The Personality of Ireland* (Cambridge, 1973), 11. See also James Edmund Harting, *British Animals Extinct within Historic Times* (1880), 115–205; Anthony Dent, *Lost Beasts of Britain* (1974), 99–134; Alain Molinier and Nicole Molinier–Meyer, 'Environnement et histoire: les loups et l'homme en France', *Revue d'histoire moderne et contemporaine*, xxxviii (1981).

17. Edward MacLysaght, *Irish Life in the Seventeenth Century* (3rd edn, Cork, 1969), 333. Cf. James Ritchie, *The Influence of Man on Animal Life in Scotland* (Cambridge, 1920), 119–21.
18. *Notes & Queries*, 3rd ser., ix (1866), 158–9; *Tenures of Land and Customs of Manors*, ed. W. C. Hazlitt (1874), 27, 285; Charles Owen, *An Essay towards a Natural History of Serpents* (1742), 144; Sir Robert Atkyns, *The Ancient and Present State of Glocestershire* (2nd edn, 1768), 202–3.
19. William Harrison, *The Description of England*, ed. Georges Edelen (Ithaca, N.Y., 1968), 325; Curtius, *European Literature*, 184.
20. *The Works of the Reverend Mr Edm. Hickeringill* (1709), i. 358.
21. 24 Hen. VIII, c. 10 (1532–3); 8 Eliz., c. 15 (1566); T. N. Brushfield, 'On the Destruction of "Vermin" in Rural Parishes', *Trans. Devon. Assoc.*, xxix (1897); J. Steele Elliott, *Bedfordshire 'Vermin' Payments* (Luton, 1936); E. L. Jones, 'The Bird Pests of British Agriculture in Recent Centuries', *AgHR*, 20 (1972).
22. Jones, *art. cit.*, 113; Charles Brears, *Lincolnshire in the 17th and 18th Centuries* (1940), 76–7; Brushfield, *art. cit.*, 325; Elliott, *op. cit.*, 58; Gilbert White, *The Natural History of Selborne* (1788), letter x to Pennant.
23. *CSPD, 1603–10*, 165; Pennant, *Zoology*, i. 142.
24. John Worlidge, *Systema Agriculturae* (1669), 223–35.
25. Ronald A. Rebholz, *The Life of Fulke Greville* (Oxford, 1971), 43; *Turner on Birds*, ed. A. H. Evans (Cambridge, 1903), 29; Sir Robert Sibbald, *A Collection of Several Treatises* (Edinburgh, 1739), ii. 47.
26. *The Hawkins' Voyages*, ed. Clements R. Markham (Hakluyt Soc., 1878), 16, 194–5, 196–7; *The Jamestown Voyages*, ed. Philip L. Barbour (Hakluyt Soc., 1969), i. 133.
27. *The Letters of John Chamberlain*, ed. Norman Egbert McLure (Philadelphia, 1939), i. 212; John Nassau Simpkinson, *The Washingtons* (1860), appendix (A) 2.
28. Ellis, vi. 128; Charles Butler, *The Feminine Monarchie* (Oxford, 1609), sig. H6; Bryan J'Anson Bromwich, *The Experienced Bee-Keeper* (2nd edn, 1783), 43–4.
29. *The Diary of Sir Simonds D'Ewes (1622–1624)*, ed. Elisabeth Bourcier (Paris, 1974), 98; Brushfield, 'Destruction of "Vermin"', 333.
30. E.g. Morton, *Northants.*, 425, 426, 431; John Hill, *An History of Animals* (1752), 322, 331, 453, 488; *Gilbert White's Journals*, 94 and *passim*.
31. *Turner on Birds*, 117; Jones, 'Bird Pests', 110.
32. On extinctions see Harting, *British Animals Extinct*; Ritchie, *Influence of Man on Animal Life in Scotland* (a model study); N. W. Moore, 'The Past and Present Status of the Buzzard in the British Isles', *British Birds*, 50 (1957); Maarten

Bijlveld, *Birds of Prey in Europe* (1974); *Book of British Birds* (Reader's Digest and A.A., 1969), 6–14.

33. Cf. Stow, *Survey*, ed. Strype, i. 29; *HMC, Portland*, ii. 309.
34. 13 Edw. I (Stat. Westmr.), c. 47 (1285); 13 Ric. II, c. 19 (1389–90); John Manwood, *A Treatise and Discourse of the Lawes of the Forrest* (1598), fol. 72; Coke, *Institutes*, iv. chap. 73; 11 Hen. VII, c. 17 (1495); 14 & 15 Hen. VIII, c. 10 (1523); 25 Hen. VIII, cc. 7, 11 (1533–4); 31 Hen. VIII, c. 12 (1539); 32 Hen. VIII, c. 11 (1540); 1 Eliz., c. 17 (1558–9); *Tudor Royal Proclamations*, i. no. 215.
35. 32 Hen. VIII, c. 8 (1540); Owen, *Description of Penbrokshire*, 149–50; *CSPV, 1617–19*, 271, 309; 7 Jac. I, c. 27 (1603–4); P. B. Munsche, *Gentlemen and Poachers* (Cambridge, 1981), 39–42.
36. William Dugdale, *The History of Imbanking and Drayning* (1662), sig. A3; John Nisbet, *Our Forests and Woodlands* (1900), 301.
37. Charles Darwin, *The Variation of Animals and Plants under Domestication* (2nd edn, 1893), i. 87; ii. 97; *The Zoologist*, 3rd ser., xv (1891), 81–7; Evelyn Philip Shirley, *Some Account of English Deer Parks* (1867), 150, 177.
38. 22 Edw. IV, c. 6 (1482–3); Goldsmith, vi. 120; Benjamin Martin, *The Natural History of England* (1759–63), 44; *The Travels through England of Dr Richard Pococke*, ed. James Joel Cartwright (Camden Soc., 1888–9), i. 94.
39. Allison, *East Riding of Yorkshire Landscape*, 124. See in general Norman F. Ticehurst, *The Mute Swan in England* (1957).
40. William Horman, *Vulgaria* (1519), fol. 192v.
41. Stow, *Survey*, ed. Strype, i. 119; Harrison, *Description of England*, 328; [E. T. Bennett], *The Tower Menagerie* (1829); *A Pleasant Funeral-Oration, at the Interment of the Three (lately deceased) Tower-Lyons* (1681).
42. Harrison, *op. cit.*, 319; *Journeys of Celia Fiennes*, 173, 229; above, p. 111.
43. George Edwards, *A Natural History of Uncommon Birds* (1743–51); *id.*, *Gleanings of Natural History* (1758–64).
44. William Chafin, *A Second Edition of the Anecdotes and History of Cranbourn Chase* (1818), 99; Basil Taylor, *Stubbs* (2nd edn, 1975), 209; Gilpin, *Observations on the High-Lands*, ii. 188–9; Hill, *History of Animals*, 348, 355, 360, 363, 419, 540, 550; *Letters of S. T. Coleridge*, i. 645.
45. Thomas Mouffet, *The Theater of Insects*, appended to Edward Topsell, *The History of Four-Footed Beasts*, rev. by J. R. (1658), preface; *VCH, Bucks.*, ii. 189; *The Flemings in Oxford*, ed. John Richard Magrath (Oxford Hist. Soc., 1904–24), i. 47; *Diary of Sir Simonds D'Ewes*, 150; North, *Lives of the Norths*, i. 366–7; *The Practical Works of Richard Baxter* (1707), i. 817; Richard D. Altick, *The Shows of London* (1978), chaps 3 and 22.
46. A. M[ullen], *An Anatomical Account of the Elephant* (1682), 4; Stuart Piggott, *William Stukeley* (Oxford, 1950), 61.
47. John Bulwer, *Anthropometamorphosis* (1653), sig. B4v.
48. Frank N. Egerton, 'Changing Concepts of the Balance of Nature', *Qtly Rev. of Biology*, 48 (1973); John Passmore, *Man's Responsibility for Nature* (1974), 22. Cf. John Locke, *Two Treatises of Government*, ed. Peter Laslett (Cambridge, 1960), 199 (i. para. 56).

49. Frank N. Egerton, 'A Bibliographical Guide to the History of General Ecology and Population Ecology', *History of Science*, xv (1977), 203. Cf. Edward Bury, *The Husbandmans Companion* (1677), sig. a6.
50. James Jennings, *Ornithologia* (1828), xxvi–xxvii, 245; Phyllis Barclay-Smith, 'The British Contribution to Bird Protection', *The Ibis*, 101 (1959), 115–16. Cf. James Bolton, *Harmonia Ruralis* (new edn, 1830), ii. 95–6.
51. *The Works, Moral and Religious, of Sir Matthew Hale*, ed. T. Thirlwall (1805), i. 273; Maurice Cranston, *John Locke* (1957), 426; Luke Booker, *Poems* (Wolverhampton, 1785), ii. 30n.
52. E.g. Ogg, *England in the Reigns of James II and William III*, 121; *The Works of Symon Patrick*, ed. Alexander Taylor (Oxford, 1858), iii. 68; Watkyns, *Flamma sine Fumo*, 48.
53. *The Bird-Fancier's Recreation* (3rd edn, 1735), sig. A3ᵛ; *The Natural History of English Song-Birds* (with figs. by Eleazar Albin) (1737), 25–6; Dagobert De Levie, *The Modern Idea of the Prevention of Cruelty to Animals and its Reflection in English Poetry* (New York, 1947), 87–8; Humphry Primatt, *A Dissertation on the Duty of Mercy and Sin of Cruelty to Brute Animals* (1776), 277; Booker, *Poems*, ii. 75; Jennings, *Ornithologia*, 400–401.
54. Cit. Bolton, *Harmonia Ruralis*, i. xx.
55. *The Collected Poems of Christopher Smart*, ed. Norman Callan (1949), i. 344; Booker, *Poems*, ii. 72–5; *The Letters of Joseph Ritson* (1833), i. 21; Samuel F. Pickering, *John Locke and Children's Books in Eighteenth-Century England* (Knoxville, 1981), 24; *The Letters of Sir William Jones*, ed. Garland Cannon (Oxford, 1970), ii. 750; De Levie, *Modern Idea of Prevention of Cruelty*, 84–8.
56. [Francis Mundy], *Needwood Forest* (Lichfield, 1776), 34.
57. [Soame Jenyns], *Disquisitions on Several Subjects* (2nd edn, 1782), 17–18.
58. Lady Newcastle, *Poems, and Fancies* (1653), 70–75; 'Philotheos Physiologus' [Thomas Tryon], *The Country-Man's Companion* (n.d. [1683]), 141–73.
59. *The Journal of John Woolman* (Secaucus, N. J., 1961), 2–3; *A Memoir of Thomas Bewick written by himself* (1979 edn), 15; *Byron's Letters and Journals*, ed. Leslie A. Marchand (1973–), iii. 253; Audrey Williamson, *Wilkes* (1974), 110.
60. Southey, cit. Jennings, *Ornithologia*, 201n; Allen, *Naturalist*, 119.
61. Allen, *ibid.*, 197–9; Barclay-Smith, 'British Contribution to Bird Protection'; Sheail, *Nature in Trust*, 11–16, 22–36.
62. C. Deering, *Catalogus Stirpium* (1738), 'To the Reader'.
63. Sir Harry Godwin, *The History of the British Flora* (2nd edn, Cambridge, 1975), 1. Allen, *Naturalist*, chaps 1–12, gives an excellent brief account.
64. Morton, *Northants.*, 450, 428, 438 and *passim*.
65. Raven, *Ray*, *passim*. A. D. Atkinson, 'William Derham', *Ann. Sci.*, 8 (1952).
66. Plot, *Staffs.*, 230; John Ray, *Synopsis Methodica Stirpium Britannicarum* (3rd edn, 1724), 35; Allen, *Naturalist*, 28–9; Henrey, ii. 243 and chap. xix; Charles Abbot, *Flora Bedfordiensis* (Bedford, 1798), vii.
67. *The Letters of the Earl of Chesterfield*, ed. Charles Strachey (1901), i. 298.
68. Henry Power, *Experimental Philosophy* (1664), 183; Pennant, *Zoology*, i. xi.
69. Daniel Mornet, *Les Sciences de la nature en France, au XVIIIᵉ siècle* (Paris, 1911),

9n, 202, 248–9; *A Selection of the Correspondence of Linnaeus*, ed. Sir James Edward Smith (1821), i. 18–19.

70. Allen, *Naturalist*, 139; *DNB*, 'Brightwen, Eliza'; Eliza Brightwen, *More about Wild Nature* (n.d.), x.
71. See Raymond Irwin, *British Bird Books* (1951); Henrey, ii; Allen, *Naturalist*, 35–6.
72. *Selection of Correspondence of Linnaeus*, i. 33.
73. S. H[ayes], *A Practical Treatise on Planting* (1794), 154n; Henrey, ii. 241–3; *The Critical Rev.*, 16 (1763), 312; *The Beauties of England* (2nd edn, 1764), 87, 55; Rashleigh Holt-White, *The Life and Letters of Gilbert White* (1901), i. 311.
74. Allen, *Naturalist*, 158–9; E. B. Ford, *Butterflies* (1945), 24–5.
75. [Dawson Turner], *Extracts from the Literary and Scientific Correspondence of Richard Richardson* (Yarmouth, 1835), 74.
76. Allen, *Naturalist*, 33; S. Peter Dance, *Shell Collecting* (1961), chap. 3.
77. Robert Henry Welker, *Birds and Men* (New York, 1966), 68.
78. *Early Letters of Robert Wodrow, 1698–1709*, ed. L. W. Sharp (Scottish Hist. Soc., 1937), xxv.
79. Ritchie, *Influence of Man on Animal Life*, 292; Edwin M. Betts and Hazlehurst Bolton Perkins, *Thomas Jefferson's Flower Garden at Monticello* (2nd edn, Charlottesville, Va., 1971), 4; Rowland Hill, *Journal of a Tour through the North of England* (1799), 87n.
80. Byng, ii. 251–2; also iii. 161–2; i. 278.
81. 'The Excursion', viii. lines 151–5.
82. Cit. Alfred Biese, *The Development of the Feeling for Nature in the Middle Ages and Modern Times* (1905), 265.
83. Gilpin, *Observations on the High-Lands*, ii. 112, 114; *id.*, *Observations on the Western Parts of England* (1798), 328; *id.*, *Observations on Cumberland and Westmoreland*, ii. 44.
84. Alison, *Essays on the Nature and Principles of Taste*, 85; William Mitford, *Principles of Design in Architecture* (2nd edn, 1824), 9. Cf. Barrell, *Idea of Landscape*, chap. 2.
85. Price, *Essays on the Picturesque*, i. 94; Richard Payne Knight, *An Analytical Inquiry into the Principles of Taste* (1805), 79–80; William Gilpin, *Three Essays* (1792), ii. 36–7; *id.*, *Remarks on Forest Scenery*, ii. 119; George Ewart Evans, *The Horse in the Furrow* (1967 edn), 174.
86. Gilpin, *Remarks on Forest Scenery*, i. 42–3; Price, *Essays on the Picturesque*, i. 269, 273–6; *Wordsworth's Guide to the Lakes*, 82–5.
87. H[umphry] Repton, *Observations on the Theory and Practice of Landscape Gardening* (1803), 94; *Man made the Land*, ed. Alan R. H. Baker and J. B. Harley (Newton Abbot, 1973), 165–6.
88. *Spectator*, 414 (25 June 1712); *Letters of William Shenstone*, 285; R. W. King, 'The Ferme Ornée', *Garden History*, ii (1974); Hussey, *Picturesque*, 128–33.
89. William Marshall, *Planting and Rural Ornament* (2nd edn, 1796), i. xxiii; Repton, *Observations*, 92. Cf. Barrell, *Idea of Landscape*, 68–83.
90. On this complex subject see Francis D. Klingender, *Art and the Industrial Revolution* (rev. edn by Arthur Elton Frogmore, 1972); Esther Moir, 'The

Industrial Revolution: a Romantic View', *History Today*, ix (1959); A. D. Harvey, *English Literature and the Great War with France* (1981), appendix 2.

91. Cf. Leo Marx, *The Machine in the Garden* (New York, 1967 edn), 364–5.
92. D. M. Palliser, *The Staffordshire Landscape* (1976), 183.

iv. Meat or Mercy?

1. 'A Gentleman', *A Description of Millennium Hall* (1762). Cf. Nathanael Homes, *ΑΠΟΚΑΛΥΨΙΣ ΑΝΑΣΤΑΣΕΩΣ. The Resurrection Revealed* (1654), 189–91, 530, 534; Mary Cary, *The Little Horns Doom and Downfall* (1651), 293–4.
2. Topsell, 285; Walter Charleton, *Physiologia Epicuro-Gassendo-Charltoniana* (1654), 362; William Kirby, *On the Power, Wisdom and Goodness of God as manifested in the Creation of Animals* (1835), 371. Cf. Isaiah, xi. 6–9.
3. Sheail, *Nature in Trust*, 22–39; Charles McDougal, *The Face of the Tiger* (1977), 163–4; *The Times*, 4 Sep. 1980.
4. Goldsmith, iv. 157 (plagiarizing Buffon).
5. George Eliot, *Middlemarch* (1872), chap. 3; Henry S. Salt, *Animals' Rights* (1892), 53; Peter Singer, *Animal Liberation* (1976), 9–10; Lord Zuckerman, foreword to *Golden Days* (1976).
6. *The Domestication and Exploitation of Plants and Animals*, ed. Peter J. Ucko and G. W. Dimbleby (1969), 526.
7. *Minor Poets of the Caroline Period*, ed. George Saintsbury (Oxford, 1968), i. 558; Arthur O. Lovejoy and George Boas, *Primitivism and Related Ideas in Antiquity* (reprint, New York, 1935), index, 'vegetarianism'.
8. Alexander Pope, 'An Essay on Man', iii. lines 152, 154.
9. *Wilson's Arte of Rhetorique 1560*, ed. G. H. Mair (Oxford, 1909), sig. Avi^{v}; also [William Alley], *The Poore Mans Librarie* (1571), fol. 96^{v}; Thomas Plume, *An Account of the Life and Death of . . . John Hacket*, ed. Mackenzie E. C. Walcott (1865), 116; Matthew Henry, *A Commentary on the Holy Bible* (1710; new edn, n.d.), i. 7.
10. Thomas Cooper, *A Briefe Exposition of Such Chapters of the Olde Testament as usually are red in the Church* (1573), fols. 106^{v}–7; Andrew Willet, *Hexapla in Genesin* (1605), 105; Benjamin Needler, *Expository Notes* (1655), 16–18; Don Cameron Allen, *The Legend of Noah* (Urbana, 1963), 144n.
11. George Boas, *Essays on Primitivism and Related Ideas in the Middle Ages* (Baltimore, 1948), 25–6; Willet, *op. cit.*, 105; Alexander Rosse, *The First Booke of Questions and Answers upon Genesis* (1620), 26–7; Thomas Muffett, *Healths Improvement*, enlarged by Christopher Bennet (1655), 30–31; James Mackenzie, *The History of Health* (Edinburgh, 1759), 44–7; Samuel Pegge, 'Whether the Antediluvians used Animal Food' (Society of Antiquaries, MS. correspondence, 13 Nov. 1788). Cf. Thomas Love Peacock, *Headlong Hall* (1816), chap. ii.
12. F. J. Powicke, 'The Reverend Richard Baxter's Last Treatise', *Bulletin of the John Rylands Lib.*, x (1926), 197.
13. Robert Hawker, *Memoirs of the Life and Writings of the late Rev. Henry Tanner* (2nd edn, 1811), 69–70.

14. Porphyry, *De Abstinentia*; Seneca, *Ad Lucilium Epistulae Morales*, viii; Boas, *Essays on Primitivism*, 17–18, 32, 114–16.
15. J. W. Gough, *The Superlative Prodigall* (Bristol, 1932), 16; *CSPD, 1639*, 467; Lodowick Muggleton, *The Acts of the Witnesses* (1764 edn), 47.
16. J. F. C. Harrison, *The Second Coming* (1979), 21–2, 140, 159.
17. *The Whole Works of ... Jeremy Taylor*, ed. Reginald Heber, rev. by Charles Page Eden (1847–54), ix. 356–62; Andrew Willet, *Hexapla in Leviticum* (1631), 57; *The Workes of John Boys* (1629), 77; William Ames, *Conscience* (1639), iv. 195; [John Dunton], *The Athenian Oracle* (1703–4), ii. 56–7, 93–6; iii. 77.
18. [P. Delany], *Revolution examin'd with Candour* (1732), ii. 1–79; Anon., *The Question about Eating of Blood Stated* (1732); [Delany] *The Doctrine of Abstinence from Blood Defended* (1734); 'A Prebendary of York', *An Enquiry about the Lawfulness of Eating Blood* (1733); *id.*, *A Defence of the Enquiry about the Lawfulness of Eating Blood* (1734); Anon., *The Apostolic Decree at Jerusalem proved to be still in force* (1734); Joseph Priestley, *Institutes of Natural and Revealed Religion* (Birmingham, 1782), ii. 447–9.
19. *Works of Jeremy Taylor*, ix. 356; Thomas Edwards, *Gangraena* (1646), ii. 2; Richard Gough, *Antiquities and Memoirs of the Parish of Myddle* (1875), 147; Laurence, *New System of Agriculture*, 87. Cf. John Rawlinson, *Mercy to a Beast* (Oxford, 1612), 34; 'A Well-Wisher to Ancient Truth' [?Thomas Barlow], *The Trial of a Black-Pudding* (1652); J[ohn] E[velyn], *Acetaria* (1699), 156–60; [John Toland], *Reasons for Naturalizing the Jews* (1714), in *Pamphlets relating to the Jews in England in the 17th and 18th Centuries*, ed. P. Radin (San Francisco, 1939), 57.
20. Pepys, *Diary*, viii. 483; *Locke Corr.*, iv. 414; Kalm, 14; *Travels of Carl Philip Moritz in England* (1924), 33–4; J.-C. Schmitt, 'Le Suicide au moyen âge', *Annales (économies, sociétés, civilisations)*, 31e (1976), 21.
21. Nathaniel Lardner, *Remarks upon the late Dr Ward's Dissertations* (1762), 132; Norbert Elias, *The Civilizing Process*, trans. Edmund Jephcott (New York, 1978), 121; John Clive, *Thomas Babington Macaulay* (1973), 53; Countess Evelyn Montenegro Cesaresco, *The Place of Animals in Human Thought* (1909), 23.
22. Edwards, *Gangraena*, i. 34, 80.
23. *The English Hermit* (1655), in *The Harleian Miscellany* (1808–11), vi. 390–405; *The Rev. Oliver Heywood ... his Autobiography, Diaries, Anecdote and Event Books*, ed. J. Horsfall Turner (Brighouse and Bingley, 1882–5), i. 361.
24. Charles Smith, *The Ancient and Present State of the County and City of Waterford* (2nd edn, 1774), 371–4. There are useful accounts of early vegetarians in Howard Williams, *The Ethics of Diet* (rev. edn, 1896), and the very rare work by Narhar Kashinath Gharpure, *Tierschutz, Vegetarismus und Konfession (eine religions-soziologische Untersuchung zum Englischen 17. und 18. Jahrhundert)* (München, 1935).
25. 'Phylotheus Physiologus' [Thomas Tryon], *Monthly Observations for the Preserving of Health* (1688), 29–30, 81–2; *id.*, *Wisdom's Dictates* (1691), 6, 14, 21, 76, 129–30; *id.*, *Country-Man's Companion*, 119–22; *id.*, *The Good House-Wife* (2nd edn, 1692), 11, 218, 268; *id.*, *Friendly Advice to the Gentleman-Planters of*

the East and West Indies (1684), 54–5; *id.*, *The Way to Make All People Rich* (1685), prefatory poem by A[phra] Behn; *The Works of Dr Benjamin Franklin* (1824), 22–3; Joseph Ritson, *An Essay on Abstinence from Animal Food* (1802), 80; Alexander Gordon, 'A Pythagorean of the Seventeenth Century', *Procs. Lit. & Phil. Soc. of Liverpool*, xxv (1871); above, pp. 155, 170, 280.

26. Tryon, *Good House-Wife*, 217–18; *id.*, *Country-Man's Companion*, 121; Gordon, *art. cit.*, 297–8.
27. William Smellie, *The Philosophy of Natural History* (Edinburgh, 1790–99), i. 60–61.
28. Plutarch, *Moralia: De Esu Carnium*; Porphyry, *De Abstinentia*; W. K. C. Guthrie, *A History of Greek Philosophy*, i (Cambridge, 1962), 186–91; Ovid, *Metamorphoses*, xv. 73–142; *The Poems of John Dryden*, ed. James Kinsley (Oxford, 1958), iv. 1736. I suspect that Professor Kinsley may have been wrong to include this interpolation among the 'links and single lines' added merely 'for better finish' (iv. 2080).
29. Pepys, *Diary*, vii. 223–4; *Philos. Trans.*, xxii (1702 for 1700–1701), 769–85; John Ray, *Historia Plantarum* (1686–1704), i. 46.
30. 'Eugenius Philalethes', *A Treatise of the Plague* (1721), 17–18. Cf. *The Educational Writings of John Locke*, ed. James L. Axtell (Cambridge, 1968), 125, 127; James Nelson, *An Essay on the Government of Children* (1763), 70.
31. Evelyn, *Acetaria*; [Edward Bancroft], *An Essay on the Natural History of Guiana* (1769), 262–3; John Small, *Biographical Sketch of Adam Ferguson* (Edinburgh, 1864), 32.
32. Charles W. Forward, *The Food of the Future* (1904), 114; E. Hare, *The Life of William Lambe, M.D.* (1873). See also James C. Whorton, '"Tempest in a Flesh-Pot": The Formulation of a Physiological Rationale for Vegetarianism', *Jnl Hist. Medicine*, xxxii (1977).
33. George Cheyne, *The English Malady* (1733), 342, 353; *id.*, *An Essay on Regimen* (1740), xv, 70.
34. William Haller, *Foxe's Book of Martyrs and the Elect Nation* (1963), 56; *Works of Sir Matthew Hale*, ii. 273–4; Powicke, 'Baxter's Last Treatise,' 197; Ray, *Historia Plantarum*, i. 46; *Œuvres complètes de Voltaire* (new edn, Paris), xxii (1879), 421–2; *Extracts from the Diary of the Rev. Robert Meeke*, ed. Henry James Morehouse (1874), 49.
35. Newcastle, *Poems, and Fancies*, 184, 112–13.
36. 'Spring', lines 336–73.
37. 'Eugenius Philalethes', *Treatise of Plague*, 16–17.
38. *The Works of Jonath^{n} Swift . . . with notes by J. Hawkesworth* (Dublin, 1774), iv. 102n; Bancroft, *Natural History of Guiana*, 261–2; Ford, *Gatherings from Spain*, 313–14.
39. More, *Utopia*, 71.
40. Philip E. Jones, *The Butchers of London* (1976), 1, 10, 78–81, 84, 95–9; E. L. Sabine, 'Butchering in Medieval London', *Speculum*, viii (1933); *VCH, Oxon.*, iv. 27; F. G. Emmison, *Elizabethan Life: Home, Work and Land* (Chelmsford, 1976),

295–6; *The Southampton Mayor's Book of 1606–1608*, ed. W. J. Connor (Southampton Recs. Ser., 1978), 64–5; *Glamorgan County History*, iv, ed. Glanmor Williams (Cardiff, 1974), 320.

41. W[illiam] Vaughan, *The Golden Grove* (2nd edn, 1608), sig. V4[v]; Muffett, *Healths Improvement*, 47.
42. Joshua Poole, *The English Parnassus* (1677 edn), 59; *Works of Hickeringill*, ii. 519.
43. John Gay, *Poetry and Prose*, ed. Vinton A. Dearing (Oxford, 1974), i. 144; David Hartley, *Observations on Man* (4th edn, 1801), ii. 222; Adam Smith, *An Inquiry into the Nature and Causes of the Wealth of Nations*, ed. R. H. Campbell and A. S. Skinner (Oxford, 1976), i. 117.
44. R. Fletcher, *A Few Notes on Cruelty to Animals* (1846), 41; Charles Booth, *Life and Labour of the People in London*, 3rd ser., v (1902), 20.
45. Rawlinson, *Mercy to a Beast*, 34; Charles George Cock, *English-Law* (1651), 155; *Harleian Miscellany*, vi. 396; Samuel Butler, *Prose Observations*, ed. Hugh de Quehen (Oxford, 1979), 262, 394; *Educational Writings of Locke*, 226; *Locke Corr.*, ii. 82–3, 112, 733; 'Prebendary of York', *Enquiry about the Lawfulness*, 8–9; *Clemency to Brutes* (1761), 19; J.-J. Rousseau, *Émile* (1762), bk ii; *Gent. Mag.*, l (1780), 462; *The Works of Jeremy Bentham*, ed. John Bowring (1843–59), vii. 61; John Oswald, *The Cry of Nature* (1791), 27; Henry Home of Kames, *Sketches of the History of Man* (new edn, Glasgow, 1817), i. 182n; Thomas Young, *An Essay on Humanity to Animals* (1798), 5–6.
46. Erasmus Darwin, *Phytologia* (1800), 467; *id.*, *Zoonomia* (1794–6), ii. 670; George Nicholson, *On the Conduct of Man to Inferior Animals* (Stourport, n.d.), 186–9; Sir Richard Phillips, *Golden Rules of Social Philosophy* (1826), 352.
47. Frederick A. Pottle, *James Boswell. The Earlier Years* (1966), 4, 33–4; *The World*, 190 (19 Aug. 1756).
48. Roberts Vaux, *Memoirs of the Lives of Benjamin Lay and Ralph Sandiford* (1816), 16; *The Memoirs of James Stephen*, ed. Merle M. Bevington (1954), 182–3.
49. For these authors see *DNB* (though the writer of the article on Ritson is very hostile, attributing his 'perverse arguments' and 'depressing diet' to 'incipient insanity'). David Lee Clark argued that Shelley's debt to Newton has been exaggerated and that he owed more to Ritson; *Studies in Philology*, xxxvi (1939).
50. *DNB*, 'Cowherd, William'; Nicholson, *Conduct of Man*, 216; Peter James Lineham, 'The English Swedenborgians 1770–1840' (Ph.D. thesis, Univ. of Sussex, 1978), chap. 5.
51. Hare, *Life of William Lambe*, 25.
52. *Letters of Joseph Ritson*, i. xlviii; *DNB*, 'Oswald, John'; 'Phillips, Sir Richard'; Phillips, *Golden Rules*, dedication, 347–56.
53. Percy Bysshe Shelley, *A Vindication of Natural Diet* (new edn, 1884), 16; William Lambe, *Additional Reports on the Effects of a Peculiar Regimen* (1815), 238–9. Cf. *Letters of Ritson*, i. 38, 41, 47; T. Forster, *Philozoia* (Brussels, 1839), 43.
54. Williams, *Ethics of Diet*, 424n.
55. Muffett, *Healths Improvement*, 56.

56. Lawrence, *Horses*, i. 122; Philip Doddridge, *A Course of Lectures on the Principal Subjects in Pneumatology, Ethics, and Divinity* (1763), 132. Cf. *Works of Bentham*, i. 142n–143n; Richard Cumberland, *A Treatise of the Laws of Nature*, trans. John Maxwell (1727), 302; William King, *An Essay on the Origin of Evil* (1731), 118–19; *The Diary of Benjamin Newton*, ed. C. P. Fendall and E. A. Crutchley (Cambridge, 1933), 35; and above, pp. 20–21.
57. *Boswell's Life of Johnson*, ed. George Birkbeck Hill, rev. by L. F. Powell (Oxford, 1934–64), iii. 53.
58. Francis Hutcheson, *A System of Moral Philosophy* (1755), i. 316; William Paley, *The Principles of Morals and Political Philosophy* (13th edn, 1801), i. 99; John Field, *The Absurdity & Falsness of Trion's Doctrine Manifested* (1685); *The Apostolic Decree at Jerusalem*, 7–8; Hartley, *Observations on Man*, ii. 223–4; William Cowper, 'The Task', vi. 450–58.
59. Hutcheson, *System of Moral Philosophy*, i. 316n–317n.
60. Benedict de Spinoza, *Ethic*, iv. prop. 37 (trans. W. Hale White, 4th edn, 1910, 209); above, p. 171.
61. *The World*, 190 (19 Aug. 1756); Hartley, *Observations on Man*, ii. 223; Darwin, *Phytologia*, 556; Hutcheson, *System of Moral Philosophy*, i. 315–16.
62. Robert Surtees, *The History and Antiquities of the County Palatine of Durham* (1816–40), iii. 193n.
63. *Remains of John Tweddell*, ed. Robert Tweddell (1815), 215.
64. Bernard Mandeville, *The Fable of the Bees*, ed. F. B. Kaye (Oxford, 1924), i. 174; Stuart Piggott, *William Stukeley* (Oxford, 1950), 146; Nathaniel Bloomfield, *An Essay on War* (1803), 21. Cf. Byng, ii. 256; *Memoirs of Thomas Bewick*, 15.
65. Gilbert White, *Garden Kalendar 1751–1771* (facsimile edn, 1975), fol. 29^{v}; *Boswell's Life of Johnson*, ed. Hill, v. 247; Muffett, *Healths Improvement*, 57; Mandeville, *op. cit.*, i. 174.
66. William Hazlitt, *The Plain Speaker* (EL, n.d.), 173; Elias, *Civilizing Process*, 118–21.

v. Conclusion

1. Oliver Goldsmith, *The Citizen of the World* (EL, 1934), 38.
2. Norman Cohn, *The Pursuit of the Millennium* (1962 edn), 342. Cf. *Diary of Thomas Burton*, i. 62; Christopher Hill, *The World Turned Upside Down* (1972), 112, 119–20, 165, 176; Rufus M. Jones, *Spiritual Reformers in the Sixteenth and Seventeenth Centuries* (1928), 247, 254–5.
3. Sir Richard Blackmore, *Creation* (1712), in *Poetical Works* (Edinburgh, 1793), 61; *The Complete Writings of William Blake*, ed. Geoffrey Keynes (1957), 160; Wordsworth, 'The Prelude' (1805–6 version), iii. lines 121–9; Leslie Stephen, *History of English Thought in the Eighteenth Century* (3rd edn, 1902), ii. 453–4.
4. G. M. Trevelyan, *Must England's Beauty Perish?* (1929), 20.
5. *Dictionary of the History of Ideas*, i. 471.
6. Barry Holstun Lopez, *Of Wolves and Men* (1978), 167, 180; and for a more moderate plea, L. David Mech, *The Wolf* (New York, 1970), esp. chap. xii.

NOTE TO PAGE 302

7. Christopher D. Stone, 'Should Trees have Standing?', *Southern Calif. Law Rev.*, 45 (1972); Laurence H. Tribe, 'Ways not to think about Plastic Trees', *The Yale Law Jnl*, 83 (1974); John Rodman, 'The Liberation of Nature', *Inquiry*, 20 (1977).

SOURCES OF ILLUSTRATIONS

INDEX

Places in England and Wales are attributed to counties as they existed before 1974. Books of the Bible are indexed under *Bible*. Saints' days, church feasts and other festivals appear under *calendar*. Individual species are listed under *animals*; *birds*; *flowers*; *fruit*; *insects*; *reptiles*; *shell-fish*; *trees and shrubs*.

About the Author

Keith Thomas was born in 1933, the son of a South Wales farmer. He was educated at Balliol College, Oxford, and is currently a Fellow and Tutor in Modern History at St. John's College, Oxford. He is also a Reader in the University of Oxford and a Fellow of the British Academy. In 1978 Thomas was a Visiting Fellow at Princeton University. His first book, *Religion and the Decline of Magic* (Penguin, 1971), won a Wolfson Literary Award for history. He is also the editor of the Oxford University Press's Past Master series. He is married and has two children.